Lecture Notes in Artificial Intelligence 13282

Subseries of Lecture Notes in Computer Science

More information about this subseries at https://link.springer.com/bookseries/1244

João Gama · Tianrui Li · Yang Yu ·
Enhong Chen · Yu Zheng · Fei Teng (Eds.)

Advances in Knowledge Discovery and Data Mining

26th Pacific-Asia Conference, PAKDD 2022
Chengdu, China, May 16–19, 2022
Proceedings, Part III

Editors
João Gama
Laboratory of Artificial Intelligence
and Decision Support
University of Porto
Porto, Portugal

Tianrui Li 🆔
School of Computing and Artificial
Intelligence
Southwest Jiaotong University
Chengdu, China

Yang Yu
National Key Laboratory for Novel
Software Technology
Nanjing University
Nanjing, China

Enhong Chen
School of Computer Science and Technology
University of Science and Technology
of China
Hefei, China

Yu Zheng
JD iCity, JD Technology & JD Intelligent
Cities Research
Beijing, China

Fei Teng
School of Computing and Artificial
Intelligence
Southwest Jiaotong University
Chengdu, China

ISSN 0302-9743   ISSN 1611-3349 (electronic)
Lecture Notes in Artificial Intelligence
ISBN 978-3-031-05980-3   ISBN 978-3-031-05981-0 (eBook)
https://doi.org/10.1007/978-3-031-05981-0

LNCS Sublibrary: SL7 – Artificial Intelligence

This Springer imprint is published by the registered company Springer Nature Switzerland AG
The registered company address is: Gewerbestrasse 11, 6330 Cham, Switzerland

General Chairs' Preface

On behalf of the Organizing Committee, it is our great pleasure to welcome you to the 26th Pacific-Asia Conference on Knowledge Discovery and Data Mining (PAKDD2022), held in Chengdu, China, during May 16–19, 2022. Starting in 1997, PAKDD has long established itself as one of the leading international conferences in data mining and knowledge discovery. PAKDD provides an international forum for researchers and industry practitioners to share their new ideas, original research results, and practical development experiences from all Knowledge Discovery and Data Mining (KDD) related areas. In response to the COVID-19 pandemic and the need for social distancing, PAKDD 2022 was held as a hybrid conference for both online and onsite attendees.

Our gratitude goes first and foremost to the researchers, who submitted their work to PAKDD 2022. We would like to deliver our sincere thanks for their efforts in research, as well as in preparing high-quality presentations. We also thank all the collaborators and sponsors for their trust and cooperation. It is our great honor that three eminent keynote speakers joined the conference: Jian Pei (Simon Fraser University, Canada), Bernhard Schölkopf (Max Planck Institute for Intelligent Systems, Germany) and Ji-Rong Wen (Renmin University, China). They were extremely professional and have high reputations in their respective areas. We enjoyed their participation and talks, which made the conference one of the best academic platforms for knowledge discovery and data mining.

We would like to express our sincere gratitude to the contributions of Steering Committee members, Organizing Committee members, Program Committee members and anonymous reviewers, led by Program Committee Co-chairs: João Gama (University of Porto), Tianrui Li (Southwest Jiaotong University), and Yang Yu (Nanjing University). We are also grateful for the hosting organization Southwest Jiaotong University which is continuously providing institutional and financial support to PAKDD 2022. We feel beholden to the PAKDD Steering Committees for their constant guidance and sponsorship of manuscripts.

Finally, our sincere thanks go to all the participants and volunteers. We hope all of you enjoyed PAKDD 2022.

April 2022

Enhong Chen
Yu Zheng

PC Chairs' Preface

It is our great pleasure to present at the 26th Pacific-Asia Conference on Knowledge Discovery and Data Mining (PAKDD 2022) as the Program Committee Chairs. PAKDD is one of the longest established and leading international conferences in the areas of data mining and knowledge discovery. It provides an international forum for researchers and industry practitioners to share their new ideas, original research results, and practical development experiences from all KDD related areas, including data mining, data warehousing, machine learning, artificial intelligence, databases, statistics, knowledge engineering, big data technologies and foundations.

This year PAKDD received 627 submissions, among which 69 submissions were rejected at a preliminarily stage due to the policy violations. There were 320 Program Committee members and 45 Senior Program Committees members involved in reviewing process. Each submission was reviewed by at least three different reviewers. Over 67% of those submissions were reviewed by four or more reviewers. Eventually, 121 submissions were accepted and recommended to be published, resulting in an acceptance rate of 19.30%. Out of these, 29 submissions were about applications, 4 submissions were related to big data technologies, 46 submissions were on data science and 42 submissions were about foundations. We would like to appreciate all PC members and reviewers, who offered a high-quality program with diligence on PAKDD 2022.

The conference program featured keynote speeches from distinguished researchers in the community, most influential paper talks, cutting-edge workshops and comprehensive tutorials.

We wish to sincerely thank all PC members and reviewers for their invaluable efforts in ensuring a timely, fair, and highly effective PAKDD 2022 program.

April 2022

João Gama
Tianrui Li
Yang Yu

Organization Committee

Honorary Co-chairs

Dan Yang
Zhi-Hua Zhou

Southwest Jiaotong University, China
Nanjing University, China

General Co-chairs

Enhong Chen

University of Science and Technology of China, China

Yu Zheng

JD.com, China

Program Committee Co-chairs

Joao Gama
Tianrui Li
Yang Yu

University of Porto, Portugal
Southwest Jiaotong University, China
Nanjing University, China

Workshop Co-chairs

Gill Dobbie
Can Wang

University of Auckland, New Zealand
Griffith University, Australia

Tutorial Co-chairs

Gang Li
Tanmoy Chakraborty

Deakin University, Australia
Indraprastha Institute of Information Technology Delhi, India

Local Arrangement Co-chairs

Yan Yang
Chuan Luo
Xin Yang

Southwest Jiaotong University, China
Sichuan University, China
Southwestern University of Finance and Economics, China

Sponsor Chair

Xiaobo Zhang   Southwest Jiaotong University, China

Publicity Co-chairs

Xiangnan Ren   Group 42, United Arab Emirates
Hao Wang   Zhejiang Lab, China
Junbo Zhang   JD.com, China
Chongshou Li   Southwest Jiaotong University, China

Proceedings Chair

Fei Teng   Southwest Jiaotong University, China

Web and Content Co-chairs

Xiaole Zhao   Southwest Jiaotong University, China
Zhen Jia   Southwest Jiaotong University, China

Registration Chairs

Hongmei Chen   Southwest Jiaotong University, China
Jie Hu   Southwest Jiaotong University, China
Yanyong Huang   Southwestern University of Finance and
                                            Economics, China

Steering Committee

Longbing Cao   University of Technology Sydney, Australia
Ming-Syan Chen   NTU
David Cheung   University of Hong Kong, China
Gill Dobbie   University of Auckland, New Zealand
Joao Gama   University of Porto, Portugal
Zhiguo Gong   University of Macau, China
Tu Bao Ho   Japan Advanced Institute of Science and
                                            Technology, Japan
Joshua Z. Huang   Shenzhen Institutes of Advanced Technology,
                                            Chinese Academy of Sciences, China
Masaru Kitsuregawa   Tokyo University, Japan
Rao Kotagiri   University of Melbourne, Australia
Jae-Gil Lee   Korea Advanced Institute of Science &
                                            Technology, South Korea

Ee-Peng Lim	Singapore Management University, Singapore
Huan Liu	Arizona State University, USA
Hiroshi Motoda	AFOSR/AOARD and Osaka University, Japan
Jian Pei	Simon Fraser University, Canada
Dinh Phung	Monash University, Australia
P. Krishna Reddy	International Institute of Information Technology, Hyderabad, India
Kyuseok Shim	Seoul National University, South Korea
Jaideep Srivastava	University of Minnesota, USA
Thanaruk Theeramunkong	Thammasat University, Thailand
Vincent S. Tseng	NCTU
Takashi Washio	Osaka University, Japan
Geoff Webb	Monash University, Australia
Kyu-Young Whang	Korea Advanced Institute of Science & Technology, South Korea
Graham Williams	Australian National University, Australia
Min-Ling Zhang	Southeast University, China
Chengqi Zhang	University of Technology Sydney, Australia
Ning Zhong	Maebashi Institute of Technology, Japan
Zhi-Hua Zhou	Nanjing University, China

Host Institute

Contents – Part III

Applications

NEWSKVQA: Knowledge-Aware News Video Question Answering

Pranay Gupta[1] and Manish Gupta[1,2(✉)]

[1] IIIT-Hyderabad, Hyderabad, India
[2] Microsoft, Hyderabad, India
pranay.gupta@research.iiit.ac.in, gmanish@microsoft.com

Abstract. Answering questions in the context of videos can be helpful in video indexing, video retrieval systems, video summarization, learning management systems and surveillance video analysis. Although there exists a large body of work on visual question answering, work on video question answering (1) is limited to domains like movies, TV shows, gameplay, or human activity, and (2) is mostly based on common sense reasoning. In this paper, we explore a new frontier in video question answering: answering knowledge-based questions in the context of *news videos*. To this end, we curate a new dataset of ∼12K news videos spanning across ∼156 h with ∼1M multiple-choice question-answer pairs covering 8263 unique entities. We make the dataset publicly available (https://tinyurl.com/videoQAData). Using this dataset, we propose a novel approach, NEWSKVQA (Knowledge-Aware News Video Question Answering) which performs multi-modal inferencing over textual multiple-choice questions, videos, their transcripts and knowledge base, and presents a strong baseline.

1 Introduction

Visual Question Answering (VQA) aims at answering a text question in the context of an image [3]. The questions can be of various types: multiple choice, binary, fill in the blanks, counting-based [1], or open-ended. Most methods for VQA use either basic multimodal fusion of language and image embeddings [11], attention-based multimodal fusion [30] or neural module networks [8]. More recently, newer problem settings have been proposed as extensions of the basic VQA framework like Text VQA [24], Video QA [34], knowledge-based VQA [23] and knowledge-based VQA for videos [5]. Extending on this rich literature, we propose a novel problem setting: entity-based question answering in the context of news videos.

Recently multiple datasets have been proposed for knowledge-based VQA and video VQA. However, many of these questions do not actually make use of the image and knowledge graph (KG) information in a holistic manner. Thus, questions in such datasets can be often answered by using just the image information, or the associated text or a limited KG of dataset-specific entities. We fill this gap by contributing a dataset where each question can be answered only by an effective combination of video understanding, and knowledge-based reasoning. Further, current domain-specific video VQA work is limited to domains like movies, TV shows, gameplay and human activity. In this

© The Author(s), under exclusive license to Springer Nature Switzerland AG 2022
J. Gama et al. (Eds.): PAKDD 2022, LNAI 13282, pp. 3–15, 2022.
https://doi.org/10.1007/978-3-031-05981-0_1

paper, we focus on the news domain since news videos are rich in real-world entities and facts, and linking such entities to Wikipedia KG and answering questions based on such associations is of immense practical value in video indexing, video retrieval systems, video summarization, learning management systems and surveillance video analysis.

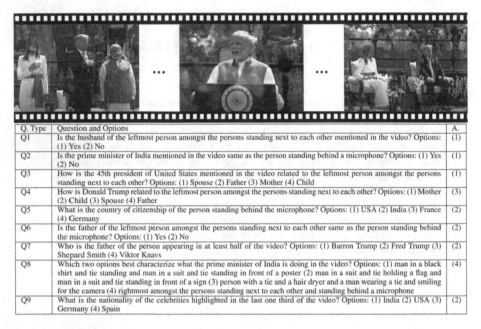

Q. Type	Question and Options	A.
Q1	Is the husband of the leftmost person amongst the persons standing next to each other mentioned in the video? Options: (1) Yes (2) No	(1)
Q2	Is the prime minister of India mentioned in the video same as the person standing behind a microphone? Options: (1) Yes (2) No	(1)
Q3	How is the 45th president of United States mentioned in the video related to the leftmost person amongst the persons standing next to each other? Options: (1) Spouse (2) Father (3) Mother (4) Child	(1)
Q4	How is Donald Trump related to the leftmost person amongst the persons standing next to each other? Options: (1) Mother (2) Child (3) Spouse (4) Father	(3)
Q5	What is the country of citizenship of the person standing behind the microphone? Options: (1) USA (2) India (3) France (4) Germany	(2)
Q6	Is the father of the leftmost person amongst the persons standing next to each other same as the person standing behind the microphone? Options: (1) Yes (2) No	(2)
Q7	Who is the father of the person appearing in at least half of the video? Options: (1) Barron Trump (2) Fred Trump (3) Shepard Smith (4) Viktor Knavs	(2)
Q8	Which two options best characterize what the prime minister of India is doing in the video? Options: (1) man in a black shirt and tie standing and man in a suit and tie standing in front of a poster (2) man in a suit and tie holding a flag and man in a suit and tie standing in front of a sign (3) person with a tie and a hair dryer and a man wearing a tie and smiling for the camera (4) rightmost amongst the persons standing next to each other and standing behind a microphone	(4)
Q9	What is the nationality of the celebrities highlighted in the last one third of the video? Options: (1) India (2) USA (3) Germany (4) Spain	(2)

Fig. 1. Examples of questions for a video on "PM Modi and President Trump attends Namaste Trump event in Ahmedabad, Gujarat" (See footnote 1). For every question, we show question type (refer Table 2), question, options and answer (A.).

Our dataset gathered from 12094 news videos spanning 156 h contains 1041352 multiple-choice question-answer pairs with links to 8263 unique Wikipedia entities. All the questions are centered around entities of person type and involve different levels of KG as well as visual reasoning complexity. Questions with high KG complexity need multiple hops of inference over KG nodes. Questions with high visual complexity require multi-frame reasoning, person-person positional relationship inference or person-object relationship inference. We have ensured that incorrect answer options are selected carefully to avoid bias. Figure 1 shows examples of questions for a video[1].

For question answering, we present results using a strong baseline which uses a combination of signals across different modes. Given the input video, there are three main sources of knowledge – text, visual and KG. From the visual part, we can retrieve object boundaries, object names, celebrities, and image representation using popular

---

[1] https://www.youtube.com/watch?v=lEQ2J_Z1DdA.

CNN models like ResNet [7]. For the text modality, the question, candidate answer (or option), transcript are used as input. Finally from KG, we can leverage KG facts for entities extracted from transcript (person type entities) or visual frames (using celebrity detection). Text as well as KG signals are processed using pretrained BERT model. Finally, we concatenate image as well as text representations and connect this to a dense layer followed by an output softmax layer to come up with the final prediction.

In this paper, we make the following main contributions. (1) We build a large dataset for knowledge-aware video QA in the news domain. We make the dataset publicly available[2]. (2) We carefully design questions and candidate answers such that (a) they need to use both visual information from videos and KG signals to be answered (b) they are diverse with respect to answer types, leverage diverse set of entities, and involve different levels of KG as well as visual reasoning complexity. (3) We experiment with a strong baseline method which leverages different combinations of signals across visual, text and KG modes to achieve good accuracy across various question types.

2 Related Work

Table 1. Description of video question answering datasets

Dataset	Domain	# Video clips	# QAs	Answer type	KG	Entity-set
Youtube2TextQA [6]	Human activity	1987	122708	MCQ	✗	none
MovieQA [25]	Movies	6771	14944	MCQ	✗	closed
PororoQA [13]	Cartoon	16066	8913	MCQ	✗	closed
MarioQA [20]	Gameplay	–	187757	MCQ	✗	none
TGIF-QA [9]	Animated GIFs	71741	165165	Open+MCQ	✗	none
Movie-FIB [18]	Movies	118507	348998	Open	✗	closed
VideoQA [34]	Generic	18100	175076	Open	✗	none
MSRVTT-QA [29]	Generic	10000	243680	Open	✗	none
MSVD-QA [29]	Generic	1970	50505	Open	✗	none
TVQA [14]	TV shows	21793	152545	MCQ	✗	closed
ActivityNetQA [31]	Human activity	5800	58000	Open	✗	none
Social-IQ [32]	Human Intent	1250	7500	MCQ	✗	none
KnowIT VQA [5]	TV shows	12087	24282	MCQ	✓	closed
TVQA+ [15]	TV shows	4198	29383	MCQ	✗	closed
NEWSKVQA (ours)	News	12094	1041352	MCQ	✓	open

Knowledge-Based Visual Question Answering: Besides typical multi-modal fusion-based methods, VQA can also be effectively done by combining image semantic representation with external information extracted from a general knowledge base. Initial systems focused on common sense-enabled VQA [2,11,33]. In this line of work, datasets like KB-VQA [26] and FVQA [27] were centered around common nouns, and

---

[2] https://tinyurl.com/videoQAData.

contain only a few images. They are far more restricted in terms of logical reasoning complexity required as compared to later datasets in this area. More recent systems leverage factual information about entities by fusing multi-modal image information with knowledge graphs (KG). Zhu et al. [35] create a specific knowledge base with image-focused data for answering questions under a certain template, whereas more generic approaches like [28] extract information from external knowledge bases, such as DBpedia, for improving VQA accuracy. R-VQA [17] has questions requiring information about relational facts to answer. KVQA [23] is a dataset where multi-entity, multi-relation, and multi-hop reasoning over KGs is needed for VQA. OK-VQA [19] has free-form questions without knowledge annotations. Knowledge-based approaches have mainly focused on fusing KG facts with image representations. We extend this line of work to leverage knowledge from KG facts for answering questions based on *videos*.

Video Question Answering: Beyond images, recently multiple video question answering datasets have been proposed as described briefly in Table 1. To the best of our knowledge, we are the first to propose a video question answering dataset focused on the news domain. Most of the datasets shown in Table 1 do not use external knowledge from KGs. Only KnowITVQA utilizes in-domain facts about entities in popular sitcoms. Unlike these datasets, we link videos with actual entities in Wikipedia and our proposed dataset contains questions which leverage multi-hop inference on the Wikipedia KG around such entities. Thus, only our dataset has questions related to an open entity set unlike other methods where the set of entities is limited to say characters in sitcoms or none. Also, like many other methods, questions in our dataset are of MCQ (multiple choice questions) form.

3   NEWSKVQA Dataset Curation

We introduce the novel NEWSKVQA dataset for exploring entity based visual QA in news videos using KG. Answering each question requires inferences over multiple modalities including video, transcripts and facts from the KG. The questions exploit the presence of multiple entities in news videos and the spatial-temporal semantics between the objects and the concerned person entities. Questions are further augmented using multi-hop facts about the entities from the KG and the transcripts.

3.1   Initial Data Collection

Video Collection. A total of 2730 English news videos are sourced from six Youtube channels of three popular news media outlets: USA today, Al Jazeera and Washington Post such that each video is less than 10 min in length. The videos belong to the following domains: entertainment, sports, politics and life. The titles, descriptions and the transcripts are downloaded along with the videos. Each video spans 3.43 min on average. These raw videos are converted into ~30 s clips by combining the consecutive shots detected via shot boundary detection[3]. Shots smaller than 30 s were combined with next shots until a clip of length $\geq$ 30 s is obtained. The aligned transcripts are also

---

[3] ffmpeg implementation of shot boundary detection.

split according to the shot boundaries. Overall, the dataset contains 12094 clips. Each clip has a transcript with $\sim$78 words on average. Code + data are publicly available (See footnote 2).

Entity Recognition. Assuming that the main person entities in video clips would be named in the transcripts, we employ a text entity linker[4] to retrieve the entities from transcripts. This led to $\sim$4.43 entities per clip on average. We retained only those entities with confidence score > 0.1. Top five entity types were human, business, film, city and country. There were a total of 8263 unique entities of which 1797 are of human type.

To temporally localize the entities within video clips, we fine-tune a face detection and recognition module[5], for all the entities retrieved from transcripts. We finetune this model using a collection of 100 images per entity obtained from Google image search. Next, keyframes are sampled with a stride of 3 s from the video clips. The bounding boxes and the faces predicted by the face detection and recognition module in each keyframe are stored. In multi-person images, the bounding boxes and the detected entities are sorted in the left-to-right order of their position. Overall, $\sim$62.5% frames had a person recognized. Of these, $\sim$39.4% frames had more than one person recognized.

Knowledge Graph Curation. For each of the 8263 recognized entities, we obtain facts as well as linkage information from Wikidata[6]. Further, we do a multi-hop traversal on the Wikidata KG to materialize up to two hop facts related to each of our entities. We store the facts as {subject entity, relation, object entity} triplets $\langle e_1, r, e_2 \rangle$, wherein the entities are marked by unique Wikidata item IDs (QIDs) and the relations are marked by unique property IDs (PIDs). Overall our KG has 42848 entities and 430985 relations.

3.2 Question Generation

Manual question generation would be slow and difficult to scale. Owing to the development of reliable image captioning systems [16], we were able to substantially bring down the human effort in generating the questions corresponding to the video clips. Automating this procedure helped us efficiently generate questions at scale. Table 2 shows a basic characterization of the 9 question types in our NEWSKVQA dataset. All the question types require inferences over data from multiple modalities like visual, textual and KG facts. Some questions need visual cues in terms of person-person while others need to understand person-object relationships from videos. Answer types could be entities, relations, binary, or person description. In addition, answering complexity varies in terms of number of KG hops across different question types. In this section, we discuss our detailed methods for generating different types of questions.

Generating Questions with Entities as Answer Type. For each video keyframe, we obtain the respective captions via Oscar caption generator [16]. The caption captures the visual relation of the person with an object or other persons in the image. E.g., 'A person standing behind the microphone', 'A group of people standing next to each other'. Thus,

---

[4] The entity linker offered by Microsoft Azure Cognitive Text Analytics Services API.
[5] https://github.com/Giphy/celeb-detection-oss.
[6] https://www.wikidata.org/wiki/Special:EntityData/$\langle$entityQID$\rangle$.json.

Table 2. Characterization of various types of questions in the NEWSKVQA dataset. V = Visual, T = Transcript, KG = Knowledge graph.

Question type	Answer evidence	Source of Q. entities	Answer type	#Questions
Q1	V+T+KG	V	Binary	29291
Q2	V+T+KG	V+T	Binary	95506
Q3	V+T+KG	V+T	Relation	23245
Q4	V+KG	V+KG	Relation	21207
Q5	V+KG	V	Entity	598762
Q6	V+KG	V	Binary	52156
Q7	V+KG	V	Entity	59536
Q8	V+KG	V	Desc.	48160
Q9	V+KG	V	Entity	113489

Fig. 2. Word cloud of answer words and phrases in our NEWSKVQA dataset

for every entity, there are 2 descriptions: surface form from transcript (which we call transcript entity) and visual description from captions (which we call visual entity).

We assume that the subject (grammatical) in the caption is the entity recognized by the celebrity detector in the keyframe. We modify the subject (grammatical) within the caption with a templatized question about the relation in the knowledge base associated with the entity. E.g., a generated caption, '*A person* standing behind a microphone', is changed to, '*Who is the mother of the person* standing behind the microphone'. Question templates for question related to every type of relation (relevant to a person entity) are manually created. In total, we have 30 such templates. Further if multiple keyframes within a video have the same caption, the question is augmented using the time of appearance of the keyframe. E.g., 'Who is the mother of the person standing behind the microphone *in the beginning one third of the video?*'. For multi-person keyframes, we leverage the knowledge of the spatial ordering of the entities in the keyframe and use positional indicators like 'leftmost', 'second to the rightmost', etc. to refer to the person being questioned about in the keyframe. E.g., 'Who is the mother of the *second from the left* amongst the persons standing next to each other in the beginning one third of the video?' All these heuristics help us obtain questions of type Q5 where the question is centered around visual entity (i.e., entity described in the video) using (1) hints from the generated captions, and (2) facts from KG. Formally, given a KG fact, $\langle e_1, r, e_2 \rangle$, Q5 questions were formed using visual entity (whose description is obtained from captions) as e_1, KG based relation as r and expected e_2 as the answer.

Generating Questions with Binary Answer Type. Given a KG fact, $\langle e_1, r, e_2 \rangle$, another way to frame questions is to check if e_2 is mentioned in the video or transcript, given visual entity e_1 and r in the question. This leads to questions of type Q1. E.g, 'Is the mother of the person standing behind the microphone mentioned in the video?' Further, more complex questions (such as of type Q6) can be designed such that given two facts $\langle e_1', r', e_2' \rangle$ and $\langle e_1'', r'', e_2'' \rangle$, visual entities e_1' and e_1'' appearing in two different keyframes in the same video, the question may contain e_1', e_1'', r' and r'' and asks to check if e_2' and e_2'' are the same. E.g, 'Is the occupation of the rightmost amongst the persons standing in a room with a clock on the wall the same as the occupation of the rightmost amongst the persons standing next to each other?'

Generating Questions with Relation as Answer Type. Questions that expect r as answer can be created by including both e_1 and e_2 in the question. While we fix e_2 to be always visual entity (to ensure that all our questions need video for answering), e_1 can be a KG entity like in questions of type Q4. E.g., '*How is Paris related to the country of citizenship of* the person standing behind the mic?'.

The choice of r for generating such questions needs to be done carefully to avoid trivial/obvious questions. Consider these examples: (1) 'What is the relation between North America and the country of the citizenship of the person behind the microphone?' with the answer '*Continent*'. (2) 'What is the relationship between Fordham University and the person standing behind the microphone?' with answer '*educated at*'. In the KG, 'North America' is only related to other entities as a 'continent', similarly 'Fordham University' is related to other person entities only as an 'alma mater'. Hence, we prune away questions where (t_1, t_2) is very frequently related to relation r (rather than other relations) where t_1 and t_2 are types of entities e_1 and e_2 resp.

Generating Questions Involving Transcript Entities. Sometimes, transcripts contain (non-person) entities which are not visualized in the video, and hence transcripts can mention novel entities. We combine such transcript entities with visual entities to create questions where answer type is binary (Q2) or relation (Q3).

The first set of questions, inquire whether the visual entities are related to the transcript entities. Thus, e_1, e_2 and r are mentioned in the question, and the answer could be Yes or No. E.g., 'Is the country of citizenship of the person standing behind the microphone mentioned in the video?' We create similar questions using multi-hop relationships of the transcript entities and visual entities. E.g., 'Is the father of the ⟨transcript entity⟩ mentioned in the video, same as the person standing behind the microphone?' However directly using the transcript entity in the question would trivialize the answering process, therefore we use a representative description of the transcript entity. For this, we leverage entity descriptions provided as part of our KG. Thus the question is converted to, 'Is the father of the 45th president of the USA same as the leftmost amongst the persons posing for a picture?'. Actor/actress entities often have very similar descriptions in the KG, hence for such entities, the description is augmented with information about their latest movies/shows. E.g., for 'Steve Carell', the generated question would be 'Is the child of the American actor who acted in the movie The Office, same as the leftmost amongst the persons standing next to each other in the video?'

The second set of questions mention transcript entity as e_1 and visual entity as e_2 in the question and seek relation r as the answer. E.g., 'How is the 45th president of the United States mentioned in the video related to the country of citizenship of the leftmost amongst the persons posing for a picture?' with the answer '*Head of Government*'. We randomly remove the questions about the most frequent transcript entities as a bias correction step.

Generating Questions Based on Video Durations. We further create 3 types of questions (Q7, Q8, Q9) about entities spanning across larger video durations. Firstly, we create questions (of type Q7) about the entities, which appear more prominently in the video, e.g., 'What is the alma mater of the celebrity appearing in more than half of the video?'.

Secondly, we create questions (of type Q8) regarding what each entity in the video is doing across different keyframes in the video. Similar to the transcript entity questions, the keyframe entity is referenced in the question using the description. E.g., 'Which statement best characterizes the American internet personality present in the video?' with answer: '*leftmost amongst the persons standing in a room with a clock on the wall and person with yellow shirt holding a mic*'. The answer is extracted from the keyframe captions, expressing the entity relationship with the object and/or other persons in the keyframe. Of course, we omit keyframes with same captions.

Finally, we create questions (of type Q9) about the most common relation among the entities appearing within a span of the video. E.g., 'What is the nationality of the celebrities highlighted in the beginning one third of the video?' The common nationality of more than 50% of the entities within the span is chosen as the answer. If less than 50% of the entities have the common entity then the question is removed.

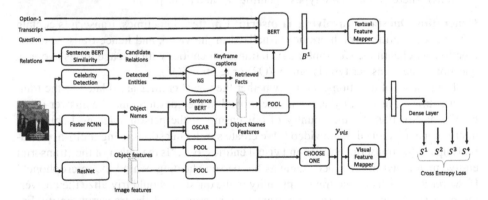

Fig. 3. NEWSKVQA system architecture

3.3 Option Generation

We select 3 options for each of the generated questions in addition to the correct answer, however the procedure for generating the options differs according to the question type. For Q5, where the answer is an entity, other entities of the same entity type as the correct answer are chosen as options. E.g., for the correct answer "United States", other countries are chosen as options. Since, randomly choosing the options adds a bias towards the more frequently occurring correct answers, entities which are approximately as frequent as the correct answer are chosen as options. The max frequency with which an entity can be a correct answer is set to 300.

For Q4, where answer type is a relation, the other relations associated with the entity whose relation is being asked about in the question are used as options. E.g., for the question "How is the 45th president of the USA mentioned in the video related to the leftmost person amongst the persons standing in front of the mic?", the options can be father, child, spouse, head of government, etc. In case less than 4 relations are present for the entity in discussion, then the remaining options are chosen randomly from the set of all possible relations. For binary questions, "yes" and "no" are the only two options. Similarly for the final person description questions the options are chosen randomly.

3.4 Dataset Analysis

On average, each clip has 86.10 questions and each question contains 24.02 words. Also, the KG contains information about 4400 different entity types. Other basic statistics related to our dataset are as follows: transcript size is 84.62 words, answers are on average 3.14 words long, average question size is 22.33 words, average number of facts per entity in our KG is 10.83 (including instance-of relationships). Table 2 reports the distribution of questions within each category. Figure 2 shows the word cloud of various answer words and phrases across all question types in our dataset.

4   Knowledge-Aware Video QA Methodology

We first delineate our approach for question answering on our proposed NEWSKVQA dataset. As shown in Fig. 3, we first extract relevant facts from the KG using the question and the entities recognized in the video. The answering module then processes these facts along with the question, transcripts and the generated captions.

Knowledge Base Fact Retrieval. Given a question and its corresponding video, relevant facts can be retrieved using the entities present in the video and the property around which the question has been framed. Thus we begin by identifying celebrities present in the video. For extracting relevant properties, we rank all the available properties based

Table 3. Main results. Q = Question, T = Transcript, C = Caption, A = Answer, OF = Object features, ON = Object names, F = facts, R = ResNet.

Info. used	Method	Q1	Q2	Q3	Q4	Q5	Q6	Q7	Q8	Q9
	Random	50	50	25	25	25	50	25	25	25
Options	Longest	50.63	50.03	32.99	25.55	24.53	50.15	26.5	27.41	25.54
	Shortest	50.65	50.03	24.82	23.39	24.92	50.15	25.68	21.06	25.34
Question + option	BERTSim	57.97	51.84	48.05	40.11	28.22	51.94	29.94	27.31	30.45
	QA	50.35	50.53	24.45	25.84	24.37	50.24	23.65	26.87	23.97
Transcript + question + option	BERTSim QTA	53.17	55.12	38.57	34.99	28.19	54.21	29.89	26.62	29.96
	QTA	50.2	50.05	28.66	24.44	25.27	50.24	25.36	26.68	25.89
Visual + transcript + question + option	QTCA	50.46	50.66	26.93	25.5	24.94	50.06	25.36	25.73	25.62
	QTONA	57.34	50.51	74.99	57.84	33.23	53.09	37	39.66	34.80
	QTOFA	56.81	53.33	33.54	23.68	25.02	56.31	25.29	25.67	26.15
	QTRA	54.65	52.12	34.84	22.58	24.78	56.85	24.51	24.55	25.16
KG facts + visual + transcript + question + option	QTCFA	53.28	**56.42**	**84.41**	**67.60**	42.71	**62.04**	51.85	47.91	45.18
	QTCONFA	50.59	51.98	64.76	52.57	**54.27**	51.41	**55.39**	**81.32**	**68.28**
	QTCOFFA	51.66	54.36	64.09	51.69	52.39	51.66	53.09	80.78	67.28

on their similarity scores with the question. The properties with the maximum similarity scores are chosen as candidate properties. The similarity scores are calculated as the cosine distance between the Sentence-BERT [21] embedding of the question and the properties. Both one and two hop facts pertaining to the detected entities and the candidate properties are extracted from the KG.

Visual Information Extraction
(1) Image Features: For each of the keyframes, we extract the image features using an ImageNet-pretrained ResNet model. The features are average-pooled along the temporal dimension to obtain the final visual embedding (y_{vis}).
(2) Object Names: A pretrained Faster-RCNN [22] outputs objects within each keyframe. For each frame, we obtain 10 objects detected with the highest confidence scores. Then we select 10 most frequent objects from these candidates. For each of these object we retrieve their Sentence-BERT [21] embeddings.
(3) Object Features: We use the final layers features of Faster-RCNN for the 10 most frequent and confidently detected objects. To add positional information about the detected objects, we concatenate the features with the normalized bounding box coordinates of the detected object. For both object names and object features, the features are first average-pooled along the temporal dimension and then concatenated along the object dimension to obtain the final embedding. Note that image features, object names or object features can be used to calculate y_{vis}.
(4) Image Captions: We generate captions of the keyframes using Oscar [16].

Answering Module. Let q denote the input question, t be the transcript, c be the concatenated captions for all keyframes, f be all the concatenated KG facts, o^i be the ith option. For $i \in (1, 2, 3, 4)$, we obtain BERT embeddings (B^i), using $y_{text}^i = [CLS]+q+t+c+[SEP]+f+o^i+[SEP]$ as input. We pass B^i through the text mapper and y_{vis} through the visual-information mapper. Both the mappers are one dense layer each. The text and visual-information mapper outputs are concatenated before being projected to a single score s^i. The score for the correct option out of the 4 (s^1, s^2, s^3, s^4) are maximized under cross entropy loss.

5   Experiments and Results

We perform extensive experimentation on our NEWSKVQA dataset using the following methods. (1) **Options**: We simply choose the longest option (Longest) or the shortest option (Shortest) as the answer. (2) **Question+Option**: In BERTSim, options with max cosine similarity with the question are selected. The embeddings are calculated using Sentence-Bert. QA is a simplification of our approach (Fig. 3) wherein only the question and the options are used to obtain $y_{text}^i = [CLS] + q + [SEP] + o^i + [SEP]$. BERT embedding B^i is directly projected to get the score s^i. (3) **Transcript+Question+Option**: In addition to the questions and the options the transcripts are also used to identify the correct answer. In BERTSim QTA, option with the maximum cosine similarity with the question and the transcripts is selected as the answer. QTA is an extension of QA wherein the question, the transcripts and the options are used to obtain $y_{text}^i = [CLS] + q + t + [SEP] + o^i + [SEP]$.

(4) **Visual+Transcript+Question+Option**: We use visual information in addition to the question, option and transcript to determine the correct answer. In QTCA, visual information is conveyed using keyframe captions. Captions are fed through BERT as $y^i_{text} = [CLS] + q + t + c + [SEP] + o^i + [SEP]$. QTONA utilizes visual information in the form of object names. y^i_{text} is obtained using the q, t and o^i, while y_{vis} is obtained using object names. Similarly in QTOFA, y_{vis} is obtained using object features. Lastly, in QTRA, y_{vis} is obtained using the image features from ResNet. (5) **KG facts+Visual+Transcript+Question+Option**: Finally, we also exploit the KG facts for answering the questions. QTCFA extends QTCA by adding KG facts to B^i generation. QTCONFA and QTCONFA follow the complete approach as illustrated in Fig. 3, the former uses object names while the latter uses object features to obtain y_{vis}.

Results. In Table 3, we report the accuracies of individual question types for each experiment. From the table, we make the following observations: (1) In both Longest and Shortest, the performance is very close to random. Although with the longest answers, the accuracy is 8% more than random accuracy, the random accuracies for all the other cases indicates that there is negligible bias due to the length of the options. (2) BERT-Sim answers Q3 with 48% accuracy. Higher performance is also observed in the Q4 type. QA performs close to random. (3) Even with the Question, transcripts and options available, performance gain is absent. Similar to the BERTsim ablation, BERTsim QTA performs better on the Q3 and Q4 types. QTA performs close to random, signifying that questions, options and the transcripts alone are not enough to predict the correct answers. (4) Unsurprisingly, just adding the captions is also not enough to lift accuracies, QTCA performs close to random in all categories. However, incorporating visual information via object names (QTONA) shows improvements. With minor improvements in other question types, major jumps are observed for the question types Q3, Q4 with 50% and 32% increment over the random accuracy. Better performance can also be seen in the Q7, Q8 and Q9 types. With object features (QTOFA) and image features (QTRA) the accuracies still lurk around random, with slight improvements in the Q3 type. (5) Finally, adding the KG facts revamps performance across the board. QTCFA improves over random by 60% for Q3 and by 42% for Q4. It shows a 17% improvement over random accuracy in Q5, 12% improvement in Q6, 26% improvement in Q7, 22% in Q8 and 20% in the Q9 type. Adding the object names (QTCONFA) information along with the captions improves on QTCFA by massive amounts (11.56, 3.54, 33.41 and 23.1% points) for Q5, Q7, Q8 and Q9 resp. Surprisingly, adding the object feature (QTCOFFA) helps but not as much as adding object names, possibly because a simple dense layer is not enough to effectively combine the visual embeddings with the BERT based textual embeddings. The relatively low performance of these methods shows that there is a large scope of work to be done for question answering on news videos.

6 Conclusions and Future Work

We studied the problem of knowledge-aware question answering on videos. We curated a large dataset with 12094 clips and associated them with more than a million questions across 9 different types. The questions have been carefully automatically generated such that each question needs video as well as a KG to be answered. Using methods which consume signals across different modes (like text, visual and KG), we establish a baseline accuracy across different question types. We observe that there is a lot of room for improvement, thus establishing that our dataset aptly captures the complexity of the video QA task. Video QA has been performed using various kinds of methods like spatio-temporal reasoning [10], multi-frame inference using memory-based methods [4], question attention-based models [12] and multimodal attention based methods [5]. Since our focus was on building a novel dataset, we presented results using a multimodal baseline method. We plan to adapt these methods for evaluation on NEWSKVQA in the future.

References

1. Acharya, M., Kafle, K., Kanan, C.: TallyQA: answering complex counting questions. In: AAAI, vol. 33, pp. 8076–8084 (2019)
2. Aditya, S., Yang, Y., Baral, C.: Explicit reasoning over end-to-end neural architectures for visual question answering. In: AAAI, vol. 32 (2018)
3. Antol, S., et al.: VQA: visual question answering. In: ICCV, pp. 2425–2433 (2015)
4. Fan, C., Zhang, X., Zhang, S., Wang, W., Zhang, C., Huang, H.: Heterogeneous memory enhanced multimodal attention model for video question answering. In: CVPR, pp. 1999–2007 (2019)
5. Garcia, N., Otani, M., Chu, C., Nakashima, Y.: KnowIT VQA: answering knowledge-based questions about videos. In: AAAI, vol. 34, pp. 10826–10834 (2020)
6. Guadarrama, S., et al.: YouTube2Text: recognizing and describing arbitrary activities using semantic hierarchies and zero-shot recognition. In: ICCV, pp. 2712–2719 (2013)
7. He, K., Zhang, X., Ren, S., Sun, J.: Deep residual learning for image recognition. In: CVPR, pp. 770–778 (2016)
8. Hu, R., Andreas, J., Rohrbach, M., Darrell, T., Saenko, K.: Learning to reason: end-to-end module networks for visual question answering. In: ICCV, pp. 804–813 (2017)
9. Jang, Y., Song, Y., Yu, Y., Kim, Y., Kim, G.: TGIF-QA: toward spatio-temporal reasoning in visual question answering. In: CVPR, pp. 2758–2766 (2017)
10. Jiang, J., Chen, Z., Lin, H., Zhao, X., Gao, Y.: Divide and conquer: question-guided spatio-temporal contextual attention for video question answering. In: AAAI, vol. 34, pp. 11101–11108 (2020)
11. Kembhavi, A., Seo, M., Schwenk, D., Choi, J., Farhadi, A., Hajishirzi, H.: Are you smarter than a sixth grader? Textbook question answering for multimodal machine comprehension. In: CVPR, pp. 4999–5007 (2017)
12. Kim, J., Ma, M., Kim, K., Kim, S., Yoo, C.D.: Progressive attention memory network for movie story question answering. In: CVPR, pp. 8337–8346 (2019)
13. Kim, K.M., Heo, M.O., Choi, S.H., Zhang, B.T.: DeepStory: video story QA by deep embedded memory networks. In: IJCAI, pp. 2016–2022 (2017)
14. Lei, J., Yu, L., Bansal, M., Berg, T.L.: TVQA: localized, compositional video question answering. arXiv:1809.01696 (2018)
15. Lei, J., Yu, L., Berg, T., Bansal, M.: TVQA+: spatio-temporal grounding for video question answering. In: ACL, pp. 8211–8225 (2020)
16. Li, X., et al.: OSCAR: object-semantics aligned pre-training for vision-language tasks. In: Vedaldi, A., Bischof, H., Brox, T., Frahm, J.-M. (eds.) ECCV 2020. LNCS, vol. 12375, pp. 121–137. Springer, Cham (2020). https://doi.org/10.1007/978-3-030-58577-8_8
17. Lu, P., Ji, L., Zhang, W., Duan, N., Zhou, M., Wang, J.: R-VQA: learning visual relation facts with semantic attention for visual question answering. In: KDD, pp. 1880–1889 (2018)

18. Maharaj, T., Ballas, N., Rohrbach, A., Courville, A., Pal, C.: A dataset and exploration of models for understanding video data through fill-in-the-blank question-answering. In: CVPR, pp. 6884–6893 (2017)
19. Marino, K., Rastegari, M., Farhadi, A., Mottaghi, R.: OK-VQA: a visual question answering benchmark requiring external knowledge. In: CVPR, pp. 3195–3204 (2019)
20. Mun, J., Hongsuck Seo, P., Jung, I., Han, B.: MarioQA: answering questions by watching gameplay videos. In: ICCV, pp. 2867–2875 (2017)
21. Reimers, N., Gurevych, I.: Sentence-BERT: sentence embeddings using Siamese BERT-networks. arXiv:1908.10084 (2019)
22. Ren, S., He, K., Girshick, R.B., Sun, J.: Faster R-CNN: towards real-time object detection with region proposal networks. In: NIPS, pp. 91–99 (2015)
23. Shah, S., Mishra, A., Yadati, N., Talukdar, P.P.: KVQA: knowledge-aware visual question answering. In: AAAI, vol. 33, pp. 8876–8884 (2019)
24. Singh, A., et al.: Towards VQA models that can read. In: CVPR, pp. 8317–8326 (2019)
25. Tapaswi, M., Zhu, Y., Stiefelhagen, R., Torralba, A., Urtasun, R., Fidler, S.: MovieQA: understanding stories in movies through question-answering. In: CVPR, pp. 4631–4640 (2016)
26. Wang, P., Wu, Q., Shen, C., Dick, A., van den Hengel, A.: Explicit knowledge-based reasoning for visual question answering. In: IJCAI, pp. 1290–1296 (2017)
27. Wang, P., Wu, Q., Shen, C., Dick, A., Van Den Hengel, A.: FVQA: fact-based visual question answering. TPAMI 40(10), 2413–2427 (2017)
28. Wu, Q., Wang, P., Shen, C., Dick, A., Van Den Hengel, A.: Ask me anything: free-form visual question answering based on knowledge from external sources. In: CVPR, pp. 4622–4630 (2016)
29. Xu, D., et al.: Video question answering via gradually refined attention over appearance and motion. In: ACM MM, pp. 1645–1653 (2017)
30. Yang, Z., He, X., Gao, J., Deng, L., Smola, A.: Stacked attention networks for image question answering. In: CVPR, pp. 21–29 (2016)
31. Yu, Z., et al.: ActivityNet-QA: a dataset for understanding complex web videos via question answering. In: AAAI, vol. 33, pp. 9127–9134 (2019)
32. Zadeh, A., Chan, M., Liang, P.P., Tong, E., Morency, L.P.: Social-IQ: a question answering benchmark for artificial social intelligence. In: CVPR, pp. 8807–8817 (2019)
33. Zellers, R., Bisk, Y., Farhadi, A., Choi, Y.: From recognition to cognition: visual commonsense reasoning. In: CVPR, pp. 6720–6731 (2019)
34. Zeng, K.H., Chen, T.H., Chuang, C.Y., Liao, Y.H., Niebles, J.C., Sun, M.: Leveraging video descriptions to learn video question answering. In: AAAI, vol. 31 (2017)
35. Zhu, Y., Zhang, C., Ré, C., Fei-Fei, L.: Building a large-scale multimodal knowledge base system for answering visual queries. arXiv:1507.05670 (2015)

Multicommunity Graph Convolution Networks with Decision Fusion for Personalized Recommendation

Shenghao Liu[1], Bang Wang[2(✉)], Bin Liu[2], and Laurence T. Yang[3]

[1] School of Cyber Science and Engineering, Huazhong University of Science and Technology (HUST), Wuhan, China
shenghao_liu@hust.edu.cn
[2] School of Electronic Information and Communications,
Huazhong University of Science and Technology (HUST), Wuhan, China
{wangbang,liubin0606}@hust.edu.cn
[3] Department of Computer Science, St. Francis Xavier University,
Antigonish, Canada
ltyang@stfx.ca

Abstract. Most graph-based recommendation approaches actually have made an implicit assumption about that representations of all users and all items can be learned in a single latent space. However, this assumption may be too strong to well describe a single use's multifaceted preferences each probably dominated by some latent type of motivations. This paper challenges this assumption and proposes a MultiGCN model (Multicommunity Graph Convolution Networks with Decision Fusion) to leverage multiple latent spaces for capturing multiple types of motivation. Specifically, we first design a community exploration module to construct multiple communities so as to explore different latent types of motivation. We next design a local recommendation module which maps the representations of entities in each community into one latent space and outputs a local recommendation list. A decision fusion module reranks the items of local lists to obtain the final recommendation list. Experiment results on three real-world datasets demonstrate that our MultiGCN outperforms the state-of-the-art algorithms.

Keywords: Recommender system · Graph convolution network · Community exploration · Random walk

1 Introduction

The core idea of graph-based recommendation approaches is to construct a user-item bipartite graph based on a user-item incidence matrix, on which representations can be learned for all entities (users and items) [3,6,8,10]. For example, Huang et al. [10] apply an attributed random walk on such a bipartite graph

Supported by National Natural Science Foundation of China (Grant No: 62172167).

J. Gama et al. (Eds.): PAKDD 2022, LNAI 13282, pp. 16–28, 2022.
https://doi.org/10.1007/978-3-031-05981-0_2

to sample some node paths and design a graph recurrent network to learn the representations of the paths for recommendation. He et al. [8] design a light graph convolution network which removes nonlinear layers and only includes the most essential component of GCN, neighborhood aggregation, for learning the representations of entities. Gao et al. [6] sample both explicit and implicit interactions in between entities from a user-item bipartite graph and uses the two kinds of interactions to learn their representations.

Most graph-based representation approaches actually have made an implicit assumption about that the representations of all entities can be learned in a single latent space. However, we argue that such an assumption is too limited to well distinguish different latent types of motivation for a user interacting with different items. For example, the user may be learned as disliking science fiction movie from his history records. But he might want to see a new science fiction movie simply owing to the recommendation from his followed celebrity. It may be difficult to represent such two different types of motivation only by a single latent space. A user who has different types of motivation may need multiple different types of representations. This motivates us to construct multiple latent spaces for *multifaceted representation learning*.

In this paper, we construct multiple communities, each of which contains entities most probably impacted by one latent type of motivation, so as to learn multifaceted representations for entities in different latent spaces. Specifically, we first select the most influential entities from the user-item bipartite graph as hub entities. We next assign the most relevant entities to each hub entity to construct a hub-centered community. Note that a user may have multiple different types of motivation, so a user needs to be assigned to multiple different communities.

This paper proposes a *Multicommunity Graph Convolution Networks with Decision Fusion* (MultiGCN) model, which consists of three components: *Community Exploration, Local Recommendation* and *Decision Fusion*. In the community exploration module, we construct multiple overlapping communities based on two kinds of random walk on graph technique. For each community, the local recommendation module constructs a subgraph based on the entities of the community and their interactions on the entire bipartite graph. We next employ a graph convolution network to learn the representations of entities on a subgraph, which we call local representations. For one user in a community, we rank items of his belonging community to output his local recommendation list. As a user may be in multiple communities, we design a decision fusion to rerank all items in his local recommendation lists by combining the contribution of an item's belonging subgraph as well as the item order in a local recommendation list. After reranking, we output the final recommendation list with the top-n items. We conduct experiments on three public datasets. Results have validated that our MultiGCN outperforms the state-of-the-art algorithms.

Our contributions are summarized as follows:

– Propose a MultiGCN model to construct multiple latent spaces based on users' different types of motivations for multifaceted representation learning;

– Propose a Community Exploration module to explore users' latent types of motivation by constructing multiple overlapping communities;
– Propose a Decision Fusion module to fuse and rerank local recommendation lists for generating the final recommendation list;
– Validate the superiority of our proposed MultiGCN compared with the state-of-the-art algorithms.

The rest of the paper is organized as follows: Sect. 2 briefly reviews the related work. The detail of our proposed MultiGCN algorithm is shown in Sect. 3. In Sect. 4, the experiments are introduced. Section 5 concludes the paper with some discussions.

2   Related Work

This section reviews mostly related work on collaborative filtering algorithm, graph based recommendation algorithm and graph embedding algorithm.

Most of collaborative filtering methods learn preference coherence among users and rating correlations among items by matrix factorization [12,25] and neural network based methods [2,14,23,26]. Those MF approaches approximate missing ratings by using the inner-product of a user representation and an item representation. Lee et al. [12] propose to extend low-rank matrix approximation by constructing several local matrices for factorization separately. Those neural network-based approaches design different neural models to learn users' and items' representations for rating predictions. Le et al. [23] aggregate preferences of each user's neighbors in a social network to learn users' representations.

Graph-based recommendation methods construct a graph based on all available entities and interactions in between them. There are two typical kind of graph-based methods. The first kind is random walk-based method which randomly walks on the graph to discover paths between each user node and each item node for discovering relations between them [13,15–17]. Lei et al. [13] propose a conversational recommendation framework which transform a session into a path in a user-item-attribute graph to mine the potential relations between entities. Mo et al. [17] propose a new reverse RWR that starts walking from an item on a heterogeneous graph for event recommendation, which solves a dangling node problem. The second kind is graph neural network-based method which exploits neural network to learn representations of each entity based on the structure of a graph [18,21,22,24]. Palumbo et al. [18] design an entity2rec algorithm which divides a knowledge graph into several subgraphs based on the properties of interactions and learns property-specific representations of interactions to predict users' ratings on items. Xu et al. [24] transform a user-item bipartite graph into a user-user homogeneous graph to learn users' representations and an item-item homogeneous graph to learn items' representations by graph convolution networks, respectively.

Graph embedding approaches learn nodes' embedding based on structure of a graph by ensuring that if two nodes are close on the graph, their representations

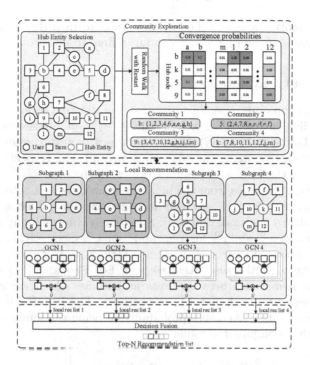

Fig. 1. Illustration of the MultiGCN framework. The community exploration module samples a user-item bipartite graph to construct multiple overlapping communities. In each community, the local recommendation module delivers a local recommendation list. The decision fusion module fuses local recommendation lists to output a final recommendation list.

are also close in the embedding space [4,5,7,11]. Grover et al. [7] propose homogeneous graph embedding algorithm, node2vec, which designs a biased random walk to discover a node's neighbor and learns the node's embedding based on a word2vec framework. Kipf et al. [11] propose a new framework of GCN on a homogeneous graph, which fuses the structure of a graph and the feature of entities for learning entities' representations. Besides applying graph embedding on homogeneous graph, Dong et al. [4] introduce a model called metapath2vec for heterogeneous graph, which designs some meta-paths to guide random walks for generating heterogeneous neighbor of each entity and learns their representation for explainable recommendation.

3   The MultiGCN Model

Figure 1 presents the framework of MultiGCN, which consists of three components: the **community exploration module** mines users' different latent types of motivation by exploring communities, the **local recommendation module** generates several local recommendation lists for a user based on his different latent types of motivations and the **decision fusion module** fuses and reranks a user's local recommendation lists to obtain a final top-n recommendation list.

3.1 Community Exploration

The community exploration module is to discover different latent types of motivation by dividing entities of a user-item bipartite graph into several overlapping communities. Let $G = (U \cup I, E)$ denote the bipartite graph, where U, I and E denote the set of users, items and edges in between them, respectively. An edge (u_m, i_n) exists, if the user u_m has interaction with the item i_n.

To explore communities from the bipartite graph, we first select top-K user $\{u_c\}_{c=1}^K$ and top-K item $\{i_c\}_{c=1}^K$ which have the greatest influences as hub entities based on their PageRank values [1]. Specifically, we randomly initialize two probability vectors $\mathbf{u}^{(0)}$ and $\mathbf{i}^{(0)}$ for users and items, respectively. Let $\mathbf{A} \in \mathbb{R}^{M \times N}$ represent the incidence matrix of G, where $\mathbf{A}(m, n) = 1$, if there is an edge in between u_m and i_n; Otherwise, $\mathbf{A}(m, n) = 0$. Let $\mathbf{P}_{UI}$ and $\mathbf{P}_{IU}$ denote probability transition matrices between users and items, which are computed by row-normalizing matrices $\mathbf{A}$ and $\mathbf{A}^T$, respectively. The PageRank value of each entity is defined as its convergence probability of being randomly walked on the graph G, which is computed as follows:

$$\mathbf{u}^{(t+1)} = (1 - \alpha) \cdot \mathbf{P}_{IU} \mathbf{i}^{(t)} + \alpha \cdot \frac{1}{|U|}, \tag{1}$$

$$\mathbf{i}^{(t+1)} = (1 - \alpha) \cdot \mathbf{P}_{UI} \mathbf{u}^{(t)} + \alpha \cdot \frac{1}{|I|}, \tag{2}$$

where $|U|$ and $|I|$ are the number of users and items, $\mathbf{u}^{(t)}$ and $\mathbf{i}^{(t)}$ are the user probability vector and item probability vector in the tth random walk iteration, and α represents randomly visit probability.

After hub entity selection, we assign other entities to their most relevant hub entities for constructing communities based on the convergence probabilities of entities computed by the random walk with restart (RWR) method. Take a hub entity u_c as an example, we set the hub entity node as a query node and its query vector is defined as $\mathbf{q}_u$, where $\mathbf{q}_u(c) = 1$ and $\mathbf{q}_u(j) = 0$ for $j \neq c$. In the process of RWR, the convergence probabilities of other user nodes and item nodes can be computed as follows:

$$\mathbf{u}^{(t+1)} = (1 - \beta) \cdot \mathbf{P}_{IU} \mathbf{i}^{(t)} + \beta \cdot \mathbf{q}_u; \tag{3}$$

$$\mathbf{i}^{(t+1)} = \mathbf{P}_{UI} \mathbf{u}^{(t)}, \tag{4}$$

where $\mathbf{u}^{(t)}$ and $\mathbf{i}^{(t)}$, respectively, represent the probability vectors of users and items in the tth generation, β is the probability of returning to the query node. For a connected graph, the RWR iteration will converge, and convergence probability vector of u_c is $\mathbf{p}_{u_c} = (\mathbf{u}^{(t)}, \mathbf{i}^{(t)}) = (p(u_1), ..., p(i_N))$. Each element $p(u_i)$ represents a convergence probability in between hub entity u_c and user u_i.

As shown in Fig. 1, we construct a relevant matrix $\mathbf{S} \in \mathbb{R}^{2K \times (M+N)}$ based on the convergence probabilities, where the cth row of $\mathbf{S}$ denotes the convergence probability vector of the cth hub entity. Take a user u_i as example, we set a scale control threshold $\mu \in (0, 1)$ to assign the user u_i to top-$2K \times \mu$ hub entities, which means the user u_i belongs to $2K \times \mu$ different communities. Specifically, these hub entities are selected based on the decreasing order of element value in the ith column of $\mathbf{S}$. Finally, $2K$ communities $\{C_k = \{U_k \bigcup I_k\}\}_{k=1}^{2K}$ are constructed, each of which consists of a hub entity and its the most relevant entities.

3.2 Local Recommendation

The local recommendation module aims to single out items that a user prefers under a latent type of motivation. For a user in a community, we construct a subgraph and design a graph convolution network (GCN) to learn local representations of entities for generating a local recommendation list. Note that this module only recommends items in the community to the user.

For example, in a community $C_k = \{U_k \bigcup I_k\}$, a subgraph $G_k = (U_k \cup I_k, E_k)$ is constructed by using the entities of C_k as nodes and their interactions on G as edges. Let $\mathbf{A}_k \in \mathbb{R}^{|U_k| \times |I_k|}$ denote the incidence matrix of G_k. If a user u_m has an edge to an item i_n on the graph G, $\mathbf{A}_k(u_m, i_n) = 1$; Otherwise, $\mathbf{A}_k(u_m, i_n) = 0$. We then construct an adjacency matrix of this subgraph as

$$\mathbf{R} = \begin{bmatrix} \mathbf{0} & \mathbf{A}_k \\ \mathbf{A}_k^T & \mathbf{0} \end{bmatrix}. \tag{5}$$

We randomly initialize an embedding matrix $\mathbf{E}^{(0)} \in \mathbb{R}^{(|U_k|+|I_k|) \times d}$ as input of GCN, where d is embedding size. Through each layer of GCN, the embedding matrix can be updated as follows:

$$\mathbf{E}^{(t+1)} = (\mathbf{D}^{-\frac{1}{2}} \mathbf{R} \mathbf{D}^{-\frac{1}{2}}) \mathbf{E}^{(t)}, \tag{6}$$

where $\mathbf{D}$ is a $(|U_k| + |I_k|) \times (|U_k| + |I_k|)$ diagonal matrix which represents degree matrix of $\mathbf{R}$. After passing through T layers of GCN, the local representations of entities of the community C_k can be computed as:

$$\mathbf{E} = meanpooling(\mathbf{E}^{(0)}, \mathbf{E}^{(1)}, \cdots, \mathbf{E}^{(T)}). \tag{7}$$

Finally, we predict a user's rating $\hat{y}$ on an item by the inner-product of their local representations and generate a top-N local recommendation list $l_u^k = \{i_1, i_2, \cdots, i_N\}$. Note that if a user u belongs to m communities, the user shall have m local recommendation lists $L_u = \{l_u^1, l_u^2, \cdots, l_u^m\}$.

3.3 Decision Fusion

The decision fusion module takes all local recommendation lists into consideration and rerank the items in these lists to construct a final recommendation list.

In this module, we propose a strategy of reranking an item by considering the contribution of its subgraph as well as its order in local recommendation list. For an item i in a local list l_u^k, we compute the contribution of its order as follows:

$$s_{u,i}^k = \frac{1}{log(order(i, l_u^k) + 1)}, \text{ if } i \in l_u^k, \tag{8}$$

where $order(\cdot)$ is a function that returns item's index in l_u^k. For example, if item i is the mth item in L_u^k, $order(i, l_u^k) = m$.

The contribution of a subgraph G_k for an item i in a local recommendation list l_u^k is defined as the ratio of node entropy of u and i in G_k and node entropy of u and i in G:

$$w_{u,i}^k = \log(\frac{N_u^k + 1}{N_u} + 1) \times \log(\frac{M_i^k + 1}{M_i} + 1), \tag{9}$$

where N_u^k and M_i^k, respectively, are the degree of user node u and item node i in G_k, and N_u and M_i are the degree of user node u and item node i in G.

A reranking score of a user-item pair (u, i) is computed by:

$$r_{u,i} = \sum_{l_u^k \in L_u} w_{u,i}^k \times s_{u,i}^k. \tag{10}$$

The final top-N recommendation list l_u is constructed by sorting items based on their reranking scores.

3.4 Optimization

In this paper, we define a user-item pair as a positive interaction, if the user has rating on the item; Otherwise, we define the user-item pair as a negative interaction. We employ the *Bayesian Personalized Ranking* (BPR) loss to learn parameters in MultiGCN algorithm, which assumes that positive interactions should get higher prediction scores than negative interactions. The loss function of BPR loss is defined as follows:

$$Loss = -\sum_{m=1}^{M} \sum_{i_k \in Q_u^+} \sum_{i_j \in Q_m^-} \ln \phi(\hat{r}(u_m, i_k) - \hat{r}(u_m, i_j)) + \lambda ||\theta||^2, \tag{11}$$

where Q_m^+ and Q_m^- denote the set of positive interactions and negative interactions of user u_m, respectively, λ is the parameter that controls the L_2 regularization strength and $\phi(\cdot)$ is the sigmoid function, θ denotes all trainable model parameters.

4 Experiments

4.1 Experiment Datasets

We conduct experiments on three widely used datasets: Yelp[1], Douban Movie[2] and MovieLens-1M[3]. Table 1 summarizes the statistics of the three datasets.

Table 1. Statistics of dataset

Dataset	Users	Items	Edges	Sparsity
Yelp	14085	14037	194255	0.00097
Douban Movie	3022	6971	195493	0.00928
MovieLens-1M	6040	3706	1000209	0.0419

4.2 Experimental Settings

Parameter Setting and Evaluation Metrics. Our proposed MultiGCN model is trained by optimizing a BPR loss function, where we sample one negative user-item pair per positive user-item pair. In the community exploration module, we set the parameters $\alpha = 0.8$, $\beta = 0.5$ and $\mu = 0.25$. In the local recommendation module, we construct a three-layer GCN with $d = 64$. To evaluate the performance of our MultiGCN algorithm on top-10 recommendation task, we adopt widely used evaluation metrics: P@N, F1, Recall, MAP and NDCG.

Comparison Algorithms. We compare the proposed MultiGCN algorithm with the following state-of-the-art algorithms:

- **BPRMF** [20]: This algorithm factorizes a user-item rating matrix to generate users' and items' representations, which is optimized by the Bayesian personalized ranking (BPR) loss.
- **RWR** [19]: This algorithm applies the RWR starting from each user on a user-item bipartite graph. After multiple walks, probability of each item node will converge, which is used as the rating predicted.
- **BiNE** [6]: This algorithm samples both explicit and implicit interactions in between users and items from a user-item bipartite graph and uses the two kinds of interactions to learn their representations.
- **NeuMF** [9]: This algorithm combines MLP with MF to learn representation of each user and each item and replaces inner product of their representations with a neural network to compute similarity in between them.
- **LightGCN** [8] designs a light graph convolution network by removing nonlinear layers for top-n recommendation.

---

[1] Yelp: https://www.yelp.com/dataset/challenge.
[2] Douban Movie: http://www.shichuan.org/HIN_dataset.html.
[3] MovieLens: https://grouplens.org/datasets/movielens/.

4.3 Experiment Results

Overall Comparison. Table 2 presents the experiment results of different competitors for top-10 recommendation task on Yelp, Douban Moive and MovieLens-1M dataset. We mark the best result of each evaluation metric in bold and use underscores to mark the second-best result. From the results, it is seen that our proposed MultiGCN outperforms the state-of-the-art algorithms in all performance metrics.

Table 2. Comparison of top-10 recommendation performance.

	Yelp				Douban				MovieLens-1M			
	MAP	Recall	F1	NDCG	MAP	Recall	F1	NDCG	MAP	Recall	F1	NDCG
BPRMF	0.1199	0.0210	0.0279	0.1719	0.0700	0.0220	0.0222	0.1011	0.4909	0.1312	0.1815	0.6143
RWR	0.1559	0.0315	0.0415	0.2247	0.0861	0.0194	0.0240	0.1287	0.4523	0.0735	0.1149	0.5752
BiNE	0.1031	0.0206	0.0271	0.1553	0.0722	0.0160	0.0197	0.1082	0.2472	0.0365	0.0552	0.3564
NeuMF	0.1528	0.0303	0.0398	0.2204	0.0875	0.0195	0.0242	0.1290	0.5818	0.1166	0.1796	0.7128
LightGCN	0.2322	0.0468	0.0612	0.3201	0.0925	0.0220	0.0271	0.1413	0.6285	0.1358	0.2072	0.7556
MultiGCN	0.2607	0.0524	0.0688	0.3581	0.1039	0.0234	0.0288	0.1541	0.6478	0.1439	0.2186	0.7729
Improvement	+12.29%	+11.95%	+12.33%	+11.86%	+12.23%	+6.16%	+6.09%	+9.11%	+3.06%	+6.02%	+5.50%	+2.29%

Specifically, we first compare BPRMF and NeuMF with our MultiGCN and observe that MultiGCN outperforms the other two algorithms. It is because BPRMF and NeuMF do not construct a bipartite graph to take implicit interactions into consideration, but only focus on the explicit interactions in between users and items. Second, although RWR, BiNE and LightGCN are all graph-based recommendation algorithms, they also perform worse than our MultiGCN. This validates the necessity of exploring community for users and items, which can mine latent types of motivation to describe more fine-grained users' preferences.

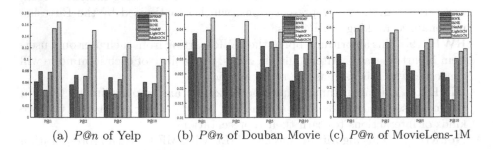

(a) P@n of Yelp   (b) P@n of Douban Movie   (c) P@n of MovieLens-1M

Fig. 2. Comparison of P@n for different recommendation list lengths.

Performance of Top-n. In Fig. 2, we evaluate the P@n on the three datasets, where the recommendation list's length n is set to $\{1, 2, 5, 10\}$. We observe that the MultiGCN conducts a comprehensive improvements over the state-of-the-art algorithms in the case of different lengths of recommendation list. Especially in the Top-1 position that users are most concerned, the MultiGCN makes improvement over the LightGCN at $P@1$ by 7.43%, 10.29% and 3.22% in the Yelp, Douban Moive and MovieLens-1M dataset, respectively. This again indicates the effectiveness of exploring users' latent motivations. Besides, it can also be observed that as the recommendation list length n increases, the P@n decreases, which suggests that MultiGCN algorithm can not only achieve high precision top-n recommendation, but also place positive items at the top of the recommendation list.

Table 3. Results of ablation study.

		MultiGCN-I	MultiGCN-E	MultiGCN
Yelp	MAP	0.2477	0.2402	0.2607
	NDCG	0.3410	0.3365	0.3581
Douban	MAP	0.0969	0.0995	0.1039
	NDCG	0.1461	0.1487	0.1541
ML-1M	MAP	0.6457	0.6145	0.6478
	NDCG	0.7715	0.7459	0.7729

Ablation Study. To prove the effectiveness of two different contributions in decision fusion module, we conduct experiments of ablation studies on Yelp, Douban Moive and MovieLens-1M dataset, where the results of the experiments are introduced in Table 3. The MultiGCN-I only leverages contributions of orders of items in each local recommendation list to rerank these items in decision fusion module, and the MultiGCN-E only considers contributions of the community of items for reranking. Note that MultiGCN-I and MultiGCN-E have the same experiment settings as MultiGCN.

We observe that MultiGCN outperforms MultiGCN-I and MultiGCN-E in terms of all evaluation metrics on the three datasets. This validates that the combination of contributions of items' orders and contribution of items' communities is effectiveness. Specifically, the contribution of an item's order represents how much a user prefer the item under a latent type of motivation and the contribution of a community denotes the significance of a latent type of motivation.

To summarize our experiment results, our MultiGCN algorithm can outperform the state-of-the-art algorithms in terms of all evaluation metrics on the three widely used datasets. This validates the necessity of taking a user's different latent types of motivation into consideration. This also indicates the superiority of community exploration module which proposes to leverage two kinds of random walk methods for construction communities. Each community can be exploited to represent a latent type of motivation.

5   Conclusion

In this paper, we propose a MultiGCN algorithm which constructs multiple latent spaces to represent users' different latent types of motivations for multifaceted representation learning. We have designed a community exploration module which mines multiple latent types of motivation by exploring multiple communities from a bipartite graph. We then propose to leverage multiple GCN to make local recommendation in each community. Furthermore, we have designed a decision fusion module to rerank items in local recommendation lists to output the final top-n recommendation list. The experiment results on three real-world public datasets have validated superiority of our MultiGCN algorithm over the state-of-the-art algorithms.

In future work, it is worth of exploring how to automatically set the number and size of communities to fit different datasets and find out the best hyper parameter settings. Besides, we shall combine the rating matrix with some side information, say for example, by sorting a user's purchased items based on purchased time and exploiting sequential prediction method to improve recommendation performance.

References

1. Berkhin, P.: A survey on PageRank computing. Internet Math. **2**(1), 73–120 (2005)
2. Chen, C., et al.: An efficient adaptive transfer neural network for social-aware recommendation. In: Proceedings of the 42nd International ACM SIGIR Conference on Research and Development in Information Retrieval, pp. 225–234 (2019)
3. Chen, Z., Zhang, Y., Li, Z.: Adversarial deep factorization for recommender systems. In: Lu, W., Zhu, K.Q. (eds.) PAKDD 2020. LNCS (LNAI), vol. 12237, pp. 63–71. Springer, Cham (2020). https://doi.org/10.1007/978-3-030-60470-7_7
4. Dong, Y., Chawla, N.V., Swami, A.: metapath2vec: scalable representation learning for heterogeneous networks. In: Proceedings of the 23rd ACM SIGKDD International Conference on Knowledge Discovery and Data Mining, pp. 135–144 (2017)
5. Fu, X., Zhang, J., Meng, Z., King, I.: MAGNN: metapath aggregated graph neural network for heterogeneous graph embedding. In: Proceedings of the 20th World Wide Web Conference, pp. 2331–2341 (2020)
6. Gao, M., Chen, L., He, X., Zhou, A.: BiNE: bipartite network embedding. In: Proceedings of the 41st International ACM SIGIR Conference on Research and Development in Information Retrieval, pp. 715–724 (2018)
7. Grover, A., Leskovec, J.: node2vec: scalable feature learning for networks. In: Proceedings of the 22nd ACM SIGKDD International Conference on Knowledge Discovery and Data Mining, pp. 855–864 (2016)
8. He, X., Deng, K., Wang, X., Li, Y., Zhang, Y., Wang, M.: LightGCN: simplifying and powering graph convolution network for recommendation. In: Proceedings of the 43rd International ACM SIGIR Conference on Research and Development in Information Retrieval, pp. 639–648 (2020)
9. He, X., Liao, L., Zhang, H., Nie, L., Hu, X., Chua, T.S.: Neural collaborative filtering. In: Proceedings of the 26th International Conference on World Wide Web, pp. 173–182 (2017)

10. Huang, X., Song, Q., Li, Y., Hu, X.: Graph recurrent networks with attributed random walks. In: Proceedings of the 25th ACM SIGKDD International Conference on Knowledge Discovery and Data Mining, pp. 732–740 (2019)
11. Kipf, T.N., Welling, M.: Semi-supervised classification with graph convolutional networks. In: Proceedings of the International Conference on Learning Representations, pp. 1–14 (2017)
12. Lee, J., Kim, S., Lebanon, G., Singer, Y.: Local low-rank matrix approximation. In: Proceedings of the 30th International Conference on Machine Learning, pp. 82–90 (2013)
13. Lei, W., et al.: Interactive path reasoning on graph for conversational recommendation. In: Proceedings of the 26th ACM SIGKDD International Conference on Knowledge Discovery & Data Mining, pp. 2073–2083 (2020)
14. Liu, H., et al.: NRPA: neural recommendation with personalized attention. In: Proceedings of the 42th International ACM SIGIR Conference on Research and Development in Information Retrieval (2019)
15. Liu, S., Wang, B., Xu, M.: Event recommendation based on graph random walking and history preference reranking. In: Proceedings of the 40th International ACM SIGIR Conference on Research and Development in Information Retrieval, pp. 861–864 (2017)
16. Liu, S., Wang, B., Xu, M., Yang, L.T.: Evolving graph construction for successive recommendation in event-based social networks. Futur. Gener. Comput. Syst. **96**, 502–514 (2019)
17. Mo, Y., Li, B., Wang, B., Yang, L.T., Xu, M.: Event recommendation in social networks based on reverse random walk and participant scale control. Futur. Gener. Comput. Syst. **79**, 383–395 (2018)
18. Palumbo, E., Rizzo, G., Troncy, R.: Entity2rec: learning user-item relatedness from knowledge graphs for top-n item recommendation. In: Proceedings of the 17th ACM Conference on Recommender Systems, pp. 32–36 (2017)
19. Pham, T.A.N., Li, X., Cong, G., Zhang, Z.: A general graph-based model for recommendation in event-based social networks. In: IEEE 31st International Conference on Data Engineering, pp. 567–578 (2015)
20. Rendle, S., Freudenthaler, C., Gantner, Z., Schmidt-Thieme, L.: BPR: Bayesian personalized ranking from implicit feedback. UAI, pp. 452–461 (2009)
21. Wang, X., Jin, H., Zhang, A., He, X., Xu, T., Chua, T.S.: Disentangled graph collaborative filtering. In: Proceedings of the 43rd International ACM SIGIR Conference on Research and Development in Information Retrieval, pp. 1001–1010 (2020)
22. Wang, X., Wang, R., Shi, C., Song, G., Li, Q.: Multi-component graph convolutional collaborative filtering. In: Proceedings of the International Conference on Artificial Intelligence (2020)
23. Wu, L., Sun, P., Fu, Y., Hong, R., Wang, X., Wang, M.: A neural influence diffusion model for social recommendation. In: Proceedings of the 42th International ACM SIGIR conference on Research and Development in Information Retrieval (2019)
24. Xu, J., Zhu, Z., Zhao, J., Liu, X., Shan, M., Guo, J.: Gemini: a novel and universal heterogeneous graph information fusing framework for online recommendations. In: Proceedings of the 26th ACM SIGKDD International Conference on Knowledge Discovery and Data Mining, pp. 3356–3365 (2020)

25. Yang, X., Wang, B.: Local matrix approximation based on graph random walk. In: Proceedings of the 42nd International ACM SIGIR Conference on Research and Development in Information Retrieval, pp. 1037–1040 (2019)
26. Zou, L., et al.: Neural interactive collaborative filtering. In: Proceedings of the 43rd International ACM SIGIR Conference on Research and Development in Information Retrieval, pp. 749–758 (2020)

Deep Learning for Prawn Farming
Forecasting and Anomaly Detection

Joel Janek Dabrowski[1]([⊠]), Ashfaqur Rahman[1], Andrew Hellicar[1],
Mashud Rana[1], and Stuart Arnold[2]

[1] Data61, CSIRO, Canberra, Australia
{Joel.Dabrowski,Ashfaqur.Rahman,Andrew.Hellicar,
Mdmashud.Rana}@data61.csiro.au
[2] Agriculture & Food, CSIRO, Canberra, Australia
Stuart.Arnold@csiro.au

Abstract. We present a decision support system for managing water quality in prawn ponds. The system uses various sources of data and deep learning models in a novel way to provide 24-h forecasting and anomaly detection of water quality parameters. It provides prawn farmers with tools to *proactively* avoid a poor growing environment, thereby optimising growth and reducing the risk of losing stock. This is a major shift for farmers who are forced to manage ponds by *reactively* correcting poor water quality conditions. To our knowledge, we are the first to apply Transformer as an anomaly detection model, and the first to apply anomaly detection in general to this aquaculture problem. Our technical contributions include adapting ForecastNet for multivariate data and adapting Transformer and the Attention model to incorporate weather forecast data into their decoders. We attain an average mean absolute percentage error of 12% for dissolved oxygen forecasts and we demonstrate two anomaly detection case studies. The system is successfully running in its second year of deployment on a commercial prawn farm.

Keywords: Forecasting · Anomaly detection · Dissolved oxygen prediction · Water quality

1 Introduction

The global trade of prawn (shrimp) is estimated at 28 billion US dollars per annum and this market continues to grow at a rate faster than any other aquaculture species [8]. The main challenge in prawn farming is to manage the highly variable water quality in prawn ponds to optimise prawn health and growth [2].

Dissolved oxygen (DO) is generally accepted as the most important water quality parameter in aquaculture [13]. Excessively low values in the diurnal cycle of DO (commonly referred to as a "DO crash") can cause the prawn to experience hypoxia, anoxia, or death. An entire crop (typically 8 to 12 tons of prawn) can be lost in a matter of hours [13].

© The Author(s), under exclusive license to Springer Nature Switzerland AG 2022
J. Gama et al. (Eds.): PAKDD 2022, LNAI 13282, pp. 29–41, 2022.
https://doi.org/10.1007/978-3-031-05981-0_3

Farmers typically monitor water quality parameters using sensors which, especially under continuous monitoring conditions, are be subject to high levels of biofouling and harsh conditions. Biofouling can reduce a sensor's accuracy or damage it (these can cost over US$25,000). Maintaining water quality sensors can thus be a challenging task.

This work is impactful as it provides forecasting and anomaly detection tools to assist a farmer in taking a *proactive* pond management approach rather than a *reactive* approach of correcting poor conditions. Forecasting gives an indication on how the temporal dynamics of a variable are expected to evolve into the future. Anomaly detection provides a means to identify any changes in the dynamics of a variable that are unusual, such as DO crashes and biofouling. With better control over water quality, animal stress can be reduced to improve growth, survivability, and consequently, production [2].

The novelty of this study is in the way that we provide a combination of forecasting and anomaly detection for decision support for this particular domain. Our novel technical contributions include (1) applying Transformer [17] for anomaly detection for the first time in the literature, (2) extending ForecastNet [4] into a multivariate model, and (3) proposing a novel approach to incorporate weather forecast data into the decoders of the Transformer [17] and Attention [1] models in a forecasting context. This work additionally provides insight into the aquaculture domain and its challenges.

Although this work is demonstrated with prawn farming, it is applicable to other domains such as other aquaculture farming industries (such as fish, molluscs, and other crustaceans), reservoir monitoring, lake monitoring, river monitoring, coastal water monitoring, and sewer monitoring.

2   Related Work

There are many challenges in precision agriculture, which have attracted various decision support tools, of which, water quality decision tools are the most common. These tools may provide decision support for determining optimal ranges of water quality [16] or provide sensing infrastructure [6]. Various physical or chemical models have also been developed, often for scenario analysis [7]. Various water quality forecasting approaches have been also developed [3]. Anomaly detection has been applied in water quality applications (e.g., [5]). However, to our knowledge, it has not been applied to aquaculture.

Our system makes use of deep learning models for forecasting and anomaly detection. Temporal deep learning models are usually based on recurrent neural networks (RNNs) and convolutional neural networks (CNNs). These include the sequence-to-sequence (seq2seq) model [15], the attention model [1], and Deep-AnT [11]. However, there are also models that are not based on RNNs or CNNs, such as the Transformer model [17] and ForecastNet [4]. To our knowledge, Transformer has not been applied for anomaly detection before. According to a recent review on anomaly detection [12], we consider "generic normality feature learning" anomaly detection approaches.

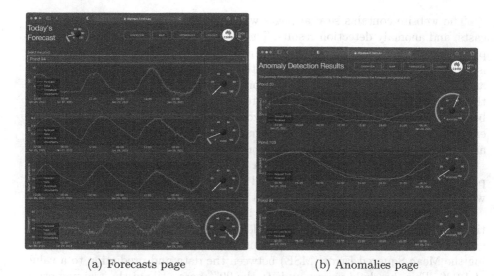

(a) Forecasts page   (b) Anomalies page

Fig. 1. Website application pages. The blue plots indicate the data and the green plots indicate the forecast/prediction. The grey filled region indicates forecast uncertainty and the red regions indicate threshold values. The gauges indicate the state of the water quality parameter (a) or the anomaly level (b). (Color figure online)

3   System Architecture and Overview

The decision support system architecture comprises 5 YSI EXO2 Multiparameter Sonde water quality sensors[1], an ATMOS41 weather station[2], the Senaps platform [10], and a server containing a website interface and the models.

Data are collected by the water quality and weather station sensors and are uploaded to the Senaps platform [10] via the local mobile (cellular) network. Senaps organises and stores this data. Weather forecasts are obtained from the Australian Bureau of Meteorology (BoM).

The server is a virtual box comprising two 2.60 GHz Intel® Xeon® CPUs and 8 GB memory. Every hour, the server collects the latest sensor data from Senaps and the latest air temperature forecasts from BoM. The data are pre-processed for each of the machine learning models and the models are run. This entire process takes approximately 6 min, which provides sufficient time to provide the hourly updates and the scope to increase the update frequency.

The website application is developed on the Plotly Dash platform[3] as it provides a seamless integration with the Python code used for model development and data processing. It also provides a website interface which is accessible from computers, tablets, and smart phones.

---

[1] https://www.ysi.com/exo2.

[2] https://www.metergroup.com/environment/products/atmos-41-weather-station/.

[3] https://plot.ly.

The website contains several pages which provide navigation, plots of forecasts, and anomaly detection results. The page displaying the forecasts for DO, pH, water temperature, and chlorophyll is illustrated in Fig. 1a. A pond "state" is indicated by a gauge to the right of the forecast plots. This guage provides an indication of how much and how long a forecast is expected to exceed a threshold. The gauge values are calculated according to the area of the region between the forecast curve and the horizontal threshold line, where the forecast exceeds the threshold. This area is normalised for a threshold of 180 min (3 h) and maximum signal values obtained from the dataset.

The gauge at the top left in Fig. 1a provides an *overall* estimation of the pond state. It is calculated as a weighted sum of the values for the other gauges where weighting is determined by expert knowledge from prawn farmers.

The page displaying the anomalies is illustrated in Fig. 1b. The plots provide the data and the anomaly detection model predictions of the data. The anomaly gauge on the right-hand side of the plot provides an anomaly score by thresholding the Mean Squared Error (MSE) between the data and prediction to a value of 1.026. This threshold corresponds to the 99% percentile of the training error produced by the anomaly detection model. That is, an MSE ≥ 1.026 saturates to an anomaly level of 100 on the gauge, which corresponds to a MSE that falls above the 99% of MSEs achieved in the training set.

4 Methods

4.1 Forecasting

With a diurnal cycle in DO, the forecasting models are designed to use 48-h of historical data to provide a 24-h, multi-step-ahead forecast. The inputs and outputs of the models are selected according to expert knowledge. The inputs include DO, pH, chlorophyll, water temperature, air temperature, and an air temperature forecast. The outputs of the models provide a forecast of DO, pH, water temperature, or chlorophyll. A different model is constructed and trained for each variable that is forecast. The forecasting models considered comprise, Transformer [17], Attention [1], and ForecastNet [4].

The Transformer and Attention models are both encoder-decoder based models. The historical input data sequence is typically provided to the model's encoder and the model's decoder outputs the forecast. In this study, the historical data are all 48-h sequences. The air temperature forecast data are however 24-h sequences, which poses a challenge in combining the input data.

To address this, we modify these models to incorporate the air temperature forecast data into the decoder rather than the encoder. This is also more natural as it provides a better separation of the data associated with the past and the future. In the Attention model, the air temperature forecasts are concatenated with the standard decoder inputs, which include the attention context and the previous decoder RNN output. In the Transformer model, the air temperature forecast is concatenated with the shifted output that is passed through the

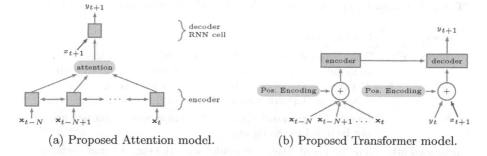

(a) Proposed Attention model.   (b) Proposed Transformer model.

Fig. 2. Model modifications to include weather forecasts in the decoder. Sensor data $x_{t-N:t}$ are provided to the encoder. Air temperature forecasts x_{t+1} are provided as inputs to the decoder. The decoder outputs the forecast y_{t+1}.

decoder's positional encoding. The proposed Attention and Transformer model architectures are illustrated in Fig. 2a and Fig. 2b respectively.

ForecastNet does not make use of the encoder-decoder architecture, but it was proposed as a univariate model [4]. The inputs to the model are in the form of a vector which contains a univariate data sequence. We extend ForecastNet into a multivariate model by concatenating the univariate sequences from all data sources into a single vector. This concatenated vector serves as the multivariate input to ForecastNet. As such, the fundamental architecture of ForecastNet is left unchanged. In the deployed model, the output layer comprises a Gaussian output to provide uncertainty in the outputs. For compatibility, a linear output is used when comparing ForecastNet with other models.

To demonstrate the effect of the model modifications proposed in this study, the standard form counterparts of the proposed models are also considered in the results. The standard forms of the models are configured with identical parameters (such as number of layers and hidden units), except the air temperature forecasts are not included in the models. The configurations for all the forecasting models are presented in Table 1. Parameters are selected by trial and error to optimise performance and memory requirements.

4.2 Anomaly Detection

We consider autoencoder and forecasting models for anomaly detection [9]. Autoencoder models make a prediction of their inputs through an encoder-decoder network, whereas forecasting models use the inputs to make a prediction into the future. Anomaly detection is achieved by comparing the predicted and observed data sequences according to the MSE. The predictive nature of the forecasting-based models encourages earlier detection of anomalies.

We consider several univariate anomaly detection models as described in Table 1. The RNN autoencoder (rnnAe) and a deep autoencoder (deepAe) provide autoencoder architectures. The remaining models are configured for

Table 1. Forecast (top) and anomaly detection (bottom) model configurations.

Model	Configuration
proposed FN	The proposed ForecastNet model [4] with hidden blocks configured with three fully connected layers, each containing 24 neurons. Outputs are linear outputs for testing, and a mixture model with a single Gaussian element in deployment
FN	Identical to *proposed FN*, but the temperature forecasts are not included in the inputs
proposed att	The proposed Attention model with the encoder and decoder configured with LSTMs, each with a hidden size of 48. Standard hyperbolic tangent alignment model is used in the attention mechanism
att	Standard Attention model [1] which is identical to *proposed att*, but the temperature forecasts are not included as inputs to the decoder
proposed trans	The proposed Transformer model is configured with a single layer in the encoder and decoder, with a model dimension of 16, 4 heads, and 24 hidden units
trans	Standard Transformer model [17] which is identical to *proposed trans*, but the temperature forecasts are not included as inputs to the decoder
rnnAe	Long Short Term Memory (LSTM) autoencoder [14] with a hidden size of 24 and sequence length of 96
deepAe	Fully connected deep autoencoder with input size of 96 and 3 hidden layers with sizes 56, 41, and 32 in the encoder and decoder
rnnAeFc	LSTM forecasting autoencoder [14] with a hidden size of 24 and sequence length of 96
seq2seq	Sequence to sequence model [15] with two LSTM layers each with a hidden size of 24 in the encoder and decoder
attention	Attention model [1] with a hidden size of 24 in the encoder and decoder, with a hyperbolic tangent alignment model in the attention mechanism
deepAnt	DeepAnT model [11] configured for 192 inputs and 96 outputs
transf.	Transformer model [17] comprising a single layer in the encoder and decoder, with a model dimension of 16, 4 heads, and 24 hidden units
forecastNet	ForecastNet model [4] with 192 inputs and 96 outputs. The hidden blocks contain two fully connected layers with 24 neurons in each layer

forecasting-based anomaly detection (see Sect. 2). Transformer is included as a new approach and is expected to perform well given its forecasting ability.

5   Dataset, Preprocessing, and Evaluation

The dataset used in this study was acquired by sensors placed in several ponds on a prominent Australian prawn farm. All sensors were configured to sample at 15-min intervals and were deployed in different ponds over two grow-out seasons covering the period of 2018-09-30 to 2020-04-19. YSI EXO2 Sensors were only in the ponds during grow-out periods (which differed between ponds). Some sensors were also removed for extended periods for servicing and repair.

The dataset used for testing the forecasting models is compiled from periods where all sensor data are present. The data are split into a training set and a test set, where both datasets contain data from all sensors that were available at the time. The training dataset contains 27 553 samples and the test set contains 1450 samples. To evaluate the forecasting models, the Mean Absolute Percentage Error (MAPE) and the Root Mean Squared Error (RMSE) metrics are used.

The anomaly detection methods are demonstrated with two case studies of known anomalous events that posed a significant risk to the farmer. The first is a DO crash event that occurred on 3 March 2020 for Sensor 1 and the second is a biofouling event that occurred over late December 2019 for Sensor 2.

During training and testing, a leave-one-out approach is adopted. The test set contains the data for the test sensor and the training dataset comprises data from all other sensors. The trained model is thus not conditioned on any form of normality or abnormality of the sensor used for testing. This also generalises the models across sensors. Additionally, the DO crash or biofouling regions in the dataset sequences are removed when these data are used in the training set.

The data are formatted such that the autoencoder anomaly detection model's inputs contain 24-h (96 sample) data sequences and the forecasting model's (both the anomaly detection and forecasting models) inputs contain 48-h (192 sample) data sequences and outputs produce 24-h (96 sample) forecasts. The result is a set of training and testing input/output sequence pairs.

Data preprocessing involves standardisation and imputation. Missing values are imputed using linear interpolation if the missing value sequence length is 8 or less samples (2 h), and removed if the missing sequence length is longer.

Both anomaly detection and forecasting models are trained using the Adam optimisation algorithm to optimise the MSE. A hyperparameter search is conducted for the learning rate over the range of 10^i, $i = [-6, \ldots, -2]$. Early stopping is used to address over-fitting.

6   Forecasting Results

The standard and proposed ForecastNet, Transformer, and Attention models are compared with each other using the forecasting dataset. The average error results for all the models are provided in Table 2 and box-whisker plots are provided in Fig. 3. Additionally, the total time for a model to process the 1450 test samples is provided in the last column of Table 2.

Comparing the difference between the standard and the proposed versions of the models, it is clear that the forecast error and its variation are consistently

Table 2. Average errors over the test dataset for the standard and proposed Forecast-Net (FN), Transformer (trans) and Attention (att) models. Proposed model results are presented in bold text.

Model	MAPE	MAPE std dev.	RMSE	RMSE std dev.	run time (s)
proposed FN	**12.259**	**5.181**	**1.665**	**0.821**	**0.84**
FN	13.248	5.665	1.722	0.842	0.83
proposed transf	**11.844**	**4.954**	**1.513**	**0.768**	**55.57**
transf	13.593	6.125	1.731	0.884	59.82
proposed att	**12.059**	**4.430**	**1.595**	**0.828**	**22.60**
att	13.663	5.786	1.713	0.821	22.24

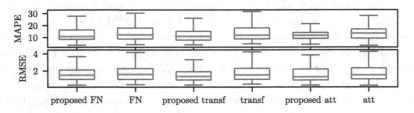

Fig. 3. Test dataset forecasting error box plots for the standard and proposed Fore-castNet (FN), Transformer (trans) and Attention (att) models.

reduced in the proposed models. Comparing the ForecastNet, Transformer, and Attention models, the error values do not differ significantly. Furthermore, the overlapping of box plots in Fig. 3 indicate that there is little statistical significance between these differences in error. ForecastNet is however significantly faster than the other models as indicated in Table 2. For example, ForecastNet is 67 times faster than the Transformer model. ForecastNet is thus deployed.

7   Anomaly Detection Case Studies

7.1   DO Crash Case Study

The anomaly detection MSE results for Sensor 1 over the 2019/2020 season are plotted in Fig. 4. As discussed in Sect. 3, the maximum anomaly detection level of 100 corresponds with a MSE of 1.026. Suppose a farmer chooses to investigate an anomaly when the anomaly detection level rises above a value of 70. This value corresponds to a MSE threshold of 0.7. A plot of the DO crash, anomaly detection MSE, and threshold value are presented in Fig. 5. Additionally, the number of hours warning that the anomaly detection provides for the DO crash is also indicated. The number of hours is measured from where the MSE exceeds the threshold to where the DO reaches its minimum.

All models detect the DO crash on 03-03-2020. The RNN autoencoder and deep autoencoder models detect the DO crash and have relatively few anomaly

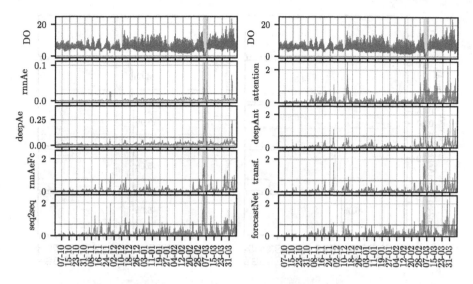

Fig. 4. Anomaly detection results for Sensor 1. The DO data (mg/L) are plotted in the top panels in blue, and the MSE for each model are plotted in the remaining panels in green. The purple horizontal lines plot the 0.7 threshold for anomaly detection. The grey region indicates the approximate period over which the DO crash occurred. Horizontal axes indicate the day and month. (Color figure online)

detections. However, these models detect the DO crash only after it occurs. As these are autoencoder models, they do not have the predictive nature that the forecasting anomaly detection models have.

The seq2seq model provides the earliest detection, however it has a relatively noisy MSE signal suggesting that it is more likely to make false positive detections. The Attention model provides similar noisy characteristics. The deepAnT and Transformer models provide strong detections of the DO crash with relatively low noise. ForecastNet provides early detection, however the detection weak. As ForecastNet is time-variant, it may be able to adapt better to the anomalous events resulting in less sensitivity to these anomalies.

If the farmer were to choose a lower threshold, earlier detections could be achieved at cost of more false positives. The choice of a threshold value is left to the discretion of the farmer as it will depend on their particular management practices, farm size, and farm resources.

7.2   Sensor Biofouling Case Study

The anomaly detection MSE results for the deployment period for the 2019/2020 season for Sensor 2 is plotted in Fig. 7. We were notified on 20-01-2020 by the farm that the sensor had (unintentionally) not been maintained. It is not clear when the sensor was last serviced. From anomaly detection model results, it appears the biofouling began to affect the sensor from late December 2020. The

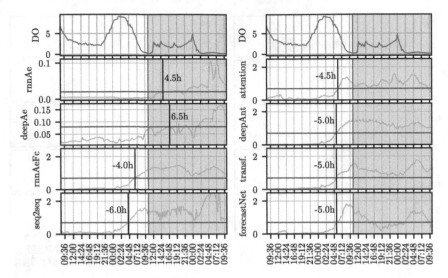

Fig. 5. DO crash for sensor 1. The DO data (mg/L) are plotted in the top panels in blue, and the MSE for each model are plotted in the remaining panels in green. The purple horizontal lines plot the 0.7 threshold for anomaly detection. The red vertical lines indicate the time at which the DO reaches its lowest point. Black vertical lines and labels indicate the time (hours) where the anomaly detection threshold is crossed. Horizontal axes indicate time. (Color figure online)

Fig. 6. Sensor 2 with severe biofouling on the left and a new sensor on the right. Notice the micro-organisms and barnacles growing on the sensor heads.

sensor was removed from the pond, cleaned, tested, and returned to the pond on 12-02-2020. A photograph of the bio-fouled sensor is provided in Fig. 6.

As indicated in Fig. 7, all anomaly detection models detect the biofouling event. Furthermore, the model's anomaly detection levels remain high throughout the duration of the biofouling event. The detections made by Transformer are strong and the noise levels in the MSE are low. The Transformer model is thus deployed in the application.

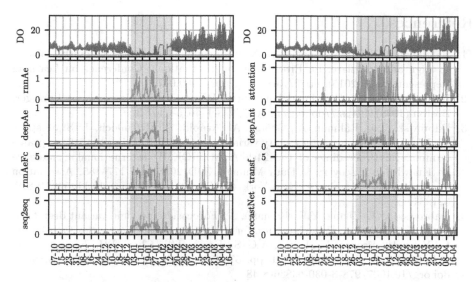

Fig. 7. Anomaly detection results for Sensor 2. The DO data (mg/L) are plotted in the top panels in blue, and the MSE for each model are plotted in the remaining panels in green. The purple horizontal lines plot the 0.7 threshold for anomaly detection. The grey region indicates the approximate period over which the sensor experienced biofouling. Horizontal axes indicate the day and month. (Color figure online)

8 Summary and Conclusion

We present a system which provides decision support through forecasting and anomaly detection. The forecasting models provide accurate forecasts with an average MAPE of 12% over a 24-h (96 step-ahead) forecast. We demonstrate that our novel approach to including weather forecast data into the models reduces both the forecast error and its variation. Two anomaly detection cases are presented demonstrating the ability to predict and identify a DO crash and sensor biofouling. The DO crash is detected 5 h before the actual event.

Our system has been deployed in various phases since 2019 and we have received positive feedback from the farm. In future work, ensembles of models may provide more accurate forecasts and more refined anomaly detection. Furthermore, in order to capture more anomalies for further validation, more sensors will need to be deployed over a longer period of time.

Acknowledgement. Thanks to Pacific Reef Fisheries for providing us with the access to their farm to conduct this study and also for assisting us in deploying and maintaining sensors. This work was supported by the CSIRO Digiscape Future Science Platform.

References

1. Bahdanau, D., Cho, K., Bengio, Y.: Neural machine translation by jointly learning to align and translate. In: Proceedings of International Conference on Learning Representations (2015). http://arxiv.org/abs/1409.0473
2. Boyd, C.E., Tucker, C.S.: Pond Aquaculture Water Quality Management. Springer, Boston (1998). https://doi.org/10.1007/978-1-4615-5407-3
3. Dabrowski, J.J., Rahman, A., George, A., Arnold, S., McCulloch, J.: State space models for forecasting water quality variables: an application in aquaculture prawn farming. In: Proceedings of the 24th ACM SIGKDD International Conference on Knowledge Discovery & Data Mining, KDD 2018, pp. 177–185. Association for Computing Machinery, New York (2018). https://doi.org/10.1145/3219819.3219841
4. Dabrowski, J.J., Zhang, Y.F., Rahman, A.: ForecastNet: a time-variant deep feed-forward neural network architecture for multi-step-ahead time-series forecasting. In: Yang, H., Pasupa, K., Leung, A.C.-S., Kwok, J.T., Chan, J.H., King, I. (eds.) ICONIP 2020. LNCS, vol. 12534, pp. 579–591. Springer, Cham (2020). https://doi.org/10.1007/978-3-030-63836-8_48
5. Dogo, E.M., Nwulu, N.I., Twala, B., Aigbavboa, C.: A survey of machine learning methods applied to anomaly detection on drinking-water quality data. Urban Water J. 16(3), 235–248 (2019). https://doi.org/10.1080/1573062X.2019.1637002
6. Encinas, C., Ruiz, E., Cortez, J., Espinoza, A.: Design and implementation of a distributed IoT system for the monitoring of water quality in aquaculture. In: 2017 Wireless Telecommunications Symposium (WTS), pp. 1–7 (2017). https://doi.org/10.1109/WTS.2017.7943540
7. Ernst, D.H., Bolte, J.P., Nath, S.S.: Aquafarm: simulation and decision support for aquaculture facility design and management planning. Aquacultural Eng. 23(1), 121–179 (2000). https://www.sciencedirect.com/science/article/pii/S0144860900000455
8. Food and Agriculture Organization of the United Nations: Towards sustainability in the shrimp industry, February 2020. http://www.fao.org/in-action/globefish/market-reports/resource-detail/en/c/1261310/
9. Gamboa, J.C.B.: Deep learning for time-series analysis. arXiv preprint arXiv:1701.01887 (2017)
10. Mac Coombea, P.N., Pasanena, J., Petersa, C., Sharmana, C., Taylora, P.: Senaps: a platform for integrating time-series with modelling systems. In: Proceedings of the 22nd International Congress on Modelling and Simulation (MODSIM 2017), Hobart, Tasmania, Australia, pp. 438–444 (2017)
11. Munir, M., Siddiqui, S.A., Dengel, A., Ahmed, S.: DeepAnT: a deep learning approach for unsupervised anomaly detection in time series. IEEE Access 7, 1991–2005 (2019). https://doi.org/10.1109/ACCESS.2018.2886457
12. Pang, G., Shen, C., Cao, L., Hengel, A.V.D.: Deep learning for anomaly detection: a review. ACM Comput. Surv. 54(2), 1–38 (2021). https://doi.org/10.1145/3439950
13. Robertson, C. (ed.): Australian Prawn Farming Manual: Health Management for Profit. Queensland Department of Primary Industries and Fisheries (QDPI&F) (2006)
14. Srivastava, N., Mansimov, E., Salakhudinov, R.: Unsupervised learning of video representations using LSTMS. In: International Conference on Machine Learning, pp. 843–852. PMLR (2015)

15. Sutskever, I., Vinyals, O., Le, Q.V.: Sequence to sequence learning with neural networks. In: Ghahramani, Z., Welling, M., Cortes, C., Lawrence, N.D., Weinberger, K.Q. (eds.) Advances in Neural Information Processing Systems 27, pp. 3104–3112. Curran Associates, Inc. (2014). http://papers.nips.cc/paper/5346-sequence-to-sequence-learning-with-neural-networks.pdf
16. Tobing, F.A.T., Dzulhaq, M.I., Overbeek, M.V., Setiawan, E.A.: Design of decision support system with C4.5 algorithm methods in determining results vaname shrimp cultivation. In: 2019 5th International Conference on New Media Studies (CONMEDIA), pp. 195–200 (2019). https://doi.org/10.1109/CONMEDIA46929.2019.8981848
17. Vaswani, A., et al.: Attention is all you need. In: Guyon, I., et al. (eds.) Advances in Neural Information Processing Systems 30, pp. 5998–6008. Curran Associates, Inc. (2017). http://papers.nips.cc/paper/7181-attention-is-all-you-need.pdf

Input Enhanced Logarithmic Factorization Network for CTR Prediction

Xianzhuang Li[1], Zhen Wang[2], Xuesong Wu[3], Bo Yuan[1], and Xueqian Wang[1(✉)]

[1] Shenzhen International Graduate School, Tsinghua University, Shenzhen, China
lixz19@mails.tsinghua.edu.cn, yuanbo@ieee.org, wang.xq@sz.tsinghua.edu.cn
[2] School of Computer Science, The University of Sydney, Sydney, Australia
zwan4121@uni.sydney.edu.au
[3] Internet Business Department, Xiaomi, Beijing, China
wuxuesong@xiaomi.com

Abstract. Factorization-based methods, which can automatically model second-order or higher-order cross features, have been the benchmark models for click-through rate (CTR) prediction. In general, they enumerate all cross features with a predetermined order and filter out useless interactions through model training. However, two significant challenges remain. First, a maximum order for feature interactions needs to be defined in advance, imposing a trade-off between computational cost and the expression ability of models; Second, enumerating all feature interactions may introduce unwanted noise. In this work, we propose a novel model called *Input Enhanced Logarithmic Factorization Network* (ILFN), which can effectively learn arbitrary-order feature interactions and identify useful cross features. More importantly, ILFN can take full advantage of the original feature distribution in a discriminative way.

The core of ILFN is the *Input Enhanced Component* (IEC), which can represent the impact of each feature in a feature interaction as a trainable coefficient to learn arbitrary-order cross features. Moreover, IEC can reduce the demand for logarithmic neurons by exploiting the essential raw information and does not need to incorporate the deep neural network (DNN) to model high-order interactions. Therefore, ILFN is more effective and efficient and can converge to satisfactory results faster. Extensive experiments on four real-world datasets demonstrate that our ILFN model can outperform start-of-the-art methods. The effectiveness of each proposed component is also verified by hyper-parameter and ablation studies.

Keywords: Factorization machines · Logarithmic neural networks · Recommender systems · Feature interactions

© The Author(s), under exclusive license to Springer Nature Switzerland AG 2022
J. Gama et al. (Eds.): PAKDD 2022, LNAI 13282, pp. 42–54, 2022.
https://doi.org/10.1007/978-3-031-05981-0_4

1   Introduction

Recommender systems are essential to many Internet companies. Among many real-world recommender systems such as news ranking, financial analysis and online advertising, Click-through rate (CTR) prediction plays a central role. The key of CTR prediction is to estimate the probability of a user clicking the recommended item, which has a direct impact on the final revenue [3].

Features play a vital role in the success of many CTR prediction tasks, and how to effectively represent features is one of the keys to obtaining optimal results. Although handcrafted feature interactions [5] are effective, data scientists need to spend great efforts on the transformation of original features to generate the best prediction systems. What is even worse, for large-scale datasets, such as the popular benchmark dataset *Criteo* with more than two million features, it is impractical to work on these features manually. The *Factorization Machine* (FM) [20] model was proposed to search for feasible transformations of the raw features. Due to its robust processing capabilities for sparse features and good scalability, the FM model has been widely adopted and has been proved effective for various tasks [20,21]. However, limited by its polynomial regression part, it only leverages first- and second-order feature interactions. In contrast, the higher-order cross features may play a more critical role in boosting the accuracy of CTR prediction [12,18].

With the success of deep neural networks (DNNs) in many fields such as natural language processing and computer vision, using deep learning for feature interactions has been a research trend in CTR prediction. Various DNN-based FM methods [3,6,13,18] have been proposed in recent years to model both low- and high-order cross features. For those models, in general, a maximum order is predefined, and then all combinatorial features within this order are enumerated while irrelevant interactions are filtered out by training. However, two significant challenges remain in the above methods. First, predefining an order, which is generally small (such as 3), may restrict the exploitation of higher-order cross features. Second, since not all useless feature combinations can be successfully removed, enumerating all cross features may introduce harmful noise and complicate the training. To this end, the logarithmic neural network (LNN) [15] was introduced to learn arbitrary-order feature interactions [4].

Unfortunately, LNN cannot handle negative or zero vectors. As a result, the AFN model [4], learning cross features via LNN, must convert negative embedding vectors to their absolute values and add noise to zero embedding vectors, leading to the lack and distortion of the initial embedding information. In this paper, we propose a novel model named *Input Enhanced Logarithmic Factorization Network* (ILFN), which can effectively and adaptively learn arbitrary-order cross features and their weights from data. The core of ILFN is the *Input Enhanced Component* (IEC), which can convert the impact of each feature in a feature interaction into a learnable coefficient with the following benefits: First, IEC can build arbitrary-order feature interactions without distortion of the initial distribution. Second, we use the attention mechanism to integrate the arbitrary-order cross features and natural embedding features, which shows the importance of these features. Besides, it reduces the need for logarithmic neurons and does not need to ensemble the deep neural network (DNN) to mine high-order feature interactions. Compared to existing methods, ILFN is more expressive yet remains efficient and simple. Third, we perform batch normalization (BN) on the output of logarithmic transformation and exponential transformation to eliminate the influence of embedding vectors close to 0. To summarize, we make the following contributions:

- We propose a novel model called *Input Enhanced Logarithmic Factorization Network* (ILFN), which can effectively learn arbitrary-order feature interactions and identify useful cross features.
- The core of ILFN is the *Input Enhanced Component* (IEC), which can make full use of the original data distribution to model feature combinations efficiently and discriminatively without destroying the initial data distribution like LNN.
- IEC reduces the demand for logarithmic neurons and does not need to ensemble DNNs, resulting in fast convergence speed.
- Extensive experiments on four real-world datasets demonstrate that ILFN outperforms several state-of-the-art methods consistently. An ablation study verifies the effectiveness of each proposed component.
- We provide a case study and share analysis of ILFN to present the impact of vital components (such as IEC and LNN) on performance in the field of CTR prediction.

2   Related Work

Logistic Regression (**LR**[1]) [9] is a linear model that only models the linear combination of features for CTR prediction. By contrast, *Factorization Machine* (**FM**) [20], which works well on large sparse data, projects features into low-dimensional vectors and learns second-order interactions by inner products. It has received significant research interest from the community due to its excellent performance on sparse data sets, low time complexity, and low memory storage. Afterwards, different variants of factorization networks have been proposed. As an extension of the standard FM, *Field-aware Factorization Machine* **FFM** [16] introduces field-aware latent vectors to model the interactions between features from different fields. **HOFM** [21] can model high-order cross features, but one major downside is that it builds all cross features, including both useful and useless interactions.

With the great success of deep neural networks (DNNs), many researchers have proposed various DNN-based methods for CTR prediction. FNN [23] uses FM to pre-train the embedding data. **PNN** [11] explores the high-order cross features via a product layer between the embedding layer and the DNN layers. **NFM** [13] introduces neural networks into the FM model to enhance its ability

---

[1] The models in bold are baselines in the experimental part.

to leverage high-order feature interactions. **CrossNet** [18] models cross features by the outer product. **Deep&Cross** [18] integrates CrossNet with DNNs. **Wide&Deep** [3] combines LR with DNNs, where LR models the linear feature combinations while DNNs learn high-order non-linear cross features. **DeepFM** [6] combines FM with DNNs, which uses FM to replace LR in Wide&Deep. **xDeepFM** [12], an improved version of DeepFM, can explicitly model higher-order feature interactions by introducing the **CIN** component.

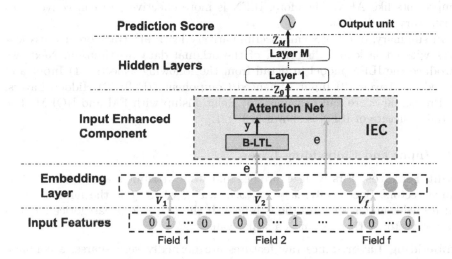

Fig. 1. The network structure of the ILFN model

However, these DNN-based models generally need to enumerate all possible cross features, introducing noise and complicating the training. To alleviate this challenge, **AFM** [14] adopts attention mechanism [1] to filter out the noise by assigning different coefficients to diverse feature interactions. Nevertheless, it can only handle second-order cross features. To this end, **AFN** [4] was proposed to learn arbitrary-order cross features with the logarithmic transformation layer, which results in the lack and distortion of the initial embedding information. In this work, we propose an effective model named *Input Enhanced Logarithmic Factorization Network* (ILFN), which can effectively and adaptively learn arbitrary-order cross features and their weights for CTR prediction. More importantly, it takes full advantage of the original information with lower time complexity.

3 Our Approach

In this section, we introduce our proposed *Input Enhanced Logarithmic Factorization Network* (ILFN). To effectively learn arbitrary-order feature interactions and take full advantage of the original information, we propose the *Input*

Enhanced Component (IEC), the core of ILFN. The *Input Enhanced Component* contains two parts: the BN-Logarithmic Transformation Layer (B-LTL) and Attention Layer. The former can represent the impact of each feature in a feature interaction as a trainable coefficient to learn arbitrary-order cross features. The latter can integrate outputs of the former and embedding layer, which handles data distortion and shows the importance of various features. In addition, the Attention Layer can fit classic DNN to learn high-order interactions by changing weights of different vectors, so ILFN no need to extra integrate DNN components like AFN. Therefore, ILFN is more effective and can converge to satisfactory results faster.

In summary, our ILFN can discriminatively learn arbitrary-order cross features without lack and distortion of the original data distribution. Next, we introduce the ILFN model in detail from the following aspects: (1) Input and Embedding Layer; (2) Input Enhanced Component (IEC); (3) Hidden Layers; (4) Prediction Score and Learning; (5) Relationship with FM and HOFM. The overall structure of ILFN is shown in Fig. 1.

3.1   Input and Embedding Layer

Input Features. Firstly, user profiles and item attributes are represented as sparse vectors. Specifically: $x = [x_1, x_2, ..., x_f]$, where f is the number of all feature fields and x_i denotes the i-th field. If the i-th field is categorical, x_i is a one-hot vector. Otherwise, it is a scalar value.

Embedding Layer. Since raw features are normally very sparse, a common practice is to project them into low-dimensional spaces by a classic embedding layer. The output of the embedding layer is a wide concatenated field vector: $e = [e_1, e_2, ..., e_f]$, where f denotes the number of all feature fields and $e_i \in \mathbb{R}^k$ is the embedding of the i-th field, and k denotes the dimension of field embedding. By introducing the embedding layer, instances of various feature lengths can be transformed into unified embedding vectors with the same size $f \times k$.

3.2   Input Enhanced Component

The Input Enhanced Component (IEC) includes two parts: the BN-Logarithmic Transformation Layer (B-LTL) and Attention Layer.

BN-Logarithmic Transformation Layer. Figure 2 shows the structure of B-LTL. It contains multiple logarithmic neurons, which can adaptively learn arbitrary-order cross features. Due to the characteristics of logarithmic neurons, two positive-value operations are introduced in the preprocessing part: First, all embedding values should be converted to their absolute values. Second, a small positive value (such as 1e−7) is added to each zero embedding vector.

In B-LTL, the output of the i-th logarithmic neuron is:

$$y_i = exp(\sum_j^f w_{ij} ln e_j) = \prod_j^f e_j{}^{w_{ij}} = e_1{}^{w_{i1}} \odot e_2{}^{w_{i2}} \odot ... \odot e_f{}^{w_{if}} \qquad (1)$$

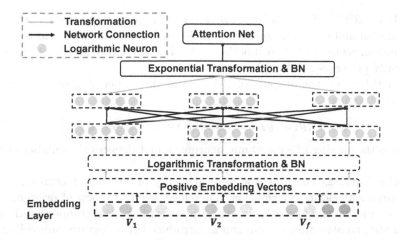

Fig. 2. The structure of the BN-Logarithmic Transformation Layer (B-LTL)

where w_{ij} denotes the power term of the i-th neuron on the j-th field, and $\odot$ denotes the bit-wise production operation. Note that the output $\boldsymbol{y}_i$ of the i-th logarithmic neuron can represent arbitrary-order cross features, and when w_i is adjusted to 0, the i-th feature will be filtered out, so it can learn useful combined features. For example, with w_{i1} and w_{i2} set to 1 and the rest set to 0 in Eq. 1, we have: $\boldsymbol{y}_i = \boldsymbol{e}_1 \odot \boldsymbol{e}_2$. This indicates that the output $\boldsymbol{y}_i$ of the i-th logarithmic neuron can represent a second-order cross-feature for the first two raw feature fields. Therefore, we can use multiple logarithmic neurons to obtain various feature combinations in any order as the output of this layer.

In addition, we perform BN on the output of logarithmic transformation and exponential transformation. We do this based on the following consideration. In real-world datasets, the feature embedding vectors are usually initialized close to zero. After logarithmic transformation, embedding vectors often involve minimal negative values, detrimental to optimizing parameters in successive layers. Since normalization can scale and convert the output to a normalized value, it is imperative to the training process of ILFN.

Attention Layer. To unify the representation, we rename vectors of the embedding layer as follow:

$$[\boldsymbol{y}_{n+1}, \boldsymbol{y}_{n+2}, ..., \boldsymbol{y}_{n+f}] = [\boldsymbol{e}_1, \boldsymbol{e}_2, ..., \boldsymbol{e}_f] \tag{2}$$

Then we use the attention mechanism to integrate the outputs of B-LTL and the embedding layer to show different importants of various features. The weights of different vectors can be defined as:

$$a_i' = \langle h, Relu(\boldsymbol{W}_a \boldsymbol{y}_i + b_a) \rangle \tag{3a}$$

$$a_i = \frac{exp(a')}{\sum_{i=1}^{n+f}(exp(a_i'))} \tag{3b}$$

where $\boldsymbol{W}_a \in \mathbb{R}^{(n+f) \times e}$, $h \in \mathbb{R}^e$, $b_a \in \mathbb{R}^e$ are model parameters, and e is the number of hidden units in attention network. Softmax function is used to normalize the attention score, and Relu function is used as an activation function, which empirically performs good performance.

After this layer, the outputs of B-LTL and the embedding vectors are concatenated together:

$$z_0 = [a_1 \boldsymbol{y}_1, a_2 \boldsymbol{y}_2, ..., a_n \boldsymbol{y}_n, a_{n+1} \boldsymbol{y}_{n+1}, ..., a_{n+f} \boldsymbol{y}_{n+f}] \tag{4}$$

where n is the number of logarithmic neurons, and f denotes the number of total fields.

In this layer, z_0 contains not only arbitrary-order cross features but also the original embedding information. Therefore, the distortion of original information caused by the logarithmic neural network has been compensated. More importantly, an attention mechanism is introduced to integrate embedding features and processed arbitrary-order cross features, which learns the importance of different parts and reduces the number of logarithmic neurons required for the model to achieve optimal results.

3.3 Hidden Layers

We stack several fully connected hidden layers upon the *Input Enhanced Component* to combine the formed feature interactions. So, z_0 is fed into M hidden layers:

$$z_1 = \sigma(\boldsymbol{W}_1 z_0 + \boldsymbol{b}_1)$$

$$...... \tag{5}$$

$$z_M = \sigma(\boldsymbol{W}_M z_{M-1} + \boldsymbol{b}_M)$$

where σ is the activation function, W_M denotes the weight matrix while b_M is the bias vector of the M-th layer. In addition, we also perform BN on the output of hidden layers. Since we use a multilayer neural network after IEC, performing BN on the hidden layer's output helps alleviate the covariance shift problem, which empirically leads to faster convergence and better model performance.

3.4 Prediction Score and Learning

Finally, we get the prediction results from the hidden layers: $\hat{y} = z_M$, where z_M is the output of the hidden layer.

For binary classification tasks, a popular objective function is the Logarithmic Loss, defined as:

$$Logloss = -\frac{1}{N} \sum_{j=1}^{N} (y_j log(\sigma(\hat{y_j})) + (1 - y_j)log(1 - \sigma(\hat{y_j}))) \tag{6}$$

where $y_j \in \{0, 1\}$ is the true value of the j-th instance while $\sigma(\hat{y_j}) \in (0, 1)$ is the predicted value (here σ is the *sigmoid* function) and N is the number of all training instances.

3.5   Relationship with FM and HOFM

Suppose we remove IEC and only use the logarithmic neurons to produce second-order cross features and approximate a sum function for the hidden layers. In that case, our ILFN model will be downgraded to the FM model. Similarly, if the logarithmic neurons are set to deliver all feature interactions within a maximum order, and the hidden layer only simulates a summation function, our ILFN model will be functionally identical to HOFM.

4   Experiments

4.1   Experimental Settings

Datasets. We conducted experiments on four public benchmark datasets: *Criteo*[2], *Avazu*[3], *Movielens*[4] and *Frappe*[5]. We split them randomly by 8:1:1 for training, validation and test, respectively. **Criteo Dataset** is a famous industry benchmarking dataset containing the click records of 45 million users, which includes 13 numerical feature fields and 26 categorical feature fields. **Avazu Dataset** has 40 million mobile user behaviours, 24 feature fields containing the user features and item attributes. **Movielens Dataset** consists of two million users' records of movies. We focused on the personalized tag recommendation by converting each tag record (user ID, movie ID, tag) into a feature vector. **Frappe Dataset** contains the application usage logs of more than 90,000 users in different contexts.

Evaluation Metrics. We adopted two evaluation metrics in our experiments: **AUC** (Area Under ROC) and **Logloss** (cross entropy) [10]. Note that **a slight improvement in AUC or a decrease in Logloss at 0.001-level is conventionally regarded as being significant** for CTR prediction because it will bring a large increase in a company's revenue if the company has a large user base. **The proposal of some crucial models in CTR prediction tasks only increases the AUC by a few thousandths** [3,4,11,14,17,19,22].

Baselines. We compared our ILFN with four classes of the existing methods: (i) first-order models that just linearly sum up raw input features; (ii) second-order approaches that model first- and second-order feature interactions; (iii) high-order methods that can capture high-order cross features; (iv) ensemble models that involve DNNs as the counterpart. These models are shown in Table 1.

Implementation Details. We implemented our methods using Tensorflow[6]. We used Adam with a learning rate of 0.001 and a mini-batch size of 4096 in all models. We also fixed the network structure (i.e., three layers, 400-400-400) in

---

[2] http://labs.criteo.com/2014/02/.
[3] https://www.kaggle.com/c/avazu-ctr-prediction.
[4] https://grouplens.org/datasets/movielens/.
[5] http://baltrunas.info/research-menu/frappe.
[6] https://www.tensorflow.org/.

all models containing DNNs. The embedding size was 10 for all approaches. The default numbers of logarithmic neurons were 1000, 800, 600, and 500 for *Criteo*, *Avazu*, *Movielens*, and *Frappe* datasets. To avoid overfitting, we also performed early-stopping according to the AUC on the validation set. The maximum order in HOFM was set to 3. All the other hyper-parameters were tuned according to the validation set to achieve the best performance. We ran the CTR prediction experiments three times for each model and reported the average value as the final result.

4.2 Performance Comparison

We compared our ILFN model with several classical models featuring linear and high-order feature interactions. The results are shown in Table 1, from which we have three critical observations:

First, feature interactions play an essential role in the improvement of the CTR model's performance. For example, LR performed worse than other models with feature interactions. Based on the result that high-order models had larger AUC and smaller Logloss than first- and second-order models, higher-order cross features play a more critical role in CTR prediction tasks. Besides, DNNs can be employed to enhance the performance of cross feature-based models; those ensembled methods usually performed better than others.

Second, learning arbitrary-order cross features is more effective than learning fixed-order feature interactions. It can be verified by the fact that ILFN performed better than methods with the preset maximum orders on *Criteo*, *Avazu*,

Table 1. The performance comparison of different CTR models

Model class	Model	Criteo		Avazu		Movielens		Frappe	
		AUC	Logloss	AUC	Logloss	AUC	Logloss	AUC	Logloss
First-order	LR [9]	0.7859	0.4635	0.7313	0.4066	0.9215	0.3080	0.9330	0.2858
Second-order	FM [20]	0.7933	0.4573	0.7496	0.3740	0.9388	0.2798	0.9641	0.2144
	FFM [16]	0.8045	0.4478	0.7501	0.3733	0.9427	0.2961	0.9687	0.2132
	AFM [14]	0.7954	0.4554	0.7454	0.3765	0.9299	0.2833	0.9640	0.2292
Ensembled	Wide&Deep [3]	0.8062	0.4453	0.7629	0.3744	0.9381	0.3309	0.9728	0.2038
	Deep&Cross [18]	0.8059	0.4463	0.7550	0.3721	0.9421	0.2789	0.9402	0.2809
	DCN V2 [2]	0.8072	0.4443	0.7557	0.3719	0.9481	0.2736	0.9772	0.1805
	DeepFM [6]	0.8026	0.4509	0.7535	0.3741	0.9424	0.3130	0.9720	0.2108
	AFN+ [4]	0.8071	0.4450	0.7555	0.3717	0.9500	0.2585	0.9783	0.1762
	xDeepFM [12]	0.8069	0.4443	0.7535	0.3738	0.9448	0.2717	0.9738	0.2098
	FiBiNET [19]	0.8067	0.4445	0.7540	0.3737	**0.9513***	**0.2564***	0.9782	0.1761
High-order	HOFM [21]	0.7960	0.4550	0.7517	0.3756	0.9409	0.3090	0.9711	0.2138
	CrossNet [18]	0.7915	0.4585	0.7498	0.3756	0.9323	0.2931	0.9395	0.2838
	PNN [11]	0.8026	0.4509	0.7526	0.3737	0.9471	0.2789	0.9735	0.2011
	NFM [13]	0.7968	0.4536	0.7531	0.3762	0.9439	0.3003	0.9725	0.2080
	AFN [4]	0.8058	0.4457	0.7512	0.3731	0.9477	0.2753	0.9759	0.1784
	AutoInt [17]	0.8039	0.4476	0.7530	0.3761	0.9492	0.2611	0.9698	0.2354
	CIN [12]	0.8040	0.4473	0.7532	0.3757	0.9494	0.2601	0.9707	0.2339
	ILFN	**0.8085***	**0.4434***	**0.7573***	**0.3701***	0.9509	0.2571	**0.9792***	**0.1749***

and *Frappe* datasets. ILFN achieved the second-best performance on *Movielens* dataset. *Movielens* only contains three feature fields, and the benefit of ILFN to mine higher-order feature interactions may be marginal. As shown in Table 1, more straightforward methods had a better prediction effect, we conjecture that the predictions on *Movielens* dataset rely more on lower-order cross features, and the advantages of ILFN are thus restricted.

At last, the introduction of the *Input Enhanced Component* (IEC) can effectively and efficiently learn arbitrary-order cross features and identify useful cross features. Beyond that, it can fully use the original data distribution to build arbitrary-order combinations without destroying the initial data distribution. More importantly, ILFN can use IEC to model high-order feature combinations in a discriminative way with a faster convergence speed. Therefore, as a non-ensembled model, ILFN had better performance than those ensemble methods on *Criteo*, *Avazu*, and *Frappe* datasets.

4.3   Hyper-parameter Study

We then studied the impact of hyper-parameters on ILFN, including (1) the number of logarithmic neurons; (2) activation functions; (3) the depth of hidden layers; (4) the number of hidden neurons. Given that the results on the four datasets were similar, we only show the results on the *Criteo* dataset here.

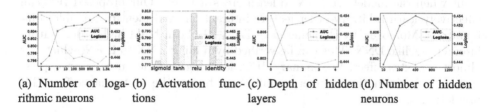

(a) Number of loga- (b) Activation func- (c) Depth of hidden (d) Number of hidden
rithmic neurons   tions   layers   neurons

Fig. 3. Impact of network hyper-parameters on the performance of ILFN on the Criteo dataset

Number Logarithmic Neurons. Figure 3(a) demonstrates the impact of the number of logarithmic neurons on *Criteo* dataset. The appropriate number of logarithmic neurons on *Criteo* is 1000, which is smaller than 1500 in AFN. Thus, preserving original data information makes the model use fewer logarithmic neurons to achieve optimal performance. This indicates that the introduction of logarithmic neural networks has led to the loss of some crucial features, and the model needs more logarithmic neurons to make up for it.

Activation Functions. We compared the performance of different activation functions in ILFN, as shown in Eq. 5. According to Fig. 3(b), *ReLU* is indeed the most appropriate for neurons in ILFN.

Depth of Hidden Layers. Figure 3(c) demonstrates the effect of the depth of hidden layers. The performance of ILFN increased when the depth of the network was varied from 0 to 3. However, when the number of hidden layers was greater than 3, the model's performance degraded.

Number of Hidden Neurons. Figure 3(d) provides the results over different numbers of hidden neurons in hidden layers. The performance of our ILFN increased when the number of neurons was increased from 10 to 400. When the parameter size of the hidden layer exceeds 400, the performance begins to decline, which is caused by overfitting.

4.4 Ablation Study

ILFN strategically integrates the BN-LTL and the Attention Layer into a unified end-to-end model for learning arbitrary-order cross features. Unlike AFN, ILFN can model high-order feature interactions by IEC; therefore, we do not involve DNN in ILFN. To gain deep insights into ILFN, we conducted a series of ablation experiments. Table 2 presents the performance of ILFN *w.r.t.* specific components. From which we can reach the following conclusions: First, introducing the logarithmic neurons for adaptively learning arbitrary-order feature interactions is effective. Second, it makes sense to introduce the attention mechanism to joint embedding vectors with the output of the B-LTL. It retains the original distribution of the data. Lastly, DNN can bring additional improvement when the model use LNN to learn cross features. Our proposed IEC can effectively learn arbitrary-order feature interactions and identify valuable combinations. Moreover, by changing a_i in Eq. 4, IEC can build high-order feature interactions like DNN. This can be verified by adding DNN to ILFN, which has almost no extra effect. In summary, IEC, learning cross features in an expressive and discriminative manner, significantly boosts the performance for CTR prediction.

Table 2. The performance of different components in ILFN

Datasets	Criteo	
Model	AUC	Logloss
FM	0.7933	0.4573
HOFM	0.7960	0.4550
ILFN-no-att	0.8063	0.4452
ILFN-no-BN	0.8075	0.4446
LNN	0.8058	0.4457
LNN-DNN	0.8071	0.4450
ILFN-DNN	0.8084	0.4436
ILFN	**0.8085***	**0.4434***

Table 3. Relative improvements (RI) of extra components on Criteo and Frappe

Method	Component	Criteo RI	Frappe RI
FM	–	–	–
FFM	Field-aware	20.8	1.9
AFM	Attention	4.2	−6.7
DeepFM	DNN	14.0	−0.3
NFM	DNN	8.1	−5.9
DCN	CrossNet	24.0	5.1
AFN	LNN	25.4	2.4
ILFN	**IEC**	**30.4***	**10.4***

4.5   Component Comparison

In ILFN and other baseline models, each extra model component has its unique functionality: DNN models high-order feature interactions; Logarithmic neural network (LNN) focuses on the arbitrary-order cross features. To more intuitively observe the different performance improvement effects brought by various components. We first train all models without using the extra parts (i.e., simulating the standard FM model). Then we freeze feature embeddings and introduce the additional components only. Upon convergence, the relative improvements (RI) of each model on *Criteo* and *Frappe* are listed in Table 3. Note that values in Table 3 are a few thousandths because the improvement at 0.001-level is conventionally regarded as significant in the CTR estimation scenario. It is clear that with the help of the IEC, ILFN achieves 30.4‰ and 10.4‰ improvement on the two datasets, respectively. Our proposed IEC outperforms other widely recognised components by the public by learning arbitrary-order cross features in a discriminative way without destroying original data distribution.

5   Conclusion

In this paper, we presented a novel model called *Input Enhanced Logarithmic Factorization Network* (ILFN) for CTR prediction. It can effectively learn arbitrary-order cross features and identify beneficial feature interactions. Furthermore, it takes full advantage of the original embedding information in a discriminative way. The key to our method is the *Input Enhanced Component* (IEC), which joints the original embedding vectors and the output of the BN-Logarithmic Transformation Layer (B-LTL) via the Attention mechanism. In addition, we also revealed that ILFN could generalize to FM and HOFM by simplifying calculations. Extensive experiments on four real-world datasets confirmed that ILFN outperforms the representative and state-of-the-art deep learning approaches such as DCN, DeepFM, xDeepFM, and FiBiNET consistently for CTR prediction. We also analyzed the impact of hyper-parameters and conducted an ablation study, which demonstrates the effectiveness of our proposed components.

References

1. Bahdanau, D., Cho, K., Bengio, Y.: Neural machine translation by jointly learning to align and translate. In: ICLR (2015)
2. Wang, R., Shivanna, R., et al.: DCN V2: improved deep & cross network and practical lessons for web-scale learning to rank systems. In: WWW, pp. 1785–1797 (2021)
3. Cheng, H.-T., Koc, L., et al.: Wide & deep learning for recommender systems. In: Proceedings of the 1st Workshop on Deep Learning for Recommender Systems, pp. 7–10 (2016)
4. Cheng, W., Shen, Y., Huang, L.: Adaptive factorization network: learning adaptive-order feature interactions. In: AAAI, pp. 3609–3616 (2020)

5. McMahan, H.B., Holt, G., et al.: Ad click prediction: a view from the trenches. In: The 19th SIGKDD, 11–14 August 2013 (2013)
6. Guo, H., Tang, R., Ye, Y., Li, Z., He, X.: DeepFM: a factorization-machine based neural network for CTR prediction. In: IJCAI, pp. 1725–1731 (2017)
7. Graepel, T., Candela, J.Q., Borchert, T., Herbrich, R.: Web-scale Bayesian click-through rate prediction for sponsored search advertising in Microsoft's Bing search engine. Omnipress (2010)
8. Wang, Z., Zhang, R., Qi, J., Yuan, B.: DBSVEC: density-based clustering using support vector expansion. In: ICDE, pp. 280–291. IEEE (2019)
9. Lee, K., Orten, B., Dasdan, A., Li, W.: Estimating conversion rate in display advertising from past erformance data. In: KDD, pp. 768–776 (2012)
10. Wang, Z., Liu, L., Tao, D.: Deep streaming label learning. In: International Conference on Machine Learning (ICML) (2020)
11. Qu, Y., et al.: Product-based neural networks for user response prediction. In: ICDM, pp. 1149–1154. IEEE (2016)
12. Lian, J., Zhou, X., Zhang, F., Chen, Z., Xie, X., Sun, G.: xDeepFM: combining explicit and implicit feature interactions for recommender systems. In: KDD, pp. 1754–1763 (2018)
13. He, X., Chua, T.-S.: Neural factorization machines for sparse predictive analytics. In: SIGIR, pp. 355–364 (2017)
14. Xiao, J., Ye, H., He, X., Zhang, H., et al.: Attentional factorization machines: learning the weight of feature interactions via attention networks. In: IJCAI (2017)
15. Hines, J.W.: A logarithmic neural network architecture for unbounded non-linear function approximation. In: IEEE ICNN 1996 (1996)
16. Juan, Y., Zhuang, Y., Chin, W.-S., Lin, C.-J.: Field-aware factorization machines for CTR prediction. In: RecSys (2016)
17. Song, W., Shi, C., et al.: AutoInt: automatic feature interaction learning via self-attentive neural networks. In: CIKM (2019)
18. Wang, R., Fu, B., Fu, G., Wang, M.: Deep & cross network for ad click predictions. In: KDD, pp. 1–7 (2017)
19. Huang, T., Zhang, Z., Zhang, J.: FiBiNET: combining feature importance and bilinear feature interaction for click-through rate prediction. In: ACM RecSys (2019)
20. Rendle, S.: Factorization machines. In: 2010 IEEE International Conference on Data Mining, pp. 995–1000. IEEE (2010)
21. Blondel, M., Fujino, A., Ueda, N., Ishihata, M.: Higher-order factorization machines. In: NIPS (2016)
22. Yu, Y., Wang, Z., Yuan, B.: An input-aware factorization machine for sparse prediction. In: IJCAI (2019)
23. Zhang, W., Du, T., Wang, J.: Deep learning over multi-field categorical data. In: Ferro, N., et al. (eds.) ECIR 2016. LNCS, vol. 9626, pp. 45–57. Springer, Cham (2016). https://doi.org/10.1007/978-3-319-30671-1_4

A Novel Bayesian Deep Learning Approach to the Downscaling of Wind Speed with Uncertainty Quantification

Firas Gerges[1]([✉]), Michel C. Boufadel[2], Elie Bou-Zeid[3], Hani Nassif[4], and Jason T. L. Wang[1]

[1] Department of Computer Science, New Jersey Institute of Technology, University Heights, Newark, NJ 07102, USA
{fg92,wangj}@njit.edu
[2] Center for Natural Resources, Department of Civil and Environmental Engineering, New Jersey Institute of Technology, University Heights, Newark, NJ 07102, USA
boufadel@njit.edu
[3] Department of Civil and Environmental Engineering, Princeton University, Princeton, NJ 08544, USA
ebouzeid@princeton.edu
[4] Department of Civil and Environmental Engineering, Rutgers University – New Brunswick, Piscataway, NJ 08854, USA
nassif@soe.rutgers.edu

Abstract. Wind plays a crucial part during adverse events, such as storms and wildfires, and is a widely leveraged source of renewable energy. Predicting long-term daily local wind speed is critical for effective monitoring and mitigation of climate change, as well as to locate suitable locations for wind farms. Long-term simulations of wind dynamics (until year 2100) are given by various general circulation models (GCMs). However, GCM simulations are at a grid with coarse spatial resolution (> 100 km), which renders spatial downscaling to a smaller scale an important prerequisite for climate-impacts studies. In this work, we propose a novel deep learning approach, named Bayesian AIG-Transformer, that consists of an attention-based input grouping (AIG), transformer, and uncertainty quantification. We use the proposed approach for the spatial downscaling of daily average wind speed (AWND), formulated as a multivariate time series forecasting problem, over four locations within New Jersey and Pennsylvania. To calibrate and evaluate our deep learning approach, we use large-scale observations extracted from NOAA's NCEP/NCAR reanalysis dataset (2.5° × 2.5° resolution), which provides a proxy for GCM data when evaluating the model. Results show that our approach is suitable for the downscaling task, outperforming related machine learning methods.

Keywords: Deep learning · Transformer · Time series forecasting · Wind speed · Climate change

© The Author(s), under exclusive license to Springer Nature Switzerland AG 2022
J. Gama et al. (Eds.): PAKDD 2022, LNAI 13282, pp. 55–66, 2022.
https://doi.org/10.1007/978-3-031-05981-0_5

1 Introduction

Spatial downscaling techniques are used to relate large-scale variables to observations over a local, smaller-scale region. Downscaling is often applied to extract high-resolution climate information (e.g., wind) from coarse outputs of general circulation models (GCMs) (e.g., wind, humidity, temperature, etc.). GCMs are used to produce simulations of multiple climate variables, decades into the future. Downscaling can be classified as dynamical or statistical. Dynamical downscaling is computationally expensive. Statistical downscaling approaches are often used instead, and they rely on inferring and extracting a relationship between the local observations and the large-scale simulated variables. Statistical approaches include bias-correction based methods, as well as machine learning regression techniques [1–4].

The majority of downscaling studies are related to temperature [3, 4] and precipitation [1, 2, 5]. Downscaling wind dynamics is crucial for local climate-impacts studies. Moreover, leveraging wind as a source of green energy is growing rapidly, and efficiently foreseeing future wind speed variabilities is crucial to overcome operational challenges, and to locate suitable areas for wind farms. Existing studies concerning downscaling wind dynamics uses statistical non-machine learning approaches to predict wind speed [6, 7] and wind gusts among other properties.

Deep learning attempts that tackle downscaling mainly cover precipitation and temperature [5, 8]. Such techniques include CNN [9–12] and LSTM [13–17]. In this paper, we develop a novel deep learning approach, dubbed Bayesian AIG-Transformer, that employs an attention-based model for implicit feature reduction (AIG), coupled with transformer [18, 19] and uncertainty quantification. We investigate the use of our approach for the statistical downscaling of daily average wind speed (AWND) over four locations within New Jersey and Pennsylvania. The main contributions of our work are summarized below:

- We develop a new transformer-based approach, named Bayesian AIG-Transformer, that employs uncertainty quantification along with an implicit feature reduction.
- Experimental results obtained from downscaling daily average wind speed show that Bayesian AIG-Transformer outperforms related machine learning methods, over each local station used.
- The use of uncertainty quantification, implemented based on the Monte-Carlo dropout sampling method, allows us to show the aleatoric and the epistemic uncertainty.

2 Problem Formulation and Data Selection

2.1 Problem Formulation

To perform statistical downscaling, one would need to collect historical local observations of the variable of interest (i.e., daily wind speed), as well as large-scale observations of multiple climate variables. We regard the downscaling task as a multivariate time series forecasting problem, and we aim to predict the daily wind speed over four locations independently (red stars in Fig. 1). In this work, we opt-in to use the NCEP/NCAR reanalysis dataset instead of GCM simulations (discussed below) to collect the large-scale climate variable values. Similar to GCMs, NCEP/NCAR data are provided as a gridded dataset, and one would need to extract the data from multiple grid points. The input of our proposed prediction model consists of sequences of multiple large-scale climate variables taken from 20 different grid points (black circles in Fig. 1). These selected grid points are referred to as the atmospheric domain. The output of the model is the daily average wind speed over a local station. More precisely, the local wind speed dataset over a local station S is a sequence of observations (ground truth values), each consisting of a timestep t, and a wind speed value $AWND_{S,t}$. To obtain the predicted value of wind speed $\widehat{AWND}_{S,t}$ at a future time t, our prediction model uses current and previous large-scale climate variables, down to a specified time lag (lag). Note that each climate variable is depicted in the input using twenty (20) feature values extracted from the twenty (20) grid points in Fig. 1. With n being the total number of features, we train the model using the sequence of attributes $A_{i,j}$, where $1 \leq i \leq n$, $(t - lag) \leq j \leq t$, over an interval from time t down to time $(t - lag)$. The default value of lag in this study is set to 7 (days).

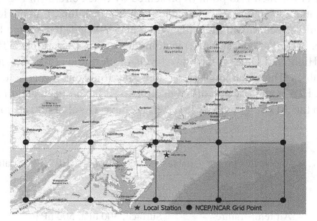

Fig. 1. Atmospheric domain of interest. The twenty black circles represent the locations of the NCEP/NCAR grid points where large-scale climate variables are extracted. The red stars represent the locations of the local stations where daily wind speed will be predicted (separately). (Color figure online)

2.2 Local Wind Data

We retrieved daily average wind speed data from the Global Historical Climatology Network (GHCN)-Daily. The GHCN dataset consists of data from thousands of stations worldwide and includes the data from the National Oceanic and Atmospheric Administration/National Climatic Data Center (NOAA/NCDC) [20]. We extracted the daily wind speed from January 1st, 1984 (earliest date where wind data is available), to December 31st, 2020 (total of 37 years). These local observations are extracted for four stations (Fig. 1): USW00014734 located at Newark Liberty International Airport (EWR), USW00014737 located at Allentown Lehigh Valley International Airport (ABE), USW00013739 located at Philadelphia International Airport (PHL), and

USW00093730 located at Atlantic City International Airport (ACY). We aim to perform the downscaling over each station/airport independently.

2.3 Reanalysis Data

The reanalysis data from the National Centers for Environmental Prediction/National Center for Atmospheric Research (NCEP/NCAR) offer a large number of climate variables at surface, and at different atmospheric levels [21]. Over the United States, the data have a $2.5° \times 2.5°$ resolution and contain daily (among other temporal scales) data since 1948. One would use this reanalysis data as a proxy of GCMs and the large-scale climate variable values when calibrating our prediction model. After calibrating and evaluating the model, one could use future large-scale variables from GCMs as input to the trained model to produce long-term projections. In this paper, we focus on the development and the calibration of the Bayesian AIG-Transformer model, and no GCM data are included. We extracted the climate variables, shown in Table 1, from NCEP/NCAR reanalysis dataset for each grid point (represented by black circles in Fig. 1) within the atmospheric domain of interest. Six variables (Air Temperature, Latent Heat Net Flux, Relative Humidity, Specific Humidity, U-Wind, and V-Wind) have three levels of elevation (surface, 300 hPa, and 500 hPa) yielding a total of 18 variables. Two variables (Geopotential Hight and Omega) have two levels of elevations (300 hPa, and 500 hPa) yielding 4 variables. The remaining five variables have only one level of elevation, yielding 5 variables. Hence, there are 25 different climate variables, each yielding 20 features (extracted from the 20 grid points). As a result, we have a total of 500 attributes/features in our study.

Table 1. Large-scale climate variables extracted from NCEP/NCAR dataset.

Climate variables	Elevation
Air temperature	Surface, 300 hPa, 500 hPa
Geopotential height	300 hPa, 500 hPa
Latent heat net flux	Surface, 300 hPa, 500 hPa
Omega (vertical velocity)	300 hPa, 500 hPa
Precipitation	Surface
Relative humidity	Surface, 300 hPa, 500 hPa
Sensible heat net flux	Surface
Specific humidity	Surface, 300 hPa, 500 hPa
Total cloud cover	Sky
Maximum temperature	Surface
Minimum temperature	Surface
U-Wind	Surface, 300 hPa, 500 hPa
V-Wind	Surface, 300 hPa, 500 hPa

3 Method

In this work, we present a new deep learning technique, named Bayesian AIG-Transformer. This technique employs three main components: attention-based input grouping (AIG), transformer, and uncertainty quantification based on the Monte-Carlo (MC) dropout method. These components are discussed in the following subsections, and the general architecture of the Bayesian AIG-Transformer is depicted in Fig. 2.

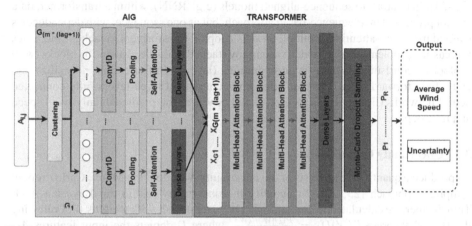

Fig. 2. Bayesian AIG-Transformer framework. For each timestep j, the input time sequences $A_{i,j}$, with $1 \leq i \leq n$, and $(t - lag) \leq j \leq t$, are clustered based on the corresponding climate variable into m clusters, each containing 20 features, corresponding to the 20 grid points in Fig. 1. In total we have $m * (lag + 1)$ clusters. Each cluster is fed into an AIG block to undergo implicit feature reduction independently, using convolution, pooling and self-attention. The red arrows in the figure depict dropout steps. The outputs of all the AIG blocks are concatenated and used as input to the transformer model. Using the Monte-Carlo dropout sampling, the model produces R predicted values (P_o for $1 \leq o \leq R$) for each timestep. The final outputs are the predicted average wind speed (mean of the R predicted values) and the uncertainty values.

3.1 Attention-Based Input Grouping

The input to the Bayesian AIG-Transformer, $A_{i,j}$, consists of a sequence of n features, from time t, down to a specified lag. We have a total of 500 features ($n = 500$) which depict 25 unique climate variables, where each variable is recorded from 20 different grid points (Fig. 1). To handle such number of features, we introduce a novel attention-based input grouping (AIG), which aims to perform implicit feature reduction, before feeding the input to the transformer. Our AIG approach consists of clustering the input features based on the climate variables they represent. Each cluster is then fed to an AIG block which performs convolution, pooling, and self-attention, producing one value per cluster. the features within each cluster represent the same variable (precipitation, air temperature, etc.), which might lead to data redundancy, and a bias model, especially if the variable is highly correlated with the output (wind speed). The rationale behind our AIG approach is to overcome noise and bias in the transformer input, by first extracting and reducing the size of the general (learning) patterns of each climate variable (via convolution and pooling), while giving selective importance to these patterns using self-attention mechanism. Note that within the AIG blocks, we opt-in to use max pooling.

3.2 Transformer

The transformer deep learning model was proposed initially for machine translation [18] in an encoder-decoder architecture, and was later incorporated within different domains [22–24]. In contrast to sequence-aligned models (e.g., RNN), within a transformer, data are not processed in a sequence-based approach, but as one sequence, with dependencies learned using self-attention. Multiple studies adopted transformers to tackle time series forecasting problems. In our model, we employ stacked multi-head attention blocks (with 4 blocks), consisting of three main operations (layers): self-attention, normalization, and a feed-forward layer. The prediction layer consists of three-layers fully connected neural networks (FNN) ($1000 \times 10 \times 1$) and produces the predicted daily wind speed at a timestep t.

3.3 Uncertainty Quantification

To produce a quantitative analysis of the uncertainty, we adopt the Monte-Carlo dropout sampling approach [25, 26]. To quantify the uncertainty within our Bayesian AIG-Transformer, we calculate the probability $P(W)$ over the network weights W. Following the Bayes' theorem $P(W|D) = \frac{P(D|W)P(W)}{P(D)}$, where D depicts the input features $A_{i,j}$ and the corresponding wind speed $AWND$, one needs to compute the exact posterior probability $P(W|D)$. Computing this probability was shown to be intractable. One could instead use variational inference [27] by learning the variational distribution $q_\theta(W)$ over the weights W parameterized by θ. This is achieved by minimizing the Kullback-Leibler divergence of said variational distribution, as well as of the posterior probability. Using dropout when training the model allows the variational approximation of the network. Moreover, minimizing the Kulback-Leibler divergence can be achieved by using the cross-entropy loss function and the Adam optimizer. The optimized wight distribution is denoted by $q_{\hat{\theta}}(W)$ with $\hat{\theta}$ being the optimized variational parameter. Dropouts are used usually to avoid overfitting in a deep learning model, allowing the model to be better generalized to perform predictions on unseen data. Dropout works by dropping out neurons in certain layers. These dropped out neurons are selected randomly following a dropout rate (which is set to 0.4 in this study). Following the Monte-Carlo dropout sampling approach, we apply dropout during testing to retrieve R Monte-Carlo samples for each test case, by running the model on each test case (test input) R times, where each time, weights are randomly drawn from $q_{\hat{\theta}}(W)$. (In this study, we set R to 100.) Subsequently, we will have R predicted values for each timestep in the testing input, allowing us to draw a mean and a variance over each R samples (each timestep). We aim to quantify two types of uncertainty, aleatoric, which captures the inherent randomness in the input data, and epistemic, which depicts the uncertainty in the model. We compute the two uncertainties following the work in [28, 29]. Note that we normalize the predicted values when capturing the aleatoric uncertainty, given that we're dealing with a regression problem.

4   Experiments and Results

4.1   Experiment Setup

We divided the dataset into training period from year 1984 to 2013 (~80%) and testing period from year 2014 to 2020. Note that 20% of the training data is used to validate the model while tuning the parameters. We adopt three evaluation metrics that are often used in downscaling studies: root mean squared error (RMSE), mean absolute error (MAE), and the Nash-Sutcliffe efficiency coefficient (NSE).

4.2   Ablation Studies

We considered multiple variants of Bayesian AIG-Transformer for downscaling the daily average wind speed. These variants include AIG-Transformer, which is the base model without uncertainty quantification; Transformer, which refers to the model without the AIG component; and AIG which represents the AIG component directly connected to fully connected layers. We show in Fig. 3 the evaluation metrics of each model over each local location. The performance of the Bayesian AIG-Transformer shows an improvement over the other variants, depicting the importance of the AIG component as well as the advantage of uncertainty quantification. Note the models were trained for 750 epochs, with a batch size of 32.

4.3   Comparative Results

In order to analyze the model performance, with respect to other machine learning techniques, we evaluated the Bayesian AIG-Transformer on each testing set (four local stations) and compared the performance with that of Random Forest (RF), Linear Regression (LR), Support Vector Regression (SVR) and Long Short-Term Memory (LSTM). Figure 4 presents the experimental results of the used techniques in terms of each evaluation metric, over each local station.

Bayesian AIG-Transformer outperformed existing machine learning methods in terms of RMSE (1.40 over EWR, 1.45 over PHL, 1.60 over ACY, and 1.50 over ABE), MAE (1.05 over EWR, 1.12 over PHL, 1.24 over ACY, and 1.18 over ABE) and NSE (0.79 over EWR, 0.77 over PHL, 0.74 over ACY, and 0.69 over ABE). We show in Fig. 5 the scatter plots between the observed and predicted values obtained by Bayesian AIG-Transformer, for each local station.

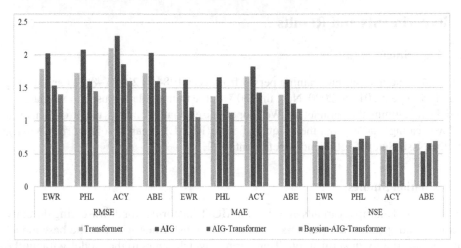

Fig. 3. Performance of the Bayesian AIG-Transformer and the three variants: Transformer, AIG, and AIG-Transformer. The figure shows the results over each dataset (EWR, PHL, ACY, and ABE) and for each evaluation metric (RMSE, MAE and NSE).

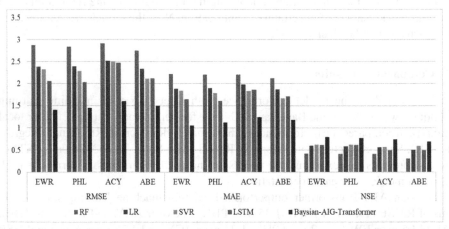

Fig. 4. Performance of the Bayesian AIG-Transformer compared to random forest (RF), linear regression (LR), support vector regression (SVR) and long short-term memory (LSTM). The figure shows the results over each dataset (EWR, PHL, ACY, and ABE) and for each evaluation metric.

In addition to improvement in performance, the use of uncertainty quantification within the AIG-Transformer model, enables us to provide interval predictions instead of single-point values. This is specifically important in this case given the uncertainties and anomalies that are usually present in reanalysis and GCM data. Moreover, the use of the Monte-Carlo dropout sampling method allows us to show the data and the model uncertainty. We display in Table 2 the average aleatoric and epistemic uncertainty values of the entire testing cases for each local station. Figure 6 shows for every local station, the observed and predicted wind speed values, along with the aleatoric and epistemic

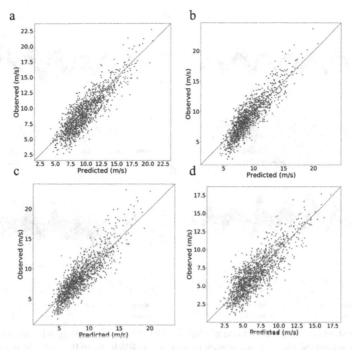

Fig. 5. Scatter plots between the observed and predicted wind speed as obtained by Bayesian AIG-Transformer over a) EWR, b) PHL, c) ACY and d) ABE.

uncertainty intervals. Note the data in Fig. 6 are averaged over each month for a more informative display. The results indicate that the model uncertainty (epistemic) is relatively low, and one could further reduce such uncertainty by using more data to train the model. The data uncertainty (aleatoric) however, depict the intrinsic uncertainty in the input data, and reducing such uncertainty requires using different data or performing noise and bias reduction on the current input.

Table 2. Average uncertainty values over each local station.

Uncertainty	EWR	PHL	ACY	ABE
Aleatoric	2.04	3.50	2.89	3.50
Epistemic	0.92	1.67	1.40	1.75

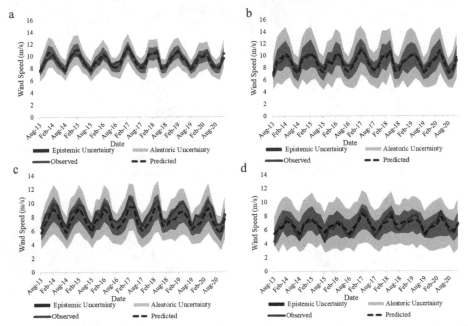

Fig. 6. Plots showing the observed and predicted wind speed values along with the aleatoric and epistemic uncertainties, averaged over each month, for a) EWR, b) PHL, c) ACY and d) ABE.

5   Conclusions

In this work, we propose a new deep learning approach, dubbed, Bayesian AIG-Transformer, for the spatial downscaling of average daily wind speed (AWND). Our proposed approach consists of an AIG model, aiming to cluster and implicitly reduce input features fed to a transformer, with convolution and attention mechanisms. Moreover, our approach implements the Monte-Carlo dropout sampling method to quantify data and model uncertainty, depicted by the aleatoric and epistemic uncertainty, respectively. We employed our Bayesian AIG-Transformer to downscale daily wind speed over 4 different locations within New Jersey and Pennsylvania. Experimental results demonstrated that our approach outperformed the related methods, in terms of each evaluation metric used (MAE, RMSE, and NSE), over each local station (EWR, PHL, ACY and ABE). The results show that our approach is suitable for the downscaling problem, and in future work, we aim to extract GCM daily climate data up to year 2100, and subsequently use our trained Bayesian AIG-Transformer model to produce daily wind speed up to year 2100, over multiple locations.

Acknowledgements. This work was supported by the Bridge Resource Program (BRP) from the New Jersey Department of Transportation.

References

1. He, X., Chaney, N.W., Schleiss, M., Sheffield, J.: Spatial downscaling of precipitation using adaptable random forests. Water Resour. Res. **52**, 8217–8237 (2016)
2. Sachindra, D., Ahmed, K., Rashid, M.M., Shahid, S., Perera, B.: Statistical downscaling of precipitation using machine learning techniques. Atmos. Res. **212**, 240–258 (2018)
3. Coulibaly, P.: Downscaling daily extreme temperatures with genetic programming. Geophys. Res. Lett. **31** (2004)
4. Li, X., Li, Z., Huang, W., Zhou, P.: Performance of statistical and machine learning ensembles for daily temperature downscaling. Theoret. Appl. Climatol. **140**(1–2), 571–588 (2020). https://doi.org/10.1007/s00704-020-03098-3
5. Misra, S., Sarkar, S., Mitra, P.: Statistical downscaling of precipitation using long short-term memory recurrent neural networks. Theoret. Appl. Climatol. **134**(3–4), 1179–1196 (2017). https://doi.org/10.1007/s00704-017-2307-2
6. Hu, W., Scholz, Y., Yeligeti, M., von Bremen, L., Schroedter-Homscheidt, M.: Statistical downscaling of wind speed time series data based on topographic variables. In: EGU General Assembly Conference Abstracts, pp. EGU21–12734 (2021)
7. Kirchmeier, M.C., Lorenz, D.J., Vimont, D.J.: Statistical downscaling of daily wind speed variations. J. Appl. Meteorol. Climatol. **53**, 660–675 (2014)
8. Sun, L., Lan, Y.: Statistical downscaling of daily temperature and precipitation over China using deep learning neural models: localization and comparison with other methods. Int. J. Climatol. **41**, 1128–1147 (2021)
9. Yang, Z., et al.: LegoNet: efficient convolutional neural networks with lego filters. In: 36th International Conference on Machine Learning, pp. 7005–7014. PMLR (2019)
10. Pan, X., Shi, J., Luo, P., Wang, X., Tang, X.: Spatial as deep: spatial CNN for traffic scene understanding. In: 32nd AAAI Conference on Artificial Intelligence (2018)
11. Jin, C., Liang, H., Chen, D., Lin, Z., Wu, M.: Identifying mobility of drug addicts with multilevel spatial-temporal convolutional neural network. In: Yang, Q., Zhou, Z.-H., Gong, Z., Zhang, M.-L., Huang, S.-J. (eds.) PAKDD 2019. LNCS (LNAI), vol. 11439, pp. 477–488. Springer, Cham (2019). https://doi.org/10.1007/978-3-030-16148-4_37
12. Liu, Z., Wan, M., Guo, S., Achan, K., Yu, P.S.: BasConv: aggregating heterogeneous interactions for basket recommendation with graph convolutional neural network. In: Proceedings of the 2020 SIAM International Conference on Data Mining, pp. 64–72. SIAM (2020)
13. Guo, T., Lin, T., Antulov-Fantulin, N.: Exploring interpretable LSTM neural networks over multi-variable data. In: 36th International Conference on Machine Learning, pp. 2494–2504. PMLR (2019)
14. Hu, Z., Turki, T., Phan, N., Wang, J.T.L.: A 3D atrous convolutional long short-term memory network for background subtraction. IEEE Access **6**, 43450–43459 (2018)
15. Liu, H., Liu, C., Wang, J.T.L., Wang, H.: Predicting solar flares using a long short-term memory network. Astrophys J. **877**, 121 (2019)
16. Segovia-Dominguez, I., Zhen, Z., Wagh, R., Lee, H., Gel, Y.R.: TLife-LSTM: forecasting future COVID-19 progression with topological signatures of atmospheric conditions. In: Karlapalem, K., et al. (eds.) PAKDD 2021. LNCS (LNAI), vol. 12712, pp. 201–212. Springer, Cham (2021). https://doi.org/10.1007/978-3-030-75762-5_17
17. Shalaby, M., Stutzki, J., Schubert, M., Günnemann, S.: An LSTM approach to patent classification based on fixed hierarchy vectors. In: Proceedings of the 2018 SIAM International Conference on Data Mining, pp. 495–503. SIAM (2018)
18. Vaswani, A., et al.: Attention is all you need. In: Advances in Neural Information Processing Systems, pp. 5998–6008 (2017)

66   F. Gerges et al.

19. Zerveas, G., Jayaraman, S., Patel, D., Bhamidipaty, A., Eickhoff, C.: A transformer-based framework for multivariate time series representation learning. In: Proceedings of the 27th ACM SIGKDD Conference on Knowledge Discovery & Data Mining, pp. 2114–2124 (2021)
20. Menne, M.J., Durre, I., Vose, R.S., Gleason, B.E., Houston, T.G.: An overview of the global historical climatology network-daily database. J. Atmos. Oceanic Tech. **29**, 897–910 (2012)
21. Kalnay, E., Kanamitsu, M., Kistler, R., Collins, W., Deaven, D., Gandin, L.: The NCEP/NCAR 40-year reanalysis project. Bull. Am. Meteor. Soc. **77**, 437–471 (1996)
22. Parmar, N., et al.: Image transformer. In: 35th International Conference on Machine Learning, pp. 4055–4064. PMLR (2018)
23. Cai, T., Shen, M., Peng, H., Jiang, L., Dai, Q.: Improving transformer with sequential context representations for abstractive text summarization. In: Tang, J., Kan, M.-Y., Zhao, D., Li, S., Zan, H. (eds.) NLPCC 2019. LNCS (LNAI), vol. 11838, pp. 512–524. Springer, Cham (2019). https://doi.org/10.1007/978-3-030-32233-5_40
24. Guo, D., Terzopoulos, D.: A transformer-based network for anisotropic 3D medical image segmentation. In: 2020 25th International Conference on Pattern Recognition (ICPR), pp. 8857–8861. IEEE (2021)
25. Gal, Y., Ghahramani, Z.: Dropout as a Bayesian approximation: representing model uncertainty in deep learning. In: 33rd International Conference on Machine Learning, pp. 1050–1059. PMLR (2016)
26. Roy, S., et al.: Deep learning for classification and localization of COVID-19 markers in point-of-care lung ultrasound. IEEE Trans. Med. Imaging **39**, 2676–2687 (2020)
27. Blei, D.M., Kucukelbir, A., McAuliffe, J.D.: Variational inference: a review for statisticians. J. Am. Stat. Assoc. **112**, 859–877 (2017)
28. Kwon, Y., Won, J.-H., Kim, B.J., Paik, M.C.: Uncertainty quantification using Bayesian neural networks in classification: application to biomedical image segmentation. Comput. Stat. Data Anal. **142**, 106816 (2020)
29. Jiang, H., et al.: Tracing Hα fibrils through Bayesian deep learning. Astrophys. J. Suppl. Ser. **256**, 20 (2021)

Bribery in Rating Systems: A Game-Theoretic Perspective

Xin Zhou[1], Shigeo Matsubara[2], Yuan Liu[3(✉)], and Qidong Liu[4(✉)]

[1] School of Computer Science and Engineering,
Nanyang Technological University, Singapore, Singapore
xin.zhou@ntu.edu.sg
[2] Center for Mathematical Modeling and Data Science,
Osaka University, Osaka, Japan
matsubara@sigmath.es.osaka-u.ac.jp
[3] Cyberspace Institute of Advanced Technology, Guangzhou University,
Guangzhou, Guangdong, China
liuyuan@swc.neu.edu.cn
[4] School of Computer and Artificial Intelligence, Zhengzhou University,
Zhengzhou, China
ieqdliu@zzu.edu.cn

Abstract. Rating systems play a vital role in the exponential growth of service-oriented markets. As highly rated online services usually receive substantial revenue in the markets, malicious sellers seek to boost their service evaluation by manipulating the rating system with fake ratings. One effective way to improve the service evaluation is to hire fake rating providers by bribery. The fake ratings given by the bribed buyers influence the evaluation of the service, which further impacts the decision-making of potential buyers. In this paper, we study the bribery of a rating system with multiple sellers and buyers via a game-theoretic perspective. In detail, we examine whether there exists an equilibrium state in the market in which the rating system is expected to be bribery-proof: no bribery strategy yields a strictly positive gain. We first collect real-world data for modeling the bribery problem in rating systems. On top of that, we analyze the problem of bribery in a rating system as a static game. From our analysis, we conclude that at least a Nash equilibrium can be reached in the bribery game of rating systems.

Keywords: Bribery · Game theory · Rating system · Nash equilibrium

1 Introduction

In e-marketplaces (*e.g.*, Amazon, app store), buyers usually select reputable services or items offered by sellers based on their direct experiences or ratings given from other buyers. Direct experience is important and trustworthy; however, it is infeasible for buyers to interact with all sellers in order to gain direct experiences in a real-world setting. As a result, a rating/reputation system evaluating the given ratings is proposed as a crucial tool for supporting buyers' decision-making [9]. In our context, we assume a seller and the owned service are one-to-one mapping. Hence, when a buyer interacts with one seller, he/she purchases the item or uses the service from that seller.

In real-world rating systems, various studies have demonstrated that high profit is attached to high reputation services or items. A study of eBay conducted by Resnick *et al.* revealed that buyers were willing to pay 8% more to the item offered by reputable sellers than that provided by new sellers [18]. Ye *et al.* [25] in their study showed that a 10% improvement in reviewers' rating on hotels could increase the number of bookings by 4.4%. Driven by profit, sellers have incentives to perform unfair rating attacks by bribing buyers and increase their aggregated rating above their competitor's, as this would increase their market share and overall revenues thereof [5,11,28]. Fake ratings have been reported on the real-world movie dataset collected by Cao *et al.* [1], and the fake ratings dominate the rating distribution in the first few days as soon as the movie is released. The fake ratings are mainly divided into fake positive ratings to boost the sellers' reputation and dishonest negative ratings to decrease their competitors' reputation. Although service attack detection frameworks or models have been proposed recently [12,13,15,20], emerging underground market (in providing fake ratings) even worsen the bribery problem [19].

We assume there are two potential reasons that entice the sellers into bribing "bot farms" or "human water armies" [23] in order to inflate service ratings. The first is the low cost of performing the bribery due to rating sparsity [26]. From the analysis on the Google Play app store, the rating ratio is as low as 2.1% based on a study of the top 20,000 apps [27]. That is, about 2 out of 100 buyers or users of the app rate their ratings on the platform. In such a situation, app owners can easily increase the positive evaluation of their apps by bribing buyers with a compensation, monetary or not. That is, sellers exchange for positive ratings by bribing a group of buyers with certain payment. The positive ratings further influence the sellers' future revenues. The second reason is the attractive revenue gained from temporal (usually on a daily basis) top app ranking. App developers resort to fraudulent means to deliberately boost up their app ranking on an App store [17,28]. In this paper, we analyze the effect of a bribing strategy by defining the utility function and strategy space for each seller. In addition, we model the problem as an economic game and examine whether an equilibrium state exists, in which the rating system is expected to be bribery-proof: no individual seller wants to perform any bribery strategy.

In single-player decision theory, the key notion is that of an optimal strategy, that is, a strategy that maximizes the player's expected payoff for a given environment in which the player operates [6,10]. Such a situation in the single-player case can be fraught with uncertainty, since the environment might be stochastic, partially observable, and spring all kinds of surprises on the player. In our work, we extend the setting into a more complex multi-players setting, in which the environment comprises other players, all of whom are also willing to maximize their payoffs. Thus, the notion of an optimal strategy for a given player is not meaningful as the best strategy also depends on the choices of others. The equilibrium state is reached when all players perform their best strategy considering others players' strategies. In the multi-player setting, we model the sellers as players and map the bribing behavior into strategy space in game theory. Based on the model, we intend to find the equilibrium state within all combinations of sellers' strategies. To simplify our analysis, we will use the terms "seller", "player" or "service" interchangeably.

In detail, we consider the static game in which the number of sellers and buyers are fixed within a time slot. The exponential increase of the strategy space with the number of sellers and buyers makes it difficult to search the optimal strategies. To find the equilibrium state in the static game, we devise a greedy algorithm to eliminate the dominated strategies.

We summarize the main contributions as follows:

- Although some researchers have studied bribery in the rating system, they modeled the problem without real-world data support and merely depended on assumptions. In this paper, we collect two real-world datasets as the backbone of our modeling and analysis.
- We extend bribery in rating systems from a single seller into multiple sellers and buyers situation and explicitly interpret the problem from a game theory perspective.
- On top of the model, we analyze the effect of bribing strategies in the case of static game and found that a Nash equilibrium exists in sellers' strategy space, with too many buyers are bribed under Nash equilibrium compared to the social optimum.

We reveal that our study can be applied to all rating-based systems in which individuals may influence the decision of one other via the rating value. Bribery can disrupt buyers to interact with fake reputable sellers with a bribed high rating value.

Related research lines: Our approach relates to several research lines in computational social choice, game theory, and reputation systems.

Computational social choice: In our studied rating system, sellers try to influence potential buyers' decisions to maximize their own expected payoff. The problem is analogous to lobbying and bribery in computational social choice, where agents manipulate or modify their outcomes to reach their objectives [2,3]. Lobbying and bribery are established concepts in computational social choice, the research ranges from the seminal contribution of [7] to recent studies such as vote-buying in [16].

Reputation systems: Dishonest ratings influence the accuracy of reputation evaluation; hence, a robust reputation system should detect the attacks and mitigate the influence caused by the attacks [24]. Otherwise, the disruptive reputation can influence other agents. In this sense, ours can be seen as a study of reputation in multi-agent systems [8]. To analyze the harm of unfair ratings, more recent studies based on information theory have been proposed in papers [21] and [22].

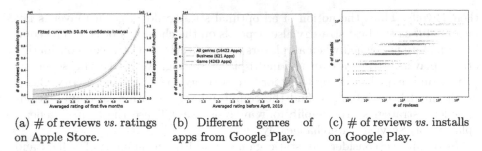

(a) # of reviews *vs.* ratings on Apple Store.

(b) Different genres of apps from Google Play.

(c) # of reviews *vs.* installs on Google Play.

Fig. 1. Snowball effect shown on apps in different stores and different genres.

2   Profit Modeling on Data Observation

We collect two real-world datasets to show how the problem can be represented as an economic game. By crawling the statistical data of 1609 apps from Apple App Store during December 2016 to May 2017 and the data of rating, app genre, reviews, and installs on 16422 apps from Google Play during April 2019 to November 2019, we first model the relation between the received rating of an app and its potential profit in the number of reviews. Although individual apps may exhibit different patterns between ratings and installs, we study whether the overall pattern is consistent between apps from different app stores or app genres. From the plotted Fig. 1(a) and (b), we find that higher-rated apps receive more reviews, although they are in different app stores and within different app genres. The data shows that the relation between the received rating r and the number of following reviews N_{re} follows a power function.

$$N_{re} = a \cdot r^n, \tag{1}$$

where $a \neq 0$ and n are real numbers, and a is a coefficient. Since not all buyers score their rating after purchase, our collected data may not accurately capture the relation between rating and the transaction volume. Seller's profile may also be evaluated in the number of installs; we further show the number of installs on a given rating in Fig. 1(c). From Fig. 1(c), we can observe that the relation between the received rating r and the number of following installs N_{in} also follows a power function. However, compared with the number of reviews, the number of installs is much higher skewed with ratings. In Google Play, the number of installs is merely an estimated value on the order of magnitude of downloads. Actually, we can obtain the relation between rating and installs through Fig. 1(b) and (c) indirectly. Figure 1(c) shows that the log scale of the number of installs N_{in} is proportional to the log scale of the number of reviews N_{re}. That is,

$$log N_{in} = b \cdot log N_{re} \Leftrightarrow N_{in} = (N_{re})^b, \tag{2}$$

where $a \neq 0$. Substituting Eq. (1) into Eq. (2), we have

$$N_{in} = k \cdot r^{\Omega}, \tag{3}$$

where $k = a^b$ and $\Omega = n \cdot b$. The number of installs based on ratings also shows a snowball effect; when a few buyers explore a good app, everyone else wants to pile it on. Following the observation, app owners competed to increase their app evaluation in order to increase their purchase revenue. For a single seller, with limited potential buyers in the platform, maximizing its own payoff will influence the other's welfare. Hence, the problem can be represented as an economic game.

3   Static Game

To present our model, we assume that, in the rating system, each seller can observe the current buyers of other sellers alongside the current evaluation from the buyers. The common knowledge in our game is as follows: *a)* The players are rational and act with the intention to maximize their revenue. *b)* In static game, all the players know their own payoffs as well as others, also all the players know that all the players know all the players' payoffs, and all the players know that all the players know that all the players know all the players' payoffs, and so on, ad infinitum. We let the number of buyers be constant in a static game, the bribery space for a seller is the buyers that have not interacted with that seller. The following analysis takes the rating system in the app store as an example; the analysis also can be applied to other numerical rating systems.

Considering that we have M *sellers* and N *buyers* in an app store within a time slot. After a seller s_i released its app on the app store, the app is evaluated by a finite set of buyers $B_i = \{b_1, b_2, ..., b_l\}, l <= N$. Low-rated app owners intend to bribe a set of buyers to increase its app evaluation. We assume one buyer can only be bribed by one seller and the apps in the market are functionally equivalent. The bribed rating is proportional to the payment, that is, more payment for higher rating. For a fair rater, based on his/her experience, he/she submits a rating to evaluate the app. The scored rating r is drawn from a set of values $r \in [0, 1]$. Although discrete assignment of 1 to 5 stars is common in online rating systems, such as Amazon, Apple Store, and Google play, the stars can be mapped into $[0, 1]$ to simplify the analysis.

We study the rating systems such as used in Apple Store; every interested buyer can see the evaluation of other buyers. Only the buyer who paid for the app can evaluate it, and the aggregation of the evaluation will influence potential buyers' decision. We represent the aggregated evaluation on the app of seller s_i as a function $\bar{r}_i = avg(R_i)$ of the received ratings $R_i = \{r_1, r_2, ..., r_l\}$, where avg is the average function across all real-valued ratings.

Suppose in a time slot, all the sellers perform bribery strategy to increase their app evaluation; as a result, the profit is based on their current rating distribution. We define the strategy of sellers in the form of an ensemble of individual strategies:

Definition 1. *An individual strategy on buyer b_j is the amount of effort $\phi(b_j)$ required to improve its current rating to a new rating. We denote: $\phi(b_j) \rightarrow [0, +\infty)$. A strategy of a seller is an ensemble of individual strategies on each buyer, such that $\sum_{j=0}^{N} b_j \in [0, +\infty)$.*

Under bribery, buyer's evaluation is closed under $min\{1, x+y\}$ for all $x, y \in [0,1]$, where x, y are two evaluations on an app. To compare with the results of [6], we adopt that the bribed cost is proportional to buyer's rating. For example, buyer b_j's evaluation is 0.5, but when bribed by seller s_i with $\phi(b_j) = 0.6$, the new evaluation will be $min\{1, x+y\} = min\{1, 1.1\} = 1$. By Definition 1, the total cost for seller s_i to perform a bribery on a set of bribed buyers B_ϕ ($|B_\phi| \leq N$) is $\sum_{b \in B_\phi} \phi(b)$. Next, we define the potential utility to a seller s_i that can be gained from bribing a set of buyers. To simplify our analysis, we assume the utility a seller can obtain is proportional to its aggregated rating. That is, we let $\Omega = 1$ in Eq. (3). The assumption does not influence our final conclusion.

Definition 2. *The expected utility of a seller s_i is a function.*

$$u(\bar{r}_i) = \mathcal{N} \cdot \bar{r}_i^{\Omega} \cdot k, \tag{4}$$

where $\mathcal{N} = N - \sum_{j=0}^{M} |B_j|$ is the potential buyers in the market, k can be viewed as the profit from an app purchase.

The utility of seller s_i amounts to the number of potential buyers in the market and the rating rated by the interacted buyers. The underlying intuition is that buyers prefer to download and use higher rated apps. Theoretically, a seller can bribe all the available buyers in the market; however, the potential profit will approach to zero because of no future interaction. Based on previous definitions, we can define the payoff of a bribery strategy.

Definition 3. *Let ϕ be a strategy and B_ϕ is the bribed buyers by ϕ. The payoff to seller s_i from ϕ is:*

$$\pi(\phi) = u^\phi - u^0 - \sum_{b \in B_\phi} \phi(b),$$

where u^0 is the initial utility and u^ϕ is the utility after execution of ϕ.

Definition 4. *Let ϕ' and ϕ'' be feasible strategies for seller s_i. Strategy ϕ' is strictly dominated by ϕ'' if for each feasible combination of the other players' strategies, s_i's payoff from playing ϕ' is less than s_i's payoff from playing ϕ'':*

$$\pi(\phi_1, ..., \phi_{j-1}, \phi', \phi_{j+1}, ..., \phi_N) < \pi(\phi_1, ..., \phi_{j-1}, \phi'', \phi_{j+1}, ..., \phi_N),$$

for each $(\phi_1, ..., \phi_{j-1}, \phi_{j+1}, ..., \phi_N)$ that can be constructed from the other players' strategy spaces $\Phi_1, ..., \Phi_{j-1}, \Phi_{j+1}, ..., \Phi_N$. Let ϕ_{-i} denote the strategy combination from players other than j. Then, if occasionally, $\pi(\phi', \phi_{-j}) = \pi(\phi'', \phi_{-j})$, we denote ϕ' is weakly dominated by ϕ''.

Given the definitions and the conclusions in paper [6], a rational seller would prefer greedy strategy to others under the same budget. In other words, a strategy of bribing non-raters always dominates others. Algorithm 1 defines the greedy bribing strategy. In the first loop, the seller bribes the history buyers who scored low ratings greedily. If the budget is still above zero, then the seller tries to bribe new buyers.

Algorithm 1: The greedy bribing strategy

Input: Seller i's evaluation set $B_i, R_i, \bar{r}_i$, budget $\mathcal{B}$;
Output: A bribing strategy ϕ;
Function *greedy* $(B_i, R_i, \bar{r}_i, \mathcal{B})$

> Sort $r_j \in R_i$ in ascending order $R_D = r_0, ..., r_l$;
> **foreach** r_j *in* R_D **do**
>> **if** $\mathcal{B} > 0$ **then**
>>> **if** $r_j < 1$ **then**
>>>> $\phi(j) = min\{1 - r_j, \mathcal{B}\}$;
>>>> $\mathcal{B} = \mathcal{B} - \phi(j)$;
>>> **end**
>> **end**
> **end**
> **if** $\mathcal{B} > 0$ **then**
>> Compute non-rater set B_{-i};
>> **foreach** j *in* B_{-i} **do**
>>> **if** $\mathcal{B} > \bar{r}_i$ **then**
>>>> $\phi(j) = min\{1, \mathcal{B}\}$;
>>>> $\mathcal{B} = \mathcal{B} - \phi(j)$;
>>> **end**
>>> **else** break
>> **end**
> **end**

end

Proposition 1. *Under same combination strategies ϕ_{-i} of other sellers', the greedy bribing strategy ϕ^o for s_i always dominates others with same budget $\mathcal{B}$.*

Under the same budget, it shows when a buyer is bribed, it is worthy to bribe it completely with the highest rating. We further analyze under what conditions the rating system is bribery-proof. Based on the above proposition, we can show that no strategy is profitable if the number of potential buyers is too low, *i.e.*, the ratio between the number of current rated buyers $|B_i|$ and potential buyers is greater than the profit of an item k.

Lemma 1. *For seller s_i, no strategy is profitable if*

$$\frac{|B_i|}{\mathcal{N}} \geq k.$$

Proof. Let ϕ be any strategy for seller s_i and s_i's current rated buyers $B_i \neq \emptyset$. Then, by Definition 3, the payoff under ϕ with bribing budget $\mathcal{B} > 0$ is

$$\pi(\phi) = u^\phi - u^0 - \mathcal{B}.$$

According to Proposition 1, the greedy strategy always dominates others. We give proof for the greedy strategy here.

$$\pi(\phi) = u^\phi - u^0 - \mathcal{B}$$
$$= \mathcal{N}\frac{\mathcal{B} + \sum_{j \in R_i} r_j}{|B_i|} \cdot k - \mathcal{N}\frac{\sum_{j \in R_i} r_j}{|B_i|} \cdot k - \mathcal{B} = \mathcal{N} \cdot k \cdot \left(\frac{\mathcal{B}}{|B_i|}\right) - \mathcal{B}.$$

If strategy ϕ is profitable, then $\mathcal{N} \cdot k \cdot \left(\frac{\mathcal{B}}{|B_i|}\right) - \mathcal{B} > 0$. Therefore, for a profitable strategy, we have $\frac{|B_i|}{\mathcal{N}} < k$.

The lemma shows for sellers that have already bribed a large amount of buyers, further bribing may result in no profit. In real world, a bad reputed seller is more willing to improve its evaluation by performing a bribery strategy. In multiplayer situations, an open question for a seller is whether the best response to other player's strategies exists. In the next section, we analyze the strategy space for each player and identify the best response in multi-player situations.

3.1 Best Response

In the previous section, we showed that bribing more buyers may not result in more profit. When a seller bribes a buyer, it is better to induce the buyer to give the highest rating because the greedy strategy always dominates other strategies. However, in real world, a buyer who has given his/her rating may refuse to improve it. Thus, in this section, we redefine the game in this situation.

Based on Lemma 1, there exists a maximum number of buyers that can be bribed by seller s_i in the market. For seller s_i, it is always profitable to bribe the interacted buyers as shown in Lemma 1. For low $\bar{r}_i$, sellers will try to bribe new buyers to get the highest rating, 1. Thus, we redefine the strategy of a seller: a strategy ϕ for seller s_i is the number of buyers that will be bribed ϕ_i. The strategy space $[0, \mathcal{N}]$ covers all the choices that could be of interest to s_i. The redefinition of the strategy implies that the optimal strategy always bribes buyers with the highest rating. The new definition focuses on buyers instead of ratings.

Given the seller s_i's current state, aggregated ratings $\bar{r}_i$, and the number of potential buyers in the market, some interacted buyers refused to improve their given rating; thus, we are curious about how many buyers should be bribed for maximum profit.

The payoff to seller s_i from bribing ϕ_i buyers when the bribery strategy combination of the other sellers are $(\phi_1, ..., \phi_{i-1}, \phi_{i+1}, ..., \phi_M)$ is

$$\pi(\phi_i, \phi_{-i}) = u^{(\phi_i, \phi_{-i})} - u^0 - \phi_i, \tag{5}$$

where $u^{(\phi_i, \phi_{-i})}$ is the utility for seller s_i after s_i's execution of ϕ_i and the others' execution of ϕ_{-i}, respectively.

Theorem 5. *The bribery game G in a rating system has at least one Nash equilibrium.*

Proof. From Nash theorem [14], every game with a finite number of players and action profiles has at least one Nash equilibrium. In a bribery game, both the numbers of sellers and the strategy space are finite. Therefore, at least one Nash equilibrium can be reached in the game.

Let $(\phi_1^*, ..., \phi_M^*)$ be a Nash equilibrium in G. Then, for each ϕ_i^*, ϕ_i^* maximizes Eq. (5) given that the other sellers choose strategy combination $(\phi_1^*, ..., \phi_{i-1}^*, \phi_{i+1}^*, ..., \phi_M^*)$. The first-order condition for this optimization problem is

$$\frac{\partial \pi(\phi_i, \phi_{-i})}{\partial \phi_i} = \frac{\partial u^{(\phi_i, \phi_{-i})}}{\partial \phi_i} - 1 = 0 \Rightarrow \frac{\partial u^{(\phi_i, \phi_{-i})}}{\partial \phi_i} = 1. \tag{6}$$

Based on Lemma 1, for seller s_i, if $\frac{|B_i|}{N} \geq k$, the optimal strategy for s_i would be $\phi_i = 0$. The best response $(\phi_1^*, ..., \phi_M^*)$ will solve

$$\frac{\partial u^{(\phi_1, \phi_{-1})}}{\partial \phi_1} = 1; \cdots = 1; \frac{\partial u^{(\phi_M, \phi_{-M})}}{\partial \phi_M} = 1. \tag{7}$$

Subject to

$$\phi_i \in \mathbb{N}; \sum \phi_i <= N \text{ and } \phi_i = 0; \, if \, \frac{|B_i|}{N} > k. \tag{8}$$

Example 1. Consider two sellers of a duopoly game in a market, in which the number of interacted buyers with seller i is $B_i = 5$ and seller j is $B_j = 2$. These buyers cannot be bribed again by other bribery strategies. The average evaluation given by those buyers on sellers i and j are $\bar{r}_i = 0.2$ and $\bar{r}_j = 0.5$, respectively. The total number of buyers in the market is $n = 20$. The profit for bribing one buyer is $k = 2$. The strategies for the sellers are ϕ_i and ϕ_j. If (ϕ_i, ϕ_j) forms a Nash equilibrium, according to the above statements, the optimal strategy should solve Eq. (6).

Substituting ϕ_i in Eq. (6) and expanding u^{ϕ_i} yields

$$k \cdot [(N - \sum \phi_i) \frac{\bar{r}_i \cdot |B_i| + \phi_i}{|B_i| + \phi_i}]' = 1. \tag{9}$$

Each ϕ_i can be solved by computing all the sellers' first-order conditions subjecting to the constraints of Eq. (8). Solving the partial differential equations considering the constraints yields $\phi_i = 2, \phi_j = 1$.

In this example, the first four strategies for both sellers dominant others; we further present the bribery problem in the accompanying bi-matrix Table 1. The above result can also be derived by iterated elimination of strictly dominated strategies. In Table 1, both seller i and j have four strategies. For seller j, strategy 1 strictly dominates the others, so a rational seller j will play strategy 1 anyway. Thus, if seller i knows that seller j is rational, then seller i can eliminate strategies 0, 2, and 3 from seller i strategy space. In the second column, seller i's strategy 2 strictly dominates the others, leaving (6.57, 12.33) as the outcome of the game. We also plot seller i's payoff in Fig. 2. In the figure, when the strategy is larger than

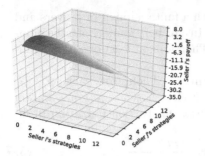

Fig. 2. Seller i's payoff.

Table 1. A bribery model of duopoly

		0	1	2	3
				j	
	0	5.20, 13.00	4.80, 15.00	4.40, 14.50	4.00, 13.00
i	1	7.00, 12.00	6.33, 13.67	5.67, 13.00	5.00, 11.40
	2	7.43, 11.00	6.57, 12.33	5.71, 11.50	4.86, 9.80
	3	7.00, 10.00	6.00, 11.00	5.00, 10.00	4.00, 8.20

4, the payoff is decreased linearly. This is consistent with Lemma 1 that bribing more buyers may not yield more profit.

In contrast to individual maximized payoff, the social optimum considers the whole social welfare and maximizes

$$\max_{0 \leq O^{**} < \infty} O^{**} = \max_{0 \leq \phi_i^{**} \leq N} \sum_{i=1}^{M} \pi(\phi_i^{**}, \phi_{-i}), \tag{10}$$

where $O = \sum_{i=1}^{M} \pi(\phi_i, \phi_{-i})$. Based on Theorem 5, we can infer this corollary.

Corollary 1. *Too many buyers are bribed in the Nash equilibrium compared to the social optimum. That is, $\mathcal{O}^* \geq \mathcal{O}^{**}$.*

Proof. We prove the corollary by contradiction and assume $\mathcal{O}^* < \mathcal{O}^{**}$. Then, there must exist at least one seller indexed by $i'(1 \leq i' \leq M)$, that $\phi_{i'}^{**} > \phi_{i'}^*$, since $\sum_{i=1}^{M} \phi_i^{**} = O^{**} > O^* = \sum_{i=1}^{M} \phi_i^*$. Because $\phi_{i'}^*$ is the number of bribed buyers by seller i' that can maximize $\pi(\phi_{i'}, \phi_{-i'})$, by definition, $\pi(\phi_{i'}^*, \phi_{-i'}) > \pi(\phi_{i'}^{**}, \phi_{-i'})$. Then, by maximizing the social welfare, we have

$$\sum_{i=1}^{i'} \pi(\phi_i, \phi_{-i}) + \pi(\phi_{i'}^*, \phi_{-i'}) + \sum_{i=i'}^{M} \pi(\phi_i, \phi_{-i}) >$$

$$\sum_{i=1}^{i'} \pi(\phi_i, \phi_{-i}) + \pi(\phi_{i'}^{**}, \phi_{-i'}) + \sum_{i=i'}^{M} \pi(\phi_i, \phi_{-i}),$$

which means there exists another $O' = \sum_{i=1}^{i'} \phi_i^* + \phi_{i'} + \sum_{i=i'}^{M} \phi_i^* < O^{**}$. This contradicts with our assumption that O^{**} is the social optimum.

The corollary illustrates the problem of commons [4], where the common resource is over-utilized because each seller considers only his or her own incentives, not the influence of his or her actions on the other sellers. In general, cooperative behavior is more desirable for the general good, but competitive behavior may bring higher individual gain.

4   Conclusion

We formally define the bribing behaviors of sellers in the rating system and analyze its effect leveraging game theory, where the sellers and their bribing behaviors are mapped into players and their corresponding strategy space, respectively. In the static game, where the sellers and buyers are static and their payoffs are common knowledge, we find Nash equilibrium existing in the system when all sellers perform their best-bribing strategy. In the equilibrium state, the system is bribery-proof and has the problem of commons [4] among sellers. For our future work, we will design a simulation framework to evaluate our previous theoretical analysis.

Acknowledgements. This work was supported in part by the National Natural Science Foundation of China under Grant No. 61906174 and 62172085, in part by the China Postdoctoral Science Foundation under Grant No. 2020M672275, and in part by JSPS KAKENHI Grant Number JP19H04170. We would like to thank Gautham Prakash for his sharing of Google Play apps dataset.

References

1. Cao, X., Huang, W., Yu, Y.: A complete and comprehensive movie review dataset (ccmr). In: Proceedings of the 39th International ACM SIGIR Conference on Research and Development in Information Retrieval, pp. 661–664. ACM (2016)
2. Faliszewski, P., Hemaspaandra, E., Hemaspaandra, L.A.: How hard is bribery in elections? J. Artif. Intell. Res. **35**, 485–532 (2009)
3. Faliszewski, P., Hemaspaandra, E., Hemaspaandra, L.A., Rothe, J.: Llull and copeland voting computationally resist bribery and constructive control. J. Artif. Intell. Res. **35**, 275–341 (2009)
4. Gibbons, R.: Game Theory for Applied Economists. Princeton University Press (1992)
5. Grandi, U., Stewart, J., Turrini, P.: Personalised rating. Auton. Agent. Multi-Agent Syst. **34**(2), 1–38 (2020). https://doi.org/10.1007/s10458-020-09479-2
6. Grandi, U., Turrini, P.: A network-based rating system and its resistance to bribery. In: Proceedings of the Twenty-Fifth International Joint Conference on Artificial Intelligence (IJCAI), pp. 301–307. AAAI Press (2016)
7. Helpman, E., Persson, T.: Lobbying and legislative bargaining. Adv. Econ. Anal. Policy **1**(1) (2001)
8. Jiang, S., Zhang, J., Ong, Y.S.: An evolutionary model for constructing robust trust networks. In: Proceedings of the 2013 International Conference on Autonomous Agents and Multi-agent Systems, pp. 813–820. International Foundation for Autonomous Agents and Multiagent Systems (2013)
9. Jøsang, A., Ismail, R., Boyd, C.: A survey of trust and reputation systems for online service provision. Decis. Supp. Syst. **43**(2), 618–644, 107441 (2007)
10. Lianju, S., Luyan, P.: Game theory analysis of the bribery behavior. Int. J. Bus. Soc. Sci. **2**(8) (2011)
11. Liu, C., et al.: Fraud transactions detection via behavior tree with local intention calibration. In: Proceedings of the 26th ACM SIGKDD International Conference on Knowledge Discovery and Data Mining, pp. 3035–3043 (2020)
12. Liu, Y., Zhou, X., Yu, H.: 3r model: a post-purchase context-aware reputation model to mitigate unfair ratings in e-commerce. Knowl.-Based Syst. **231**, 107441 (2021)

13. Manzoor, S., Luna, J., Suri, N.: Attackdive: diving deep into the cloud ecosystem to explore attack surfaces. In: 2017 IEEE International Conference on Services Computing (SCC), pp. 499–502. IEEE (2017)
14. Nash, J.F.: Equilibrium points in n-person games. Proc. Natl. Acad. Sci. USA **36**(1), 48, 107441 (1950)
15. Ouffoué, G.L., Zaïdi, F., Cavalli, A.R., Lallali, M.: An attack-tolerant framework for web services. In: 2017 IEEE International Conference on Services Computing (SCC), pp. 503–506. IEEE (2017)
16. Parkes, D.C., Tylkin, P., Xia, L.: Thwarting vote buying through decoy ballots. In: Proceedings of the 16th Conference on Autonomous Agents and MultiAgent Systems (AAMAS), pp. 1679–1681 (2017)
17. Ramos, G., Boratto, L., Caleiro, C.: On the negative impact of social influence in recommender systems: a study of bribery in collaborative hybrid algorithms. Inf. Process. Manage. **57**(2), 102058 (2020)
18. Resnick, P., Zeckhauser, R., Swanson, J., Lockwood, K.: The value of reputation on ebay: a controlled experiment. Exp. Econ. **9**(2), 79–101, 102058 (2006)
19. Sampath, V.S., Gardberg, N.A., Rahman, N.: Corporate reputation's invisible hand: bribery, rational choice, and market penalties. J. Bus. Ethics **151**(3), 743–760, 102058 (2018)
20. Saúde, J., Ramos, G., Boratto, L., Caleiro, C.: A robust reputation-based group ranking system and its resistance to bribery. ACM Trans. Knowl. Discov. Data **16**(2), 1–35, 102058 (2021)
21. Wang, D., Muller, T., Zhang, J., Liu, Y.: Quantifying robustness of trust systems against collusive unfair rating attacks using information theory. In: IJCAI, pp. 111–117 (2015)
22. Wang, D., Muller, T., Zhang, J., Liu, Y.: Is it harmful when advisors only pretend to be honest? In: AAAI, pp. 2551–2557 (2016)
23. Wang, G., et al.: Serf and turf: crowdturfing for fun and profit. In: Proceedings of the 21st International Conference on World Wide Web, pp. 679–688. ACM (2012)
24. Xu, C., Zhang, J., Sun, Z.: Online reputation fraud campaign detection in user ratings. In: Proceedings of the Twenty-Sixth International Joint Conference on Artificial Intelligence, (IJCAI), pp. 3873–3879 (2017)
25. Ye, Q., Law, R., Gu, B.: The impact of online user reviews on hotel room sales. Int. J. Hosp. Manag. **28**(1), 180–182 (2009)
26. Zhou, X., Lin, D., Ishida, T.: Evaluating reputation of web services under rating scarcity. In: 2016 IEEE International Conference on Services Computing (SCC), pp. 211–218. IEEE (2016)
27. Zhou, X., Murakami, Y., Ishida, T., Liu, X., Huang, G.: Arm: toward adaptive and robust model for reputation aggregation. IEEE Transac. Autom. Sci. Eng. **17**(1) (2019)
28. Zhu, H., Xiong, H., Ge, Y., Chen, E.: Discovery of ranking fraud for mobile apps. IEEE Trans. Knowl. Data Eng. **27**(1), 74–87 (2014)

IDSGAN: Generative Adversarial Networks for Attack Generation Against Intrusion Detection

Zilong Lin[1,2], Yong Shi[1], and Zhi Xue[1(✉)]

[1] Shanghai Jiao Tong University, Shanghai, China
{shiyong,zxue}@sjtu.edu.cn
[2] Indiana University Bloomington, Bloomington, IN, USA
zillin@indiana.edu

Abstract. As an essential tool in security, the intrusion detection system bears the responsibility of the defense to network attacks performed by malicious traffic. Nowadays, with the help of machine learning algorithms, intrusion detection systems develop rapidly. However, the robustness of this system is questionable when it faces adversarial attacks. For the robustness of detection systems, more potential attack approaches are under research. In this paper, a framework of the generative adversarial networks, called IDSGAN, is proposed to generate the adversarial malicious traffic records aiming to attack intrusion detection systems by deceiving and evading the detection. Given that the internal structure and parameters of the detection system are unknown to attackers, the adversarial attack examples perform the black-box attacks against the detection system. IDSGAN leverages a generator to transform original malicious traffic records into adversarial malicious ones. A discriminator classifies traffic examples and dynamically learns the real-time black-box detection system. More significantly, the restricted modification mechanism is designed for the adversarial generation to preserve original attack functionalities of adversarial traffic records. The effectiveness of the model is indicated by attacking multiple algorithm-based detection models with different attack categories. The robustness is verified by changing the number of the modified features. A comparative experiment with adversarial attack baselines demonstrates the superiority of our model.

Keywords: Generative adversarial networks · Intrusion detection · Adversarial examples · Black-box attack

1 Introduction

With the spread of security threats on the Internet, the intrusion detection system (IDS) has become the essential tool to detect and defend network attacks in

Z. Lin—This work was done when the author was at Shanghai Jiao Tong University.

J. Gama et al. (Eds.): PAKDD 2022, LNAI 13282, pp. 79–91, 2022.
https://doi.org/10.1007/978-3-031-05981-0_7

the form of malicious network traffic. The IDS monitors the network by analyzing the features extracted from the network traffic and raises the alarm if unsafe
traffic is identified. The main aim of IDS is to audit and classify the network traffic records between normal ones and malicious ones. As a classification issue, the
IDS has widely leveraged machine learning algorithms to classify traffic based
on the feature records, including KNN, SVM, Decision Tree, etc. [13]. In recent
years, deep learning algorithms further contributed with an improvement to IDS
in accuracy and simplification [7].

However, the classification algorithms expose the vulnerability under the
adversarial examples in the recent work, in which the designed adversarial inputs
would cause the classifiers' misclassification [2]. For such an attack, the generative adversarial network (GAN) [3] is the potential method for such adversarial
example generation. GAN has been implemented in attacks within information
security, like malware generation, author attribute anonymity, and password
guessing [4,5,9,12]. Although some previous work has applied adversarial learning methods to attack target IDS models [14,15], we still hardly know GAN's
capability of dynamically attacking multiple IDS models with adversarial malicious traffic with its functionality preserved.

In this paper, we proposed a new framework of GAN, named IDSGAN, for
the adversarial attack generation against intrusion detection systems. The goal of
the model is to generate malicious feature records of the attack traffic, which can
deceive and bypass the detection of the defense systems and, finally, to guide the
evasion attack in real networks. Following adversarial malicious traffic records,
the attackers can design traffic to evade the detection in real-world attacks. In
the model, we designed and improved the generator and the discriminator based
on Wasserstein GAN [1]. The generator generates adversarial malicious traffic
records. Given that the internal structure and parameters of the black-box IDS
are unknown for attackers, the discriminator learns the IDS by its real-time
outputs and provides feedback for the training of the generator. We assumed that
the outputs of the black-box IDS models can be obtained by querying IDS with
traffic records. In the experiment, we modeled multiple IDS models powered by
different baseline machine learning algorithms, simulating the intrusion detection
systems in reality. In summary, the following contributions are made in this work:

- An improved framework named IDSGAN is proposed to generate adversarial
  malicious traffic records to guide the evasion attack against IDS. To preserve
  the functionality of real traffic generated by IDSGAN, we designed a mechanism that restricts the modification to functional features of original malicious
  traffic records in adversarial example generation. By dynamically learning the
  real-time results from IDS models in adversarial training, IDSGAN can attack
  the updated IDS models powered by different algorithms.
- We attacked different variants of IDS models to evaluate the effectiveness
  and robustness of IDSGAN. We measured the attack effectiveness with quantitative evaluations. To verify the robustness, we analyzed the attacks under
  different numbers of the modified features.

– We demonstrated that our model outperformed other adversarial attack baselines of intrusion detection.

2 Related Work

With the rapid development of adversarial learning, adversarial examples generation has attracted researchers' interests and been applied in intrusion detection.

Rigaki leveraged FGSM and JSMA methods to generate the adversarial traffic records, which can evade the detection of IDS, based on the NSL-KDD
dataset [10]. Wang further proposed to apply more adversarial attack algorithms
(including JSMA, Targeted FGSM, DeepFool, CW, etc.) to craft adversarial traffic records [14]. Assuming that the attackers have knowledge about the target
victim models, both works focused on the white-box adversarial attacks.

In the research on black-box attacks, Yang proposed zeroth-order optimization and generative adversarial networks to attack IDS [15]. However, in this
work, the traffic record features were manipulated without the discrimination of
features' function, leading to the ineffectiveness of the traffic's attack functionality. Additionally, without learning the latest knowledge of IDS like querying,
the discriminator of GAN, a pretrained classifier, cannot dynamically adapt the
generation for the attack against the updated IDS models. To solve the above
issues, we designed a mechanism to preserve the original attack functionality
in different types of malicious traffic, and leveraged the real-time query results
from the target IDS to let our model dynamically learn and adapt the target
IDS model.

3 Methodology

IDSGAN, an improved GAN model, aims to perform the evasion attacks by
generating adversarial malicious traffic records, which enable fooling the black-
box IDS models. In IDSGAN, the generator and the discriminator engage in
a two-player minimax game for adversarial example generation. Based on the
characteristics of the benchmark dataset NSL-KDD, it is necessary to preprocess
the dataset to fit the model.

3.1 Dataset: NSL-KDD Dataset Description

NSL-KDD is used as a benchmark dataset to evaluate IDS today. In NSL-KDD,
the dataset comprises the training set KDDTrain+ and the test set KDDTest+.
Extracted from the real network environment, the data contains the normal
traffic and four main categories of malicious traffic, including Probing (Probe),
Denial of Service (DoS), User to Root (U2R), and Root to Local (R2L).

The traffic records in NSL-KDD are extracted into the feature sequences,
as the abstract description of the normal and malicious network traffic.

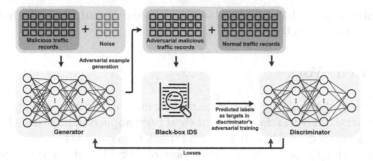

Fig. 1. The training of IDSGAN. The training set is divided into normal traffic records and malicious traffic records. After adding noise, the malicious records are sent into the generator and modified as the adversarial ones. The black-box IDS predicts the adversarial malicious records and normal ones. The predicted labels are used in the discriminator to learn the black-box IDS. The losses of the generator and discriminator are calculated based on the results of the discriminator and the predicted labels of IDS.

Each element in the sequence represents one feature of the traffic. There are 9 features in discrete values and 32 features in continuous values, a total of 41 features. Based on the meanings of each feature, these features consists of four sets including "intrinsic", "content", "time-based traffic", and "host-based traffic".

3.2 Data Preprocessing

Given multiple feature types and ranges in NSL-KDD, the numeric conversion and the normalization are leveraged for preprocessing data before being fed into models. Three nonnumeric features (including protocol_type, service, flag) are embedded. For instance, "protocol_type", including three attributes: TCP, UDP, and ICMP, will be converted into one-hot vectors. To eliminate the range impact between feature values in the input vectors, the min-max normalization method transforms all original numeric features of the data into the range of $[0, 1]$.

3.3 Structure of IDSGAN

Many GAN structures have been designed for different requests. To prevent the non-convergence and instability of GAN, IDSGAN is proposed based on Wasserstein GAN [1]. For the evasion attack against IDS, the generator modifies features to generate adversarial malicious traffic records. Complying with the proposed mechanism of restricted modification, the generation preserves original malicious functionalities of adversarial traffic. The discriminator is trained to learn the black-box IDS and feedbacks the training of the generator. Different machine learning algorithms power the black-box IDS to simulate the IDS models in the real world. The framework of IDSGAN is delineated in Fig. 1.

Restricted Modification Mechanism. Although the aim of IDSGAN's generating adversarial attack examples is to evade IDS, the premise is that this generation should retain the original attack functionality of malicious traffic so that such traffic generated based on the adversarial records from IDSGAN can be reproduced and launch network attacks in reality.

Based on the attack principles and purposes, it is evident that each category of attacks has its specific functional features representing the basic functionality of this attack. It means that, in adversarial example generations, the attack attribute would remain unaltered if we solely fine-tuned nonfunctional features, not functional features. Thus, in our mechanism, the functional features of each attack should be kept unchanged to preserve malicious functionalities. The mechanism allows the fine-tuning or retention of nonfunctional features that do not represent the functionality relevant to that attack. These retained features, including functional features, are named the unmodified features in our work. The functional features of each attack category in NSL-KDD are in Table 1.

Table 1. The functional features of each attack category.

Attack	Intrinsic	Content	Time-based traffic	Host-based traffic
Probe	✓		✓	✓
DoS	✓		✓	
U2R	✓	✓		
R2L	✓	✓		

Generator. As a crucial part of the model, the generator plays the role of generating adversarial malicious traffic records for the evasion attack to IDS.

Aiming to transform an original example into an adversarial one, the initial noise perturbation is added to the original traffic record example before the generation. We concatenated the m-dimensional original example vector M and the n-dimensional noise vector N as an input vector fed into the generator. As the original example part, M has been preprocessed. To be consistent with the normalized vector M, the elements of the noise part are randomized in a uniform distribution within the range of $[0, 1]$.

Our proposed structure of the generator has a neural network structure with five linear layers. The ReLU non-linearity $F = \max(0, x)$ is utilized to activate the outputs of former four linear layers. To ensure that the adversarial examples share the same dimension as the original example vector M, the output layer has m units. The loss of the generator is calculated based on the classification results from the discriminator (see Eq. 1).

In addition, some tricks exist in the processing of the modified features. To restrict the output elements into the range of $[0, 1]$, we set the element above 1 as 1 and the element below 0 as 0. Concerning that "`intrinsic`" features are

---

Algorithm 1. IDSGAN

Require:
    Original normal and malicious traffic records S_{normal}, S_{attack};
    The noise N for the adversarial generation;
    The pretrained black-box IDS B;
Ensure:
    The optimization of the generator G and the discriminator D;
1: Initialize the generator G and the discriminator D;
2: **for** number of training iterations **do**
3:    **for** G-steps **do**
4:      G generates the adversarial malicious traffic examples based on S_{attack};
5:      Update the parameters of G according to Eq. 1;
6:    **end for**
7:    **for** D-steps **do**
8:      B classifies the training set $(S_{normal}, G(S_{attack}, N))$, getting predicted labels;
9:      D classifies the same training set $(S_{normal}, G(S_{attack}, N))$;
10:     Update the parameters of D according to Eq. 2;
11:    **end for**
12: **end for**

---

the functional features in all the attacks of NSL-KDD, the nonnumeric features will not be modified. For the binary features in the feature vector, the values of those modified binary features will be transformed into binary values with 0.5 as the threshold after the generator's processing. We transformed the values above the threshold into 1 and those below the threshold into 0.

Discriminator. Without the knowledge of the structure and parameters in black-box IDS models, we assumed that the real-time classification results of the black-box IDS models can be obtained by querying. The discriminator is a multi-layer neural network to classify malicious records and normal ones. Also, the discriminator is responsible for learning and imitating the black-box IDS based on the detected samples and their latest predictions from the target IDS. In adversarial training, the normal traffic records and the adversarial malicious traffic records are first classified by the black-box IDS. Then, for the imitation to the black-box IDS, the same dataset labeled by the target IDS is shared to the discriminator as the training set whose current labels are the real-time predictions from this IDS, shown in Fig. 1. Reflecting the dynamic optimization of the structure and parameters in the black-box IDS, the IDS's real-time predictions are leveraged for the discriminator to learn the IDS dynamically.

Additionally, the discriminator helps the training of the generator, whose gradient is back-propagated from the discriminator. Based on the loss calculated by the real-time results of the discriminator and the black-box IDS, the generator dynamically optimizes the evasion strategy to fine-tune malicious records, enabling adversarial examples from IDSGAN to attack the real-time IDS models.

Training Algorithms. In the generator's training, the discriminator's detection results to the adversarial examples provide the gradient information for the generator. The loss function of the generator is defined in Eq. 1.

$$L_G = \mathbb{E}_{M \in S_{attack}, N} D(G(M, N)) \tag{1}$$

where S_{attack} is the original malicious traffic records; G and D represent the generator and the discriminator, respectively. To train and optimize the generator for fooling the black-box IDS, we need to minimize L_G.

The training set of the discriminator includes adversarial malicious records and normal records. With the aim of learning the black-box IDS, the discriminator gets the loss calculated by the output labels of the discriminator and the predicted labels achieved from the black-box IDS. Thus, the loss function for the discriminator's optimization is in Eq. 2.

$$L_D = \mathbb{E}_{s \in B_{normal}} D(s) - \mathbb{E}_{s \in B_{attack}} D(s) \tag{2}$$

where s means the training set of the discriminator; B_{normal} and B_{attack} mean the normal traffic records and the adversarial traffic records with the predicted labels from the black-box IDS as the ground truth, respectively.

Based on Wasserstein GAN, RMSProp is the optimizer of IDSGAN to optimize the parameters in the model. The outline of the IDSGAN adversarial training is shown in Algorithm 1.

4 Empirical Evaluation

4.1 Experimental Setup

In our implementation, we utilized PyTorch as the deep learning framework to construct the IDSGAN model. We built black-box IDS models by scikit-learn. The proposed model was evaluated on a Linux PC with Intel Core i7-2600.

IDSGAN is trained with a 64 batch size for 100 epochs. The learning rates of the generator and discriminator are both 0.0001. The weight clipping threshold in the discriminator's training is 0.01. The dimension of the noise vector is 9.

Based on the relevant researches in intrusion detection, multiple machine learning algorithms have been used in IDS. To evaluate the capacity and generalization of our purposed model against IDS comprehensively, we built seven types of algorithm-based black-box IDS models which are commonly applied as the baseline approaches to validate improved intrusion detection systems [13]. The algorithms of the black-box IDS models adopted in the evaluation include Support Vector Machine (SVM), Naive Bayes (NB), Multilayer Perceptrons (MLP), Logistic Regression (LR), Decision Tree (DT), Random Forest (RF), and K-Nearest Neighbors (KNN). In the evaluation, the black-box IDS models have been pretrained with their training set before generating adversarial samples.

The training and test sets are designed based on the NSL-KDD dataset (containing KDDTrain+ and KDDTest+). The training set for the black-box IDS consists of one-half of the records in KDDTrain+, containing normal and malicious traffic records. The training set of the discriminator includes the normal traffic records in the other half of KDDTrain+ and the adversarial malicious records from the generator. Given that the nonfunctional features to be modified in each attack category are different, IDSGAN generates the adversarial examples for solely one attack category each time. So, the training set of the generator for one attack category is the records of that category in the other half of KDDTrain+. The records of one attack category in KDDTest+ are the test set for the generator, which aims for the adversarial samples of that category.

For the experimental metrics, the detection rate and the evasion increase rate are calculated, presenting the performance of IDSGAN directly and comparatively. The detection rate (i.e., DR) is the proportion of correctly detected malicious traffic records by the black-box IDS to all of those attack records, directly showing the evasion ability of the adversarial examples and the robustness of the black-box IDS. The original and adversarial detection rates represent the detection rate of the original malicious traffic records and that of the adversarial malicious traffic records, respectively. In addition, we defined the evasion increase rate (i.e., EIR) as the rate of the increase in the undetected malicious examples by IDS between original malicious examples and adversarial examples, evaluating the evasion attack of IDSGAN. These metrics are calculated as follows.

$$DR = \frac{Num.\ of\ correctly\ detected\ attacks}{Num.\ of\ all\ the\ attacks} \tag{3}$$

$$EIR = 1 - \frac{Adversarial\ detection\ rate}{Original\ detection\ rate} \tag{4}$$

A lower detection rate means more malicious traffic records evade the detection of the black-box IDS, revealing a higher adversarial attack effectiveness. On the contrary, a lower evasion increase rate reflects a low increase rate on the number of adversarial examples evading the black-box IDS, meaning a low improvement in the evasion attack capacity of adversarial examples compared with the original ones. So, the goal for the IDSGAN optimization is to obtain a lower detection rate and a higher evasion increase rate.

4.2   Effectiveness in Different Attack Categories

To evaluate the model comprehensively, the trained IDSGAN was tested to generate the adversarial malicious samples based on KDDTest+. Given that DoS and Probe are both attacks based on network, we only tested on DoS to show the performance of IDSGAN on such kinds of attacks. Also, the attacks based on the traffic content like U2R and R2L were tested. Due to the similar characteristics shared among U2R and R2L—leading to the same functional features among two categories—and their small data amount in the dataset, we gathered U2R and R2L as one attack group in our work.

Before the generation by IDSGAN, the original detection rates to DoS, U2R, and R2L were calculated on the trained black-box IDS, shown in Table 2. By reason of the small number of U2R and R2L records in the training set, the insufficient learning makes the low original detection rates to U2R and R2L. A similar performance has been reported in the previous work [6,11].

First, to evaluate the effectiveness of IDSGAN, we tested the capacity of IDSGAN in different attacks with only the functional features unmodified. In

Table 2. The performance of IDSGAN under DoS, U2R and R2L. The first two lines are the black-box IDS's original detection rates to the original test set. In Column "Add", "×" means the adversarial generation with only the functional features unmodified, while "✓" means that the unmodified features are added in the generation.

Add	Attack	Metric (%)	SVM	NB	MLP	LR	DT	RF	KNN
–	DoS	Original DR	82.37	84.94	82.70	79.85	75.13	73.28	77.22
–	U2R & R2L	Original DR	0.68	6.19	4.54	0.64	12.66	2.44	5.69
×	DoS	Adversarial DR	0.46	0.01	0.72	0.36	0.20	0.35	0.37
×	DoS	EIR	99.44	99.99	99.13	99.55	99.73	99.52	99.52
×	U2R & R2L	Adversarial DR	0.00	0.01	0.00	0.00	0.02	0.00	0.00
×	U2R & R2L	EIR	100.00	99.84	100.00	100.00	99.84	100.00	100.00
✓	DoS	Adversarial DR	1.03	1.26	1.21	0.97	0.36	0.77	1.16
✓	DoS	EIR	98.75	98.52	98.54	98.79	99.52	98.95	98.50
✓	U2R & R2L	Adversarial DR	0.01	0.08	0.01	0.00	0.07	0.00	0.00
✓	U2R & R2L	EIR	98.53	98.71	99.78	100.00	99.45	100.00	100.00

the experiment results shown in Table 2 and Fig. 2, all of the adversarial detection rates to DoS, U2R, and R2L under different IDS algorithms decline compared with the original detection rates and reach around 0, indicating that the IDS models are almost incapable of classifying any adversarial examples.

As shown in Fig. 2(a), the adversarial detection rates to DoS under all detection algorithms remarkably decrease from around 80% to less than 1%. Although Multilayer Perceptrons shows the best robustness in the list of all IDS models, its adversarial detection rate to DoS is only 0.72%. More than 99.0% of the adversarial DoS traffic examples evade the detection of the black-box IDS model in each test. The evasion increase rate in each DoS test is above 99.0%. The results indicate the excellent effectiveness of IDSGAN in DoS.

For U2R and R2L in Fig. 2(b), while the difference of the original detection rates between algorithms is noticeable, all of the adversarial detection rates are equal to or close to 0, indicating almost all of the original detectable examples of U2R and R2L can fool and evade the IDS after the adversarial generation. With a significant increase in the adversarial malicious examples fooling the IDS, the evasion increase rates are also high, all of which are above 99.5%.

The low adversarial detection rates and high evasion increase rates obtained under various attack categories and multiple IDS algorithms indicate IDSGAN's great effectiveness and generalization in adversarial attacks of evading various IDS models. Besides, some tiny differences still exist in the performance under different attack categories and different IDS models.

4.3   Robustness with Different Numbers of Modified Features

The number of the modified features is an influential factor that significantly affects our proposed model's success in adversarial attacks. To evaluate its robustness, we conducted the contrast tests on DoS, U2R, and R2L, altering

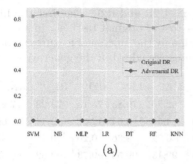 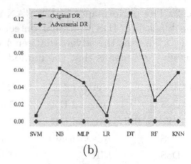

(a)   (b)

Fig. 2. The comparisons of the adversarial detection rates and the original detection rates under different black-box IDS models with only the functional features unmodified. (a) is the results of DoS and (b) is the results of U2R and R2L.

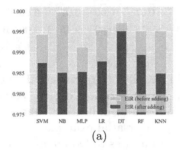 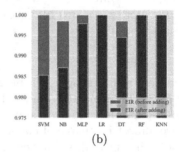

(a)   (b)

Fig. 3. The comparisons of the evasion increase rates under various algorithms of black-box IDS models before adding unmodified features and after adding unmodified features. (a) is the results of DoS and (b) is the results of U2R and R2L.

the number of the modified features and measuring such impact. Given that the attack with the fewest unmodified features is the scenario with only retaining this attack's functional features representing the attack function, the way to alter the number of modified features is adding some nonfunctional features into the group of the unmodified features. All of the nonfunctional features were modified features in Sect. 4.2. We randomly picked 50% of features from each nonfunctional feature set as the added unmodified features in the contrast tests.

In Table 2 and Fig. 3, the evasion increase rates decrease slightly or maintain in contrast experiments, compared with the results in Sect. 4.2 with only the functional features unmodified. In Fig. 3, the minor changes in metric results indicate IDSGAN's strong robustness under fewer modified features. The strong perturbation to modified features still causes high evasion increase rates for all IDS algorithms, even if lowering the number of modified features. For instance, the evasion increase rates keep unchanged after adding unmodified features when Logistic Regression detects U2R and R2L.

However, with the increase of unmodified features, more original information of a traffic record is retained in the adversarial generation, leading to the rise of accuracy in IDS detection. Besides, the improvements in detection results are different under various attack categories and IDS algorithms. Considering the detection improvement's relatedness with attack categories, the evasion increase rate of DoS under K-Nearest Neighbors decreases about 1.00% after changing the modified features in Fig. 3(a). However, no variation occurs in results when that algorithm detects U2R and R2L in Fig. 3(b). Also, for the relatedness with IDS algorithms, the evasion increase rates under Naive Bayes in two attacks change more remarkably than others, revealing the smaller number of modified features has a more significant impact on the evasion against Naive Bayes.

Thus, although the number reduction of the modified features slightly causes more failure in adversarial malicious records bypassing the IDS detection, the sustained high evasion increase rates verify the strong robustness of IDSGAN in the evasion attack. Also, the result shows that the robustness performance of adversarial evasion attacks relies on attack categories and IDS algorithms.

4.4 Baseline Comparisons

Besides GAN, different adversarial learning methods have been proposed and applied to help traffic records evade the detection of IDS [8, 10, 14]. We compared our proposed approach with the following competitive baseline attack models: JSMA Attack, Targeted FGSM Attack, DeepFool Attack, and CW Attack.

For a fair comparison of the baselines, the intrusion detection system in all the experiments is Multilayer Perceptrons. The detection system and attack models share the same architecture and hyper-parameters as the setting in the previous work [14], while the proposed restricted modification mechanism takes action to keep traffic functionality. Leveraging the dataset described in Sect. 4.1, we trained and tested the models in the experiments. Table 3 summarizes the malicious traffic record detection results of adversarial attacks. Our proposed model outperforms all the baselines by a wide margin, and the attack results are notable in both malicious traffic categories. In addition, the results have observable differences between various adversarial attack models, in which JSMA Attack and CW Attack are less devastating than other attack approaches.

We also compared with the GAN baselines in [15], which statically attacked the target IDS without the restriction on the feature modification. The restricted modification mechanism leads the GAN's capacity of evading IDS in [15] to drop quickly. The detection rates rise from 24.34% to 32.45% in DoS and from 2.23% to 3.39% in U2R and R2L. Also, IDSGAN's dynamical imitation strategy by querying the target IDS strengthens evasion attacks remarkably. Compared with [15], IDSGAN outperforms in evasion effectiveness and attack functionality preservation significantly.

Table 3. Detection rates on different adversarial attack approaches

Attack	Original	JSMA	FGSM	DeepFool	CW	Unrestricted static GAN	Static GAN	IDSGAN
DoS	79.12%	20.75%	7.19%	15.86%	24.67%	24.34%	32.45%	0.61%
U2R & R2L	4.78%	3.06%	0.15%	1.04%	3.20%	2.23%	3.39%	0.00%

5 Conclusion and Future Work

IDSGAN is a novel framework of generative adversarial networks aiming to generate adversarial attacks that can evade IDS. The model design and the restricted modification mechanism enable IDSGAN to attack against real-time black-box IDS models powered by multiple machine learning algorithms and preserve traffic's malicious functionalities, respectively. In the evaluation, IDS-GAN shows its effectiveness in generating adversarial malicious traffic records of different attacks, lowering the detection rates of various IDS models to around 0%. In the robustness evaluation, the evasion capacity of adversarial malicious examples maintains or slightly reduces after limiting the number of modified features, indicating the model's strong robustness. Also, the comparisons with other adversarial attack methods demonstrate its better performance.

We focused on generating the adversarial malicious traffic records capable of evading the target IDS in this work. In the next step, depending on such adversarial examples, we would produce the malicious network traffic, whose features match the adversarial traffic records, to experimentally attack the running IDS after being approved by our institutional Ethics Board.

References

1. Arjovsky, M., Chintala, S., Bottou, L.: Wasserstein generative adversarial networks. In: International Conference on Machine Learning, pp. 214–223. PMLR (2017)
2. Carlini, N., Wagner, D.: Adversarial examples are not easily detected: bypassing ten detection methods. In: Proceedings of the 10th ACM Workshop on Artificial Intelligence and Security, pp. 3–14. ACM, Dallas, TX, USA (2017)
3. Goodfellow, I., et al.: Generative adversarial nets. In: Advances in Neural Information Processing Systems, pp. 2672–2680 (2014)
4. Hitaj, B., Gasti, P., Ateniese, G., Perez-Cruz, F.: PassGAN: a deep learning approach for password guessing. In: Deng, R.H., Gauthier-Umaña, V., Ochoa, M., Yung, M. (eds.) ACNS 2019. LNCS, vol. 11464, pp. 217–237. Springer, Cham (2019). https://doi.org/10.1007/978-3-030-21568-2_11
5. Hu, W., Tan, Y.: Generating adversarial malware examples for black-box attacks based on gan. arXiv:1702.05983 (2017)
6. Ingre, B., Yadav, A., Soni, A.K.: Decision tree based intrusion detection system for NSL-KDD dataset. In: Satapathy, S.C., Joshi, A. (eds.) ICTIS 2017. SIST, vol. 84, pp. 207–218. Springer, Cham (2018). https://doi.org/10.1007/978-3-319-63645-0_23

7. Li, Z., Qin, Z., Huang, K., Yang, X., Ye, S.: Intrusion detection using convolutional neural networks for representation learning. In: Liu, D., Xie, S., Li, Y., Zhao, D., El-Alfy, E.-S.M. (eds.) ICONIP 2017. LNCS, vol. 10638, pp. 858–866. Springer, Cham (2017). https://doi.org/10.1007/978-3-319-70139-4_87
8. Pacheco, Y., Sun, W.: Adversarial machine learning: a comparative study on contemporary intrusion detection datasets. In: ICISSP, pp. 160–171 (2021)
9. Pasquini, D., Gangwal, A., Ateniese, G., Bernaschi, M., Conti, M.: Improving password guessing via representation learning. In: 2021 IEEE Symposium on Security and Privacy (SP), pp. 1382–1399. IEEE (2021)
10. Rigaki, M.: Adversarial deep learning against intrusion detection classifiers (2017)
11. Sapre, S., Ahmadi, P., Islam, K.: A robust comparison of the kddcup99 and NSL-KDD IoT network intrusion detection datasets through various machine learning algorithms. arXiv:1912.13204 (2019)
12. Shetty, R., Schiele, B., Fritz, M.: A4nt: author attribute anonymity by adversarial training of neural machine translation. In: 27th USENIX Security Symposium (USENIX Security 2018), pp. 1633–1650 (2018)
13. Tsai, C.F., Hsu, Y.F., Lin, C.Y., Lin, W.Y.: Intrusion detection by machine learning: a review. Expert Syst. Appl. 36(10), 11994–12000 (2009)
14. Wang, Z.: Deep learning-based intrusion detection with adversaries. IEEE Access 6, 38367–38384 (2018)
15. Yang, K., Liu, J., Zhang, C., Fang, Y.: Adversarial examples against the deep learning based network intrusion detection systems. In: MILCOM 2018–2018 IEEE Military Communications Conference (MILCOM), pp. 559–564. IEEE (2018)

Recommending Personalized Interventions to Increase Employability of Disabled Jobseekers

Ha Xuan Tran[1]([✉]), Thuc Duy Le[1], Jiuyong Li[1], Lin Liu[1], Jixue Liu[1], Yanchang Zhao[2], and Tony Waters[3]

[1] University of South Australia, Adelaide, Australia
ha.tran@mymail.unisa.edu.au
[2] Data61, CSIRO, Canberra, Australia
[3] Maxima Training Group (Aust) Ltd., Adelaide, Australia

Abstract. An emerging problem in Disability Employment Services (DES) is recommending to disabled jobseekers the right skill to upgrade and the right upgrade level to achieve maximum increase in their employment potential. This problem involves causal reasoning to estimate the causal effect on employment status to determine the most effective personalized intervention. In this paper, we propose a causal graph based method to solve the intervention recommendation problem. Personalized causal graphs of individual training samples are reverse engineered from a population-level causal graph using linear interpolation. A prediction model is built from these personalized graphs to recommend interventions. Experiments with a case study from an Australian DES provider show that by adopting interventions recommended by our method, disabled jobseekers would increase their employability by up to 24%. Evaluations with public datasets also show its advantages in other applications.

Keywords: Causal inference · Intervention recommendation

1 Introduction

Disabled people experience more long-term unemployment than those without disability [1]. DES providers give employment support to disabled jobseekers by providing expert consultation and upskilling programs. A problem confronting the providers is which factor (or skill) jobseekers should intervene (or upgrade) to increase their employability. Given a jobseeker's profile, there are many options for intervention. The key challenge is to determine the most effective intervention for the jobseeker: not only *which factor* to intervene but also its *intervention level* such that this person can achieve maximum employability increase.

Supplementary Information The online version contains supplementary material available at https://doi.org/10.1007/978-3-031-05981-0_8.

J. Gama et al. (Eds.): PAKDD 2022, LNAI 13282, pp. 92–104, 2022.
https://doi.org/10.1007/978-3-031-05981-0_8

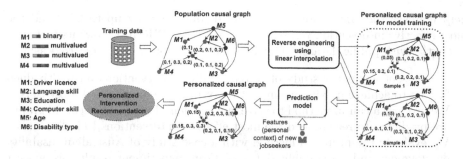

Fig. 1. Overview of personalized intervention recommendation (PIR). Different weights of an edge directed to the outcome (orange node) denote the effects of a causal factor (blue node) on employment for different intervention levels. (Color figure online)

For example, a 30 year-old jobseeker with physical disability wants to increase his employability. He has a trade certificate, basic computer skill, basic language skill, and no driver's licence. Among possible interventions such as obtaining a driver's licence, enhancing education and so forth, the jobseeker wants to know which is the most effective intervention in his context. This involves answering causal questions, e.g. how would his employment chance increase if he intervened his education (the factor) to Bachelor (the level)? The increase in employment chance, i.e. the causal effect, is the difference between the counterfactual employment chance if this person intervened a factor and the factual employment chance before the intervention. A factor together with its intervention level that would create maximum causal effect is recommended to the jobseeker.

Most employment recommendation methods are based on association to recommend factors with high correlation with employment [7,10,19]. Our problem focuses on recommending interventions that would create maximum employability *increase*, which involves causal inference. Other related works are treatment effect heterogeneity and uplift modelling where heterogeneous effects are estimated for a treatment. Most methods designed for observational data deal with binary factors [3,13,23]. Our problem involves binary and multivalued factors.

Figure 1 shows the overview of our method where a predicted personalized causal graph is used to make personalized recommendations. In this graph, the weights of an edge directed to the outcome represent causal effects of a relevant factor for different intervention levels. The intervention with the largest causal effect on the outcome is recommended. The challenge is how to obtain personalized causal graphs for individual training samples to train the prediction model. It is impossible to directly compute a personalized causal graph of a training sample using only this sample. We tackle this challenge by developing a linear interpolation technique to reverse engineer a population causal graph, which is estimated for the whole training data, to achieve personalized causal graphs for training data. These graphs are then used as ground truth for model training.

Our method (PIR) is evaluated using a case study of 4697 Australian disabled jobseekers and three public datasets. The case study indicates that PIR would help increase the employability of disabled jobseekers by up to 24% and the most effective intervention varies among jobseekers. PIR also shows its technical strengths with public datasets. Our main contributions are summarized as:

– We present a pioneer study of recommending interventions on multivalued factors to achieve maximum employability increase for disabled jobseekers.
– We propose a causal method to solve our problem. Personalized causal graphs are predicted to recommend the most effective interventions to individuals.
– We evaluate the proposed method with a case study of Australian disability employment and three public datasets. From experiment results, we answer DES domain questions and introduce practical implications of our method.

2   Problem Setup

2.1   Preliminaries

Let a jobseeker be described by features $\mathbf{X} = \{X_1, ..., X_H\}$. $\mathbf{X}$ includes manipulable attributes $\mathbf{M}$, also called factors, and non-manipulable attributes $\mathbf{I}$. Factor M can be binary or multivalued, $M \in \mathbf{J} = \{m_0, ..., m_K\}$ where m_0 is a baseline level. Outcome $Y \in \{0, 1\}$ is the employment status of the jobseeker.

Causal Graph. $\mathbf{G} = (\mathbf{V}, \mathbf{E})$ is a causal graph where $\mathbf{V}$ is nodes, $\mathbf{V} = \mathbf{X} \cup \{Y\}$, and $\mathbf{E}$ is edges. A directed edge $V_i \rightarrow V_j$ denotes a causal relationship. V_i is a cause (parent) of effect V_j (child). We adopt a directed acyclic graph (DAG) to represent the causal mechanism, and standard causal assumptions [20] (Markov, faithfulness, and causal sufficiency conditions) to link a DAG with distribution.

The DAG concept is extended with edge weights $\mathbf{W}$. The weight of edge $V_i \rightarrow V_j$ denotes causal effect of V_i on V_j. If V_i is multivalued, we have multiple weights for this edge, and each weight represents the causal effect for an intervention level. In our method, we build two types of causal graphs. A *population causal graph* is built for an entire population, where weights denote average causal effects. A *personalized causal graph* is built for a sample (a jobseeker) and the weights denote individual causal effects specific to this person (Fig. 1).

Causal Effect. Let $\{Y(m_0), ..., Y(m_K)\}$ be potential outcomes for intervening factor M to different levels. Individual causal effect on Y, also called individual treatment effect (ITE), for intervention level m_t is $Y(m_t) - Y(m_0)$. In personalized causal graphs, the weights are estimated ITE values, i.e. $\widehat{ITE}$. For a new jobseeker, we need to predict a personalized causal graph for this person. Basing on the weights of edges directed to Y, i.e. $\mathbf{W}_Y$, in the personalized causal graph, we can identify and recommend the most effective intervention.

2.2   Problem Statement

Our objective is to recommend personalized interventions to achieve maximum employability increase for disabled jobseekers. This requires predicting personalized causal graphs. Then, interventions are ranked based on $\widehat{ITE}$, i.e. the weights of edges directed to the outcome. Interventions with the highest ranks, i.e. the largest causal effect, are recommended to jobseekers. Factors considered for intervention can be binary, multivalued or mix of the two types. An intervention recommendation for a multivalued factor includes the factor to be intervened and its intervention level that would generate maximum employability increase.

3   Method

3.1   Overview of the Proposed Method (PIR)

PIR, presented in Fig. 1, uses personalized causal graphs to make recommendations. To build the model for predicting personalized graphs, we need to have personalized graphs of individual training samples for model training. However, these graphs are not available in data. A personalized causal graph has two components: 1) *causal structure* which shows causal factors of the outcome, 2) *weights of edges directed to the outcome*, i.e. $\mathbf{W}_Y$. Thus, we cannot directly compute a personalized graph for a training sample based on solely data of this sample. To tackle this challenge, we first assume a causal structure is persistent [14], but the causal effect of a factor on employment varies among jobseekers. Given a causal structure, then the task of estimating a personalized causal graph becomes estimating weights $\mathbf{W}_Y$. We develop a technique to reverse engineer a population causal graph to achieve personalized causal graphs for training samples.

3.2   Building a Population Causal Graph

Estimating Causal Structure. We use the PC algorithm [20] to infer a population causal structure from data. Detected causal relationships, represented by a DAG, are validated and refined using domain knowledge. This is to encode human belief into the DAG, making it consistent with both data and domain knowledge. From the detected causal structure, we can determine manipulable causal factors of employment status $\mathbf{M}_c \subset \mathbf{M}$ and adjustment variables for unbiased causal estimation. This causal structure is applied to all personalized causal graphs as the causal structure is assumed to be persistent.

Estimating the Weights of Edges Directed to the Outcome. We use the average causal effect, also known as the average treatment effect (ATE), to denote $w \in \mathbf{W}_Y$ in a population causal graph. Given causal factor M intervened to level m_t, its ATE is defined as $E[Y(m_t) - Y(m_0)]$, where $m_0 \equiv 0$ and $m_t \equiv 1$ for a binary factor. We adopt standard assumptions in causal inference [12,18]. Unbiased ATE estimation for level m_t is as:

$$\widehat{ATE}(m_t) = \frac{\sum_{i=1}^{N} \frac{\mathbb{1}(M_i=m_t)Y_i}{r(m_t,\mathbf{c}_i)}}{\sum_{i=1}^{N} \frac{\mathbb{1}(M_i=m_t)}{r(m_t,\mathbf{c}_i)}} - \frac{\sum_{i=1}^{N} \frac{\mathbb{1}(M_i=m_0)Y_i}{r(m_0,\mathbf{c}_i)}}{\sum_{i=1}^{N} \frac{\mathbb{1}(M_i=m_0)}{r(m_0,\mathbf{c}_i)}}, \tag{1}$$

where $r(m_t, \mathbf{c}_i) = Pr(M_i = m_t | \mathbf{C} = \mathbf{c}_i)$ is a propensity score (PS), probability of receiving an intervention at level m_t for person i. $\mathbf{C}$ is a adjustment set.

We use the back-door criterion [16] to determine the adjustment variables $\mathbf{C}$ based on the causal structure. Since all causal relationships can be learned, parents of M are sufficient to block all back-door paths between M and Y. We select these variables for adjustment. To estimate PS for a multivalued factor, we assume underlying multinomial distribution for intervention assignment. Softmax function is used for PS estimation: $Pr(M_i = m_t|\mathbf{c}_i; \boldsymbol{\Theta}) = \dfrac{e^{\theta_t \cdot \mathbf{c}_i}}{\sum_{j=0}^{K} e^{\theta_j \cdot \mathbf{c}_i}}$, where $\boldsymbol{\Theta} = (\theta_0, ..., \theta_K)$ is estimated by maximizing the following likelihood function:

$$\mathcal{L}(\boldsymbol{\Theta}) = \prod_{i=1}^{N} \prod_{t=0}^{K} \left(Pr(M_i = m_t|\mathbf{c}_i; \boldsymbol{\Theta}) \right)^{1(M_i = m_t)} = \prod_{i=1}^{N} \prod_{t=0}^{K} \left(\frac{e^{\theta_t \cdot \mathbf{c}_i}}{\sum_{j=0}^{K} e^{\theta_j \cdot \mathbf{c}_i}} \right)^{1(M_i = m_t)}.$$

For a binary factor, we assume underlying binomial distribution for intervention assignment. Sigmoid function is used for the estimation: $Pr(M = 1|\mathbf{C} = \mathbf{c}) = \dfrac{1}{1 + e^{-\theta \cdot \mathbf{c}}}$, where θ model parameters. The likelihood to be maximized is:

$$\mathcal{L}(\theta) = \prod_{i=1}^{N} Pr(m_i|\mathbf{c}_i; \theta) = \prod_{i=1}^{N} \left(\frac{1}{1 + e^{-\theta \cdot \mathbf{c}_i}} \right)^{m_i} \left(1 - \frac{1}{1 + e^{-\theta \cdot \mathbf{c}_i}} \right)^{1-m_i}.$$

3.3 Reverse Engineering Using Linear Interpolation

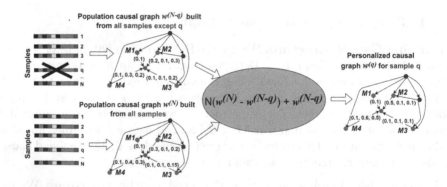

Fig. 2. Estimating a personalized causal graph for sample q in training data.

We assume each jobseeker has his/her own causal graph to reflect individual heterogeneity. Then, we propose that a population (aggregate) causal graph built for all N samples is the average of personalized causal graphs built for individual samples. This means the weight $w^{(N)}$ of an edge in the aggregate graph can be modelled as linear aggregation of the weights of that edge across personalized graphs, i.e. $w^{(N)} = \sum_i^N s_i^{(N)} w^{(i)}$, where $\sum_i^N s_i^{(N)} = 1$ and $s_i^{(N)}$ is the relative contribution of sample i. Similarly, in an aggregate graph built from all samples except q, its weight is $w^{(N-q)} = \sum_{i \neq q}^N s_i^{(N-q)} w^{(i)}$, where $\sum_{i \neq q}^N s_i^{(N-q)} = 1$.

Assuming the ratio of relative contribution s_i of a sample to aggregate graphs with and without sample q is the same for all samples, i.e. $s_i^{(N)} = \alpha s_i^{(N-q)}, \forall i$ where α is a constant, we have: $1 = \sum_{i \neq q}^{N} s_i^{(N-q)} = \sum_i^N s_i^{(N)} = s_q^{(N)} + \sum_{i \neq q}^{N} s_i^{(N)}$.

$$\Leftrightarrow s_q^{(N)} = 1 - \sum_{i \neq q}^{N} s_i^{(N)} = 1 - \frac{\sum_{i \neq q}^{N} s_i^{(N)}}{\sum_{i \neq q}^{N} s_i^{(N-q)}} = 1 - \alpha. \tag{2}$$

We also have: $w^{(N)} - w^{(N-q)} = s_q^{(N)} w^{(q)} + \sum_{i \neq q}^{N} (s_i^{(N)} - s_i^{(N-q)}) w^{(i)}. \tag{3}$

From Eq. 2 and 3, we obtain: $w^{(N)} - w^{(N-q)} = s_q^{(N)} w^{(q)} - s_q^{(N)} \sum_{i \neq q}^{N} s_i^{(N-q)} w^{(i)}$.

$$\Leftrightarrow w^{(q)} = \frac{1}{s_q^{(N)}} \left(w^{(N)} - w^{(N-q)} \right) + w^{(N-q)}. \tag{4}$$

Assume all samples have the same contribution, i.e. $s_i^{(N)} = N^{-1}$, we have:

$$w^{(q)} = N \left(w^{(N)} - w^{(N-q)} \right) + w^{(N-q)}. \tag{5}$$

Equation 5 is used to estimate a personalized causal graph for sample q where $w^{(q)} \equiv \widehat{ITE}^{(q)}$, $w^{(N)} \equiv \widehat{ATE}^{(N)}$, and $w^{(N-q)} \equiv \widehat{ATE}^{(N-q)}$ as showed in Fig. 2. Next, we verify Eq. 5 with the application to a multivalued intervention. Using Eq. 5, $\widehat{ITE}^{(q)}(m_t)$ of intervention level m_t for sample q is:

$$\widehat{ITE}^{(q)}(m_t) = N \left(\widehat{ATE}^{(N)} - \widehat{ATE}^{(N-q)} \right) + \widehat{ATE}^{(N-q)} \tag{6}$$

Use Eq. 1 for the ATE estimation, and average over all samples, we achieve:

$$\frac{1}{N} \sum_q^N \widehat{ITE}^{(q)}(m_t) = \left(\frac{\sum_q^N \frac{1(M_q=m_t)Y_q}{r(m_t, c_q)}}{\sum_i^N \frac{1(M_i=m_t)}{r(m_t, c_i)}} - \frac{\sum_q^N \frac{1(M_q=m_0)Y_q}{r(m_0, c_q)}}{\sum_i^N \frac{1(M_i=m_0)}{r(m_0, c_i)}} \right)$$
$$+ \left(\frac{\sum_q^N \sum_{i \neq q}^N \frac{1(M_i=m_t)Y_i}{r(m_t, c_i)}}{\sum_i^N \frac{1(M_i=m_t)}{r(m_t, c_i)}} - \frac{N-1}{N} \frac{\sum_q^N \sum_{i \neq q}^N \frac{1(M_i=m_t)Y_i}{r(m_t, c_i)}}{\sum_{i \neq q}^N \frac{1(M_i=m_t)}{r(m_t, c_i)}} \right) \tag{7}$$
$$+ \left(\frac{\sum_q^N \sum_{i \neq q}^N \frac{1(M_i=m_0)Y_i}{r(m_0, c_i)}}{\sum_i^N \frac{1(M_i=m_0)}{r(m_0, c_i)}} - \frac{N-1}{N} \frac{\sum_q^N \sum_{i \neq q}^N \frac{1(M_i=m_0)Y_i}{r(m_0, c_i)}}{\sum_{i \neq q}^N \frac{1(M_i, m_0)}{r(m_0, c_i)}} \right).$$

In the limit of a large number of samples $N \to \infty$:

$$\lim_{N \to \infty} \frac{1}{N} \sum_q^N \widehat{ITE}^{(q)}(m_t) = \frac{\sum_q^N \frac{1(M_q=m_t)Y_q}{r(m_t, c_q)}}{\sum_i^N \frac{1(M_i=m_t)}{r(m_t, c_i)}} - \frac{\sum_q^N \frac{1(M_q=m_0)Y_q}{r(m_0, c_q)}}{\sum_i^N \frac{1(M_i=m_0)}{r(m_0, c_i)}}. \tag{8}$$

Equation 8 is the same as Eq. 1, reassuring the linear assumption made by Eq. 5.

Building Personalized Causal Graphs for Model Training. A personalized causal graph for a training sample is built using: *1) population causal structure, 2) Eq. 5 to compute the weights of edges directed to the outcome*. By sequentially leaving out each sample, we build personalized causal graphs for all samples in training data. These graphs are used to train the prediction model.

3.4 Building the Prediction Model

To recommend the most effective interventions, we train the prediction model using personalized causal graphs achieved from the reverse-engineering process. The model takes feature set $\mathbf{S} = \mathbf{X} \setminus \mathbf{M}_c$ as jobseeker context to predict personalized causal graphs. Basing on the weights (ITE) of the predicted graphs, we recommend the most effective interventions to jobseekers. The prediction model is denoted as $\Psi(\mathbf{S}) = \{\Psi_1(\mathbf{S}), ..., \Psi_d(\mathbf{S})\}$, where $\Psi_i(\mathbf{S})$ is a sub-model for predicting weight $w \in \mathbf{W}_Y$. Gradient boosting tree [9] are used to build a sub-model. It is presented by $\Psi(\mathbf{S}) = \sum_i^T f_i(\mathbf{S})$, where $f_i(\mathbf{S})$ is a Classification And Regression Tree. Function Ψ is estimated by minimizing the following objective:

$$\mathcal{L}(\boldsymbol{\theta}) = \sum_i^N l\Big(w_i, \hat{w}_i\Big) + \sum_j^T \Omega\Big(f_j(\mathbf{S})\Big), \qquad (9)$$

where $l = (w_i - \hat{w}_i)^2$, $\Omega(f) = \gamma L + \frac{1}{2}\lambda||c||^2$ is a regularization term to control model complexity. L is the number of leaves and c is the leaf weight.

4 Setup for Empirical Evaluation

Datasets. One case study and three public datasets are used for evaluation. Datasets are split into 70% for model training and 30% for evaluation.

Seven Baselines. The baselines for comparison are: Causal Tree (CT) [3], Causal Forest (CF) [5], X-Learner (XL) [13], Doubly robust learner (DRL) [22], Transformed Outcome Tree (TOT) [4], Squared T-Statistic Tree (ST) [21], and Fit-based Tree (FT) [24]. These methods only work with binary factors. We perform pair-wise comparison to recommend the best intervention.

Three Metrics. 1) We adopt Area Under Uplift Curve (AUUC) [11] to evaluate how methods help increase in the outcome. 2) We evaluate how jobseekers are separated into groups with different degrees of outcome increase. Jobseekers are sorted by $\widehat{ITE}$ in descending order, and partitioned into segments. Observed outcome increase (OC Inc.) of segment top $\phi\%$ is $(\bar{Y}_{R\phi} - \bar{Y}_{NR\phi}) \cdot 100$, where $\bar{Y}_{R\phi}$ and $\bar{Y}_{NR\phi}$ are the average of outcome among those following recommendations and among those not. Spearman and Kendall coefficients are used to assess the consistency between orderings by causal effect and by outcome increase. 3) We use Kaplan-Meier curves [17] for evaluation on datasets with job retention time.

5   Australian Disability Employment Case Study

1) Preliminaries. A dataset obtained from a provider records information of 4697 Australian disabled jobseekers with over 40 attributes. Table 1 shows 12 relevant attributes for building our model. We want to achieve three objectives:

Table 1. Attributes of Australian jobseekers with disability.

Attribute	Description	Attribute	Description
Age	Age of workers	Gender	Gender of workers
Allowance	Allowance types	Licence	Driving licence
Disability	Disability type	Motivation	Motivation of workers
E_Comm	Email communication	Funding level	Support fee level
Education	Education level	Program	Employment support programs
Employment	Employment status	Work cap	Number of weekly hours

- *To evaluate model performance with real-world data.* We want to evaluate how methods would help increase employability of disabled jobseekers.
- *To evaluate causal factors for intervention.* We want to examine reasonableness of the causal factors recommended for intervention.
- *To answer DES domain questions.* Our DES partner wants to know: a) how much our method would help increase employability of jobseekers; b) whether the same intervention is applicable to all jobseekers; c) whether our method can distinguish jobseekers with different degrees of employability increase.

1) Employability improvement. Figure 3(a) shows our method (PIR) has the highest performance: its AUUC is 29% higher than the best baseline. This implies that interventions recommended by PIR would help jobseekers achieve the highest employability increase among evaluated methods.

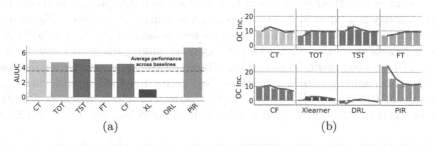

(a)   (b)

Fig. 3. Performance with Australian disability employment: (a) AUUC, (b) Observed outcome increase (OC Inc.) ranked by causal effects in descending order.

2) Heterogeneity discovery. The rank correlation coefficients in Table 2 show PIR achieves a good consistency between ranking jobseekers by causal effect and by employability increase. This means PIR has a capability of distinguishing jobseekers with different degrees of employability increase. Figure 3(b) shows the outcome increase of top groups ranked by descending causal effect: top 20%, 40% and so forth. PIR captures the heterogeneity well for accuracy as denoted by a consistent declining trendline, and for the diversity of outcome increase as denoted by decreased amounts between two consecutive groups. Another aspect of heterogeneity discovery is discussed in Section *Reasonable causal factors*.

Table 2. Kendall and Spearman correlation coefficients.

Coefficient	CT	TOT	TST	FT	CF	XL	DRL	PIR
Kendall	0.40	0.00	0.40	−0.60	**0.80**	0.20	0.00	**0.80**
Spearman	0.60	−0.10	0.60	−0.70	**0.90**	0.00	−0.10	**0.90**

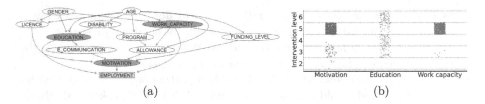

(a)   (b)

Fig. 4. (a) Causal relationships inferred from the data. (b) Heterogeneous recommendations. Each dot represents an intervention recommended to a jobseeker.

3) Reasonable causal factors. Figure 4(a) shows that causal factors to be intervened to increase employability are motivation, work capacity and education. This aligns well with previous findings in social science: motivation significantly affects employment of disabled people [2], while education is believed to have an impact on their employment rate [6]. Our industry partner also confirms work capacity is an important factor to make recruitment decisions. Figure 4(b) shows that the most effective interventions recommended by our method differ among jobseekers. This reflects the heterogeneity among disabled individuals.

4) Answering domain questions. Based on the above results, we see that:

a) Intervention recommended by our method would increase employability. Figure 3(b) denotes that if jobseekers adopt personalized interventions generated by PIR, their employability would increase by from 12% to 24%.

b) Heterogeneous personalized intervention. Although we found that three causal factors can be intervened to increase the employability of disabled jobseekers, the most effective intervention varies among individuals.

c) Degree of potential employability increase. As shown in Fig. 3 (b), our method can distinguish jobseekers with different degrees of employability increase via predicted causal effects.

6   Experiments with Public Datasets

Three public datasets are used for additional evaluations: Adult census income (ACI) with 48842 instances [8], employee turnover (TO) with 1129 instances[1], and HR with 15000 instances[2]. Outcome of the first dataset is employment status for a professional job. Outcome of others is employment retention time.

Table 3 shows PIR achieves the best performance with the highest AUUC. The Spearman metric denotes that averaging over three datasets, PIR has the highest consistency between ranking by causal effect and by outcome increase.

Table 3. Performance for three public datasets.

AUUC								
Dataset	CT	TOT	TST	FT	CF	XL	DRL	PIR
ACI	20.67	20.05	20.90	20.79	11.06	13.60	16.96	**22.05**
TO	1.13	1.14	1.23	1.13	1.53	5.25	9.87	**13.69**
HR	51.10	41.50	51.10	26.66	62.23	67.12	40.62	**76.28**
Spearman correlation coefficient								
ACI	1.00	1.00	1.00	0.80	−1.00	−1.00	1.00	0.80
TO	0.40	0.40	0.40	0.40	−0.40	1.00	1.00	1.00
HR	−0.80	−0.80	−0.80	−0.80	−1.00	−0.20	−1.0	0.80
Average	0.20	0.20	0.20	0.13	−0.80	−0.07	0.33	**0.87**

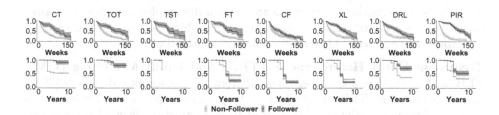

Fig. 5. Survival curves for TO (upper) and HR (lower).

Figure 5 shows PIR has the largest separation between the survival curves of those following recommendations and those not following recommendations for the TO dataset. With the HR dataset, PIR is also among the methods creating the largest separation. This means if people adopt recommendations generated by PIR, their employment retention would increase higher than other methods.

---

[1] https://www.analyticsinhr.com/blog/hr-data-sets-people-analytics.
[2] https://www.kaggle.com/liujiaqi/hr-comma-sepcsv.

7   Practical Implication

Our method has been deployed for an Australian DES provider[3]. Our work would make positive impacts on jobseekers with disability and the DES sector:

1) Helping disabled jobseekers increase their employment potential. The case study shows that the employability of disabled jobseekers who follow our intervention recommendations would increase by up to 24%.

2) Personalized employment assistance programs. The most effective intervention varies among heterogeneous jobseekers. DES providers can design assistance programs tailored to individual jobseekers. This enhances the effectiveness of the DES programs, and avoid the costs of ineffective interventions.

3) Resource allocation for employment assistance. Our model can identify possible factors to increase the employability of jobseekers. This helps providers focus solely on the effective factors in their support programs. PIR can distinguish jobseekers with different degrees of employability increase. Thus, providers can properly allocate suitable resources to support different jobseekers.

8   Related Work

Disability Employment. A growing literature in social science focuses on factors that promote or hinder disabled people in work participation. Association-based methods are used to detect the factors that affect the employability of jobseekers with data collected from studies designed for research [2,15]. Our work supplements this literature by developing a causality-based approach for recommending the most effective intervention using observational data.

Employment Recommendation. Popular methods for job, candidate, and skill recommendations [7,10,19] are based on content-based, collaborative filtering or hybrid approaches. Their goals are to optimize employment based on association between factors and the outcome. Our problem focuses on the *increase* in employment chance. This requires causation instead of association.

Treatment Effect Heterogeneity (TEH) and Uplift Modelling (UM). TEH is the fact that a treatment has different effects across individuals [3]. In precision medicine, TEH allows doctors to prescribe the most effective treatment. In business analytics, UM is used to estimate the effect of promotions. Most methods designed for observational data deal with binary factors [3,13,23]. In our problem, interventions involve binary and multivalued factors. Methods with multivalued factors are also proposed but for randomized trials [25]. Most methods assume known causal factors, which are unknown in our context.

9   Conclusion and Future Work

We present a pioneer study of recommending interventions on multivalued factors to achieve maximum employability increase for disabled jobseekers. In our

---

[3] Codes and public datasets are at https://github.com/trxuanha/pir.

method, personalized causal graphs are predicted to generate recommendations. The application of our method to a case study shows the promising performance of increasing employability of jobseekers by up to 24%. Experiments with public datasets show the promise of our method in different settings.

References

1. ABS: Disability, ageing and carers: summary of findings (2019)
2. Achterberg, T., Wind, H., De Boer, A., Frings-Dresen, M.: Factors that promote or hinder young disabled people in work participation: a systematic review. J. Occup. Rehabil. **19**(2), 129–141 (2009)
3. Athey, S., Imbens, G.: Recursive partitioning for heterogeneous causal effects. Proc. Natl. Acad. Sci. **113**(27), 7353–7360 (2016)
4. Athey, S., Imbens, G.W., et al.: Machine learning for estimating heterogeneous causal effects. Technical report (2015)
5. Athey, S., Wager, S.: Estimating treatment effects with causal forests: an application. arXiv:1902.07409 (2019)
6. Bengtsson, S., Datta Gupta, N.: Identifying the effects of education on the ability to cope with a disability among individuals with disabilities. PLoS ONE **12**(3), e0173659 (2017)
7. Dave, V.S., Zhang, B., Al Hasan, M., AlJadda, K., Korayem, M.: A combined representation learning approach for better job and skill recommendation. In: Proceedings of the 27th ACM International Conference on Information and Knowledge Management, pp. 1997–2005 (2018)
8. Dua, D., Graff, C.: UCI machine learning repository (2017). http://archive.ics.uci.edu/ml
9. Friedman, J.H.: Greedy function approximation: a gradient boosting machine. Ann. stat. **29**, 1189–1232 (2001)
10. Guo, S., Alamudun, F., Hammond, T.: Résumatcher: a personalized résumé-job matching system. Expert Syst. Appl. **60**, 169–182 (2016)
11. Gutierrez, P., Gérardy, J.Y.: Causal inference and uplift modelling: a review of the literature. In: International Conference on Predictive Applications and APIs, pp. 1–13 (2017)
12. Imbens, G.W.: The role of the propensity score in estimating dose-response functions. Biometrika **87**(3), 706–710 (2000)
13. Künzel, S.R., Sekhon, J.S., Bickel, P.J., Yu, B.: Metalearners for estimating heterogeneous treatment effects using machine learning. Proc. Natl. Acad. Sci. **116**(10), 4156–4165 (2019)
14. Li, J., Liu, L., Le, T.D.: Practical Approaches to Causal Relationship Exploration. Springer, Cham, (2015). https://doi.org/10.1007/978-3-319-14433-7
15. Lindsay, S.: Discrimination and other barriers to employment for teens and young adults with disabilities. Disabil. Rehabil. **33**(15–16), 1340–1350 (2011)
16. Pearl, J.: Causality: Models, Reasoning, and Inference, 2nd edn. (2009). https://doi.org/10.1017/CBO9780511803161
17. Rich, J.T., Neely, J.G., Paniello, R.C., Voelker, C.C.J., Nussenbaum, B., Wang, E.W.: A practical guide to understanding kaplan-meier curves. Otolaryngol. Head Neck Surg. **143**(3), 331–336 (2010). https://doi.org/10.1016/j.otohns.2010.05.007
18. Rubin, D.B.: Estimating causal effects of treatments in randomized and nonrandomized studies. J. Educ. Psychol. **66**(5), 688 (1974)

19. Shalaby, W., et al.: Help me find a job: a graph-based approach for job recommendation at scale. In: 2017 IEEE International Conference on Big Data (Big Data), pp. 1544–1553. IEEE (2017)
20. Spirtes, P., Glymour, C., Scheines, R.: Causation, Prediction, and Search. MIT Press (2000)
21. Su, X., Tsai, C.L., Wang, H., Nickerson, D.M., Li, B.: Subgroup analysis via recursive partitioning. J. Mach. Learn. Res. **10**(2) (2009)
22. Syrgkanis, V., Lei, V., Oprescu, M., Hei, M., Battocchi, K., Lewis, G.: Machine learning estimation of heterogeneous treatment effects with instruments. arXiv:1905.10176 (2019)
23. Tran, H.X., et al.: Intervention recommendation for improving disability employment. In: 2020 IEEE International Conference on Big Data (Big Data), pp. 1671–1680. IEEE (2020)
24. Zeileis, A., Hothorn, T., Hornik, K.: Model-based recursive partitioning. J. Comput. Graph. Stat. **17**(2), 492–514 (2008)
25. Zhao, Y., Fang, X., Simchi-Levi, D.: Uplift modeling with multiple treatments and general response types. In: Proceedings of the 2017 SIAM International Conference on Data Mining, pp. 588–596. SIAM (2017)

Estimating Skill Proficiency from Resumes

Anindita Sinha Banerjee, Sachin Pawar[✉], Girish K. Palshikar,
Devavrat Thosar, Jyoti Bhat, and Payodhi Mandloi

TCS Research, Tata Consultancy Services, Mumbai, India
{anindita.sinha2,sachin7.p,gk.palshikar,d.thosar,jyoti.bhat1}@tcs.com

Abstract. This paper attempts to solve a real-life problem of estimating proficiency levels of skills held by a person from her resume. This is a challenging problem because no other source of information than resumes is available and skill proficiency is a complex function of various aspects of a person's experience. A resume often mentions various skills held by a person and her experience in applying these skills as part of work or project experience. We extract skills and other relevant information automatically from resumes and capture skill related information in terms of a feature vector. We propose two techniques to automatically learn a skill proficiency estimation function using these feature vectors – (i) supervised neural network based technique where we introduce a novel loss function to combine label information with domain-specific constraints and (ii) weakly supervised clustering based technique. We evaluate these techniques along with two competent baselines on a dataset of Information Technology (IT) resumes focusing on 5 major skills – Java, Python, Databases, Embedded Systems, and Machine Learning.

Keywords: Skill proficiency estimation · Resume information extraction · Neural networks · Weakly supervised learning

1 Introduction

In many practical situations, there is a need to quickly and accurately assess the proficiency level of a person in a particular skill. For example, just before an interview, interviewers often identify a few major skills of a candidate from her resume and informally (and subjectively) assess her proficiency level in each of them, usually using mental markers like *beginner*, *intermediate*, *expert*. An internal employees-to-projects assignment system would benefit by having an automated skill-wise proficiency assessment system from employees' resumes [8]. A job portal may want to automatically assign proficiency levels to various skills for millions of resumes in their repository for suitable job recommendations.

Proficiency level (or just *proficiency*) of a person in a skill is a numerical score that measures the extent of knowledge, understanding, experience, and ability to apply these to solve a practical problem. For the purposes of this paper, we define a *skill* as a broad technical subject or area of knowledge; examples: Relational Database, Java, Machine Learning etc. The subjective assessments

© The Author(s), under exclusive license to Springer Nature Switzerland AG 2022
J. Gama et al. (Eds.): PAKDD 2022, LNAI 13282, pp. 105–118, 2022.
https://doi.org/10.1007/978-3-031-05981-0_9

of proficiency in a skill are opaque (not easily explainable), error-prone, and vary from person to person. To overcome these issues, we formulate and address the *skill-proficiency estimation* problem: *how to automatically and quantitatively estimate the proficiency of a person C in a given skill s, based on her resume R_C.* For concreteness, we discuss IT domain skills and resumes in this paper; however, the techniques are general and can apply to any other domains, provided appropriate activities and features are identified.

There are several complications in a realistic formulation of this skill proficiency estimation problem. First, a skill includes many *skill markers* (SM) such as sub-skills, tools, and concepts in that subject; e.g., skill Machine Learning includes sub-skills like Reinforcement Learning, Deep Learning, tools like Keras, Weka and concepts like activation function, backpropagation. A resume may explicitly only mention some skill markers related to a skill, without mentioning the skill itself. For example, for the skill Database, we may find that a resume only mentions skill markers like PL/SQL, Oracle, JDBC. We use a list of skill markers corresponding to a particular skill which is created semi-automatically using DBPedia. Secondly, multiple skills; e.g., Database, Java, Linux, are usually present in a resume and proficiencies among them may be inter-related. To avoid this complication, we estimate proficiencies of multiple skills independently, assuming them to be independent of each other. Thirdly, skill markers for a given skill have complex relationships among them, like IS_A, PART_OF etc. We ignore such relationships among skill markers, and make the simplifying assumption that all skill markers are at the same level of abstraction and complexity. Even after these simplifying assumptions, the problem of estimating proficiency in a skill from the resume text remains challenging, because (i) it is non-trivial to identify which text indicates proficiency in that skill, (ii) decide quantitatively how much a piece of text contributes to the proficiency in the skill; and (iii) how to aggregate the proficiency estimated from various text segments.

Although, there have been several attempts to estimate skill proficiencies in various settings [1,5,6,10–12,14], these are not applicable directly for estimating skill proficiencies from resumes. We propose two techniques to automatically learn a skill proficiency estimation function – (i) supervised neural network based technique where we introduce a novel loss function to combine label information with domain-specific constraints, and (ii) weakly supervised clustering based technique. We briefly describe how the information relevant to skill proficiencies is automatically extracted from resumes.

2   Related Work

Extraction of Skills: Gugnani et al. [2] proposed an approach for the extraction of skills mentioned in candidate profiles like resumes, using several skill ontologies. The extracted skills are further used to create a personalized skill graph which is used for career path recommendations. Gugnani and Misra [3] also described an approach to extract skills from resumes including inferring implicit skills which are not explicitly mentioned. However, both of these papers do not address the problem of estimating proficiencies of the extracted skills.

Proficiency Estimation: There has been a line of research in estimating student proficiency using Item Response Theory (IRT) [6], knowledge tracing [1], and deep knowledge tracing [10]. However, these approaches rely on *interactions* with students whereas we propose to estimate skill proficiencies in non-interactive way using only the information available in resumes. There have been attempts [12,14] to estimate skill proficiency levels using the Elo rating system[1]. Horesh et al. [5] described an ordinal regression clustering approach to classify employees into five different proficiency levels as per their expertise in a broad subject area. Stojkovic et al. [11] described ranking-based learning for scoring functions where the model is trained from pairwise comparisons.

To the best of our knowledge, this is the first attempt to predict proficiency levels of skills directly from resumes without using any other source of information like internal employee assessment, internal social media activity within an organization [5], or exam scores [11].

3   Problem Definition

3.1   Resume Information Model

Given a set of resumes $R = \{R_1, \ldots, R_N\}$ where each R_i is a free-form textual document, we extract and map the information in the given resume to a structured model (Sect. 5 describes the extraction technique). We now describe this model. We assume that we are given a particular skill s, for which we want to estimate the proficiency level from any given resume. Let S be the set of all possible *skill markers* for skill s. Let AT be the set of activity types that involve the use of the given skill s; e.g., $AT = \{certification, course, degree, project, employment, paper, patent, award\}$. Let T be the set of *tasks* in the domain, each of which involves the use of the given skill s. We define a *task* to be a well-defined knowledge-based volitional action with a specific goal, e.g., prepare test plan, database design, coding, bug fixing, produce flowcharts. Let R denote the set of all *roles*; e.g., Developer, Test Engineer, Project Leader, GUI Designer etc.

We extract necessary information from a given resume R_i and create an temporally ordered sequence $\sigma_R = \langle A_1, \ldots, A_m \rangle$, where each A_j is an activity related to the given skill s. In general, each *activity* is a tuple $A = (a_t, a_s, a_{dur}, a_T, a_R)$, where $a_t \in AT$ is the activity type, $a_s \subset S$ is the set of skill markers associated with this activity, a_{dur} is the duration of the activity, $a_T \subset T$ is the set of tasks executed as part of the activity, and $a_R \subset R$ are the set of roles performed in the activity. Let $S_{R_i} \subset S$ denote the set of skill markers occurring in resume R_i i.e., $S_{R_i} = \bigcup_{j=1}^m A_j.a_s$ where $A_j.a_s$ denotes the set of skill markers in the activity A_j. See Table 1 for an example.

3.2   Skill Proficiency Function

Given this structured representation σ_R of a resume R, the task is to estimate the proficiency for each skill $s \in S_R$ using the information in this sequence.

Table 1. Activities of type *project* for skill Java extracted from a resume.

Dur	Skill markers	Roles	Tasks
24	Java, Struts, J2EE, Tomcat, Log4J, Junit, Web services	Backend developer	System Architecture Design, Design data models
24	Java, J2EE, Spring, Hibernate, Jboss, SQL, Oracle 10g	Backend developer, Web developer	Implement a web friendly web UI, Implement modules, fast tracking and closure for essential change requirements, led the team
5.5	Java Restful web services, Angular, Typescript, Javascript, Jasmine/karma	Web developer	Creating and rendering Backbase widgets, Using Jasmine/karma for unit tests, Work with Agile process, Implement complex business logic

Table 2. Set of features for a resume R for skill s

Features	Values for the example in Table 1
x_1: Total duration of all projects related to skill s in months	53.5 (sum of durations of the 3 projects related to Java)
x_2: #projects related to s	3 (there are 3 projects related to Java)
x_3: #distinct roles in projects related to s	2 (Backend developer & Web developer)
x_4: #distinct skill markers related to s mentioned within projects	15 (all skill markers except SQL and Oracle 10g are related to Java)
x_5: #distinct tasks mentioned in projects related to s	10 ($2 + 4 + 4$ tasks in the 3 projects)
x_6: ratio of x_4 to total #skill markers (whether or not related to s) mentioned in projects related to s	0.88 ($\frac{15}{17}$ where there are 17 distinct skill markers within these projects)
x_7: #distinct certifications or courses related to s	1 (not shown in Table 1, a certification mentioned elsewhere in the resume)
x_8: #distinct skill markers related to s mentioned anywhere in the resume	25 (not shown in Table 1, 10 more skill markers mentioned outside projects)

For this, we further process the structured representation σ_R as follows. We need to know a mapping that associates each skill marker found in the resume to the appropriate skill. Thus if a resume mentions J2EE, SQL, JDBC then we need to map the first to Java, the second to Database, and the third to both Java and Database. Let s be a given skill for which we want to compute the proficiency level of any resume. We map the sequence σ_R of activities extracted from a resume R to a simple set of features described in Table 2. The example in Table 1 maps to $\mathbf{x} = (53.5, 3, 2, 15, 10, 0.88, 1, 25)$. The current set of features is mainly based on the *project* and *certification* activity types, although the extension to other activity types is straightforward. The features are neutral with respect to demographic details like gender to avoid unfair or biased estimations.

Our goal is to learn a function that maps any $\mathbf{x} \in \mathbb{R}^8$ to one of the five proficiency levels - from E_0 (lowest proficiency level) to E_4 (highest proficiency level) – $f_s : \mathbb{R}^8 \mapsto \{E_0, E_1, E_2, E_3, E_4\}$ E.g., $f_{Java}(\mathbf{x} = (53.5, 3, 2, 15, 10, 0.88, 1, 25)) = E_2$, indicates that the resume in Table 1 has proficiency level of

E_2 in Java. Please note that we have chosen 5 proficiency levels because it is consistent with how HR of our organization manually assigns and maintains the skill proficiency levels for the associates. However, all the proposed techniques would apply for a different number of proficiency levels than 5 without loss of generality.

4   Skill Proficiency Estimation Techniques

We explore two baseline techniques (B_1 and B_2) based on the prior literature and propose two new techniques (M_1 and M_2).

4.1 B_1: Ordinal Regression Clustering

B_1 is based on the ordinal regression clustering technique proposed in Horesh et al. [5]. Clusters resulting from standard clustering algorithms (such as k-means) in N-dim space, have no natural ordering. Whereas, the cluster boundaries from ordinal regression clustering are a set of parallel hyperplanes dividing the space in ordered clusters. The slope vector and the intercepts of the hyperplanes are the set of weights to be learned. The linear SVM algorithm is used as a component to find the set of hyperplanes in the multi-dimensional space that best separates the candidate resumes into different proficiency levels [13]. Cluster size of 5 is considered and resumes are categorized into five different levels of proficiency - E_0 to E_4. The advantage of this method is that it is completely unsupervised.

4.2 B_2: Supervised Linear Model

B_2 is based on the approach proposed by Stojkovic et al. [11] which learns the following linear function.

$$f(\mathbf{x}) = 5.0 \times \sigma(\mathbf{w}^T \cdot \mathbf{x} + b) \tag{1}$$

Here, $\mathbf{x} \in \mathbb{R}^8$ is a feature vector for a resume-skill combination. σ is the sigmoid function whose output is in the range $[0, 1]$ and hence the estimated proficiency score is in the range $[0, 5]$. The parameters to be learned are $\mathbf{w} \in \mathbb{R}^8$ and $b \in \mathbb{R}$. We used annotated pairs of resumes where each resume pair is labeled with a relative proficiency label $\in \{1, -1, 0\}$, to train our model to learn the scoring function (see Sect. 6.1 for details of the training dataset). We used a neural network model with one linear layer and trained it to optimize for the following loss functions (at most 1 loss is non-zero at a time, depending on the label):

- **Margin loss 1**: For a pair of resumes $\langle \mathbf{x}, \mathbf{y} \rangle$, if its label indicates that $\mathbf{x}$ has a better proficiency than $\mathbf{y}$, then $f(\mathbf{x})$ should be greater than $f(\mathbf{y})$ by the margin of 1.

$$loss_1 = max(0, 1 - (f(\mathbf{x}) - f(\mathbf{y}))) \tag{2}$$

- **Margin loss 2**: For a pair of resumes $\langle \mathbf{x}, \mathbf{y} \rangle$, if its label indicates that $\mathbf{x}$ has comparable proficiency as that of $\mathbf{y}$, then the absolute difference between $f(\mathbf{x})$ and $f(\mathbf{y})$ should not be more than the margin of 1.

$$loss_2 = max(0, |f(\mathbf{x}) - f(\mathbf{y})| - 1) \tag{3}$$

We use Stochastic Gradient Descent (SGD) to optimize for the above loss functions and also use L2 regularization to avoid over-fitting. Once, the model is trained, we can use the learned weights ($\mathbf{w}$ and b) to estimate the proficiency score, given the feature vector for any new resume. In order to obtain the discrete proficiency level (E_0, E_1, E_2, E_3, E_4) corresponding to the predicted score, we simply use the floor of the predicted proficiency score ($\lfloor f(\mathbf{x}) \rfloor$). Relative proficiency label l_R for any new pair of resumes $\langle \mathbf{x}, \mathbf{y} \rangle$ is predicted as follows:
If $\lfloor f(\mathbf{x}) \rfloor > \lfloor f(\mathbf{y}) \rfloor$, then $l_R = 1$
Else If $\lfloor f(\mathbf{y}) \rfloor > \lfloor f(\mathbf{x}) \rfloor$ then $l_R = -1$
Else If $f(\mathbf{x}) - f(\mathbf{y}) \geq 0.5$ then $l_R = 1$
Else If $f(\mathbf{y}) - f(\mathbf{x}) \geq 0.5$ then $l_R = -1$ Else $l_R = 0$

4.3 M_1: Supervised Model with Constraints

We observed that the proficiency levels obtained using the above supervised linear model tend to overestimate proficiency. Moreover, the estimated levels do not satisfy several constraints that are considered key by subject matter experts. E.g., the experts usually do not assign any particular proficiency level to a candidate for a skill, unless the candidate has some minimum experience of using the skill. Hence, it is important to integrate such domain constraints in the model training which specify the *upper bound* on the skill proficiency level of a candidate. Given a set of such constraints, it is straightforward to calculate $UB(\mathbf{x})$ which is the maximum allowable proficiency level for $\mathbf{x}$ considering all the constraints. In other words, $UB(\mathbf{x})$ is the highest possible proficiency level for $\mathbf{x}$ above which at least one constraint gets violated. As these are not *hard* constraints, we integrated them in our model training by introducing another loss function $loss_{cons}$ which incurs a loss proportional to the extent to which the constraint is violated.

$$loss_{cons} = (max(0, f(\mathbf{x}) - UB(\mathbf{x})) + max(0, f(\mathbf{y}) - UB(\mathbf{y})))/2 \tag{4}$$

As the constraints can be based on any arbitrary characteristic, we considered it better to model them through this constraints loss, rather than adding them as additional features. We integrated the following domain constraints in B_2 model:
If $x_1 < 12$ then $f(\mathbf{x}) < 1$; If $x_1 < 36$ then $f(\mathbf{x}) < 2$; If $x_1 < 72$ then $f(\mathbf{x}) < 3$; If $x_1 < 120$ then $f(\mathbf{x}) < 4$ (A person needs to have at least 12 months of project experience in a skill to be qualified for E_1 proficiency level, at least 36 months of project experience for E_2 and so on.) In practice, we observed that most constraints specified an upper bound on the proficiency level. The proposed loss function can be easily extended to consider lower bounds as well.

The overall loss is the weighted average of margin loss functions (Eq. 2 and 3) and the constraints loss.

$$loss_{final} = \lambda \cdot (loss_1 + loss_2) + (1 - \lambda) \cdot loss_{cons} \tag{5}$$

To improve the model capacity, we learn a non-linear function instead of the simple linear function used in B_2.

$$f(\mathbf{x}) = 5.0 \times \sigma(\mathbf{w}^T \cdot ReLU(W_h \cdot \mathbf{x} + \mathbf{b}_h) + b) \tag{6}$$

Here, we introduce a hidden layer of 10 units. σ is the sigmoid activation function and $ReLU$ is the rectified linear unit activation function. $W_h \in \mathbb{R}^{10 \times 8}$, $\mathbf{b}_h \in \mathbb{R}^{10}$, $\mathbf{w} \in \mathbb{R}^{10}$ and $b \in \mathbb{R}$ (bias) are the learnable parameters of the model. We used $\lambda = 0.7$ based on 5-fold cross-validation.

4.4 M_2: Weak Supervision Using Clustering

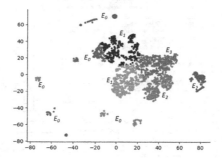

Fig. 1. Clusters obtained for Java by M_2^{tSNE} along with proficiency levels assigned to each cluster

Although the method M_1 performed well, we developed another method that required lower annotation efforts. The idea behind this method is quite simple and is based on clustering unlabelled instances. Each cluster is then represented with an 8-dim feature vector corresponding to the cluster center and an actual representative resume whose feature vector is closest to the center. A human expert then assigns proficiency level to each cluster by inspecting the cluster center and its representative resume. We use K-means clustering and have found that $K = 10$ works well empirically. Hence, the manual annotation efforts are reduced to identifying proficiency levels of just 10 clusters. Further, we also explored dimensionality reduction using using t-Distributed Stochastic Neighbor Embedding (t-SNE) [7] followed by clustering in reduced dimensional space ($d = 2$). We refer to this variant as M_2^{tSNE} method (see example in Fig. 1). To determine proficiency level for any new feature vector, we simply find the nearest cluster center and assign the proficiency level associated with that cluster.

4.5 Handling New Skills

Both the methods M_1 and M_2 need some manual annotation effort – to annotate resume pairs (in case of M_1) and to assign labels to cluster centers (in case of M_2). We can predict the skill proficiency level for any new resume only for those skills for which we have such annotated data. We explore one simple idea to extend our techniques for predicting proficiency levels for new skills. In case of M_1 (and B_2), we combined training instances for all 5 skills (from Table 3) and trained a single combined model. Similarly, in case of M_2, we pooled together unlabelled instances for all the 5 skills, clustered these pooled instances, and then labelled the cluster centers with appropriate proficiency levels. Table 4 shows the performance of such combined models in case of 2 new skills - SAP and Computer Vision (CV), as well as the 5 existing skills.

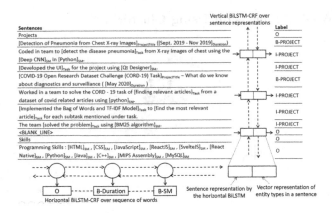

Fig. 2. Hierarchical BiLSTM based extraction model

5 Resume Information Extraction

We model the key information to be extracted from resumes in terms of *entity* and *relation* types. As described in Sect. 3.1, for skill proficiency estimation, we need to extract skill markers, certifications, and project details. We extended the joint entity and relation extraction technique by Pawar et al. [9] to extract project details by modeling it as a PROJECT relation type. It captures the relation between entity types ProjectTitle, Duration, Role, Skill_marker, and Task. Each mention of this relation type is similar to the tuples shown in the rows of Table 1.

We need to extract the following entities which are needed for computing our feature values – SM (Skill marker), Duration, Role, Task, and Certification. Similar to Pawar et al. [9], we also use 3 different techniques for entity extraction and consider an ensemble of their predictions – Linguistic patterns, CRF based on manually engineered features, and BiLSTM-based entity extractor. We also need to extract PROJECT relation type for computing project-related feature values. This relation is extracted by identifying a section of a resume (i.e., sequence of sentences) such that all the sentences in this section correspond to a single project. The corresponding PROJECT relation tuple is then constructed by linking together the entity mentions occurring within this section. Figure 2 shows an example of a resume where 2 project sections are identified, where each project leads to a single PROJECT relation tuple. Such sections are realized by labeling each sentence in a resume with an appropriate label in BIO encoding– (i) B-PROJECT indicates that the current sentence is the first sentence of a project section, (ii) I-PROJECT indicates that the current sentence is part of the same project section as the previous sentence, and (iii) O indicates that the sentence does not belong to any project. Overall model architecture is based on a hierarchical BiLSTM model as shown in Fig. 2 where – (i) the horizontal BiLSTM-CRF layer accepts a sequence of words as input and predicts entity type labels, and (ii) the vertical BiLSTM-CRF layer accepts a sequence of sentences as input and predicts relation type labels. A sentence is represented by concatenating sentence representation by the horizontal BiLSTM and a vector representing the presence of entity types in the sentence. We train this model in the pipeline fashion by sequentially training the horizontal BiLSTM-CRF layer and then the vertical BiLSTM-CRF layer. During inference for a new resume, the entity extraction output consists of an ensemble of extractions by linguistic patterns, CRF-based entity extractor and the horizontal BiLSTM-CRF layer. The PROJECT relation tuples for are constructed using this entity extraction output and sentence labels predicted by the vertical BiLSTM-CRF layer.

6 Experimental Analysis

6.1 Dataset

We use a set of 10290 resumes of IT professionals. The project information along with skill markers and certifications is extracted automatically from these resumes using the technique described in Sect. 5. We first convert the resumes in PDF format to plain text using Apache PDFBox[2]. Resumes in Microsoft Word format are first converted to HTML format using LibreOffice unoconv utility[3] and text is extracted from HTML using BeautifulSoup[4]. We then use spaCy [4] for NLP pre-processing such as sentence splitting, tokenization, part-of-speech tagging, dependency parsing etc. We retain information of blank lines in the resumes by introducing a place-holder sentence <BLANK_LINE> as shown in Fig. 2 as it is useful to capture project boundaries in some cases. The extracted information is further processed for facilitating features generation for skill proficiency estimation. Extracted durations are automatically converted to months, e.g., Sept. 2019 - Nov 2019 is converted to 3 months. The duration is assumed to be 6 months for projects which do not explicitly mention it.

---

[2] https://pypi.org/project/python-pdfbox/.
[3] https://github.com/unoconv/unoconv.
[4] https://pypi.org/project/beautifulsoup4/.

Table 3. Evaluation results where each skill is evaluated using the skill-specific models (DB: Database, ES: Embedded Systems, ML: Machine Learning)

Skill	Relative ordering accuracy					Average absolute error					Constraints violation factor				
	B_1	B_2	M_1	M_2	M_2^{tSNE}	B_1	B_2	M_1	M_2	M_2^{tSNE}	B_1	B_2	M_1	M_2	M_2^{tSNE}
Java	0.43	0.63	**0.66**	0.41	0.44	0.80	1.50	**0.68**	0.74	0.82	**0.04**	0.58	0.08	0.12	0.18
Python	0.54	**0.70**	0.61	0.45	0.45	2.08	2.48	0.58	0.52	**0.36**	1.00	1.00	0.56	0.36	**0.24**
DB	0.56	**0.57**	0.53	0.23	0.51	1.16	1.76	**0.78**	1.10	0.96	0.66	0.78	0.18	**0.10**	0.54
ES	0.44	**0.52**	0.46	0.46	0.29	2.06	2.46	**0.48**	0.58	0.54	0.9	1.00	0.14	**0.02**	0.22
ML	0.59	**0.73**	0.63	0.52	0.56	1.88	2.22	**0.51**	0.51	0.56	0.86	0.96	0.40	**0.31**	0.40
Avg.	0.51	**0.63**	0.58	0.41	0.45	1.60	2.08	**0.61**	0.69	0.65	0.69	0.86	0.27	**0.18**	0.32

Table 4. Evaluation for new skills using combined models for existing skills

Skill	Relative ordering accuracy					Average absolute error					Constraints violation factor				
	B_1	B_2	M_1	M_2	M_2^{tSNE}	B_1	B_2	M_1	M_2	M_2^{tSNE}	B_1	B_2	M_1	M_2	M_2^{tSNE}
Java	0.45	0.55	**0.63**	0.51	0.44	1.2	1.54	**0.72**	0.72	0.86	0.4	0.58	**0.06**	0.14	0.16
Python	0.47	**0.64**	0.55	0.35	0.47	1.2	1.60	**0.30**	0.68	0.60	0.74	1.00	**0.26**	0.58	0.44
DB	0.51	**0.56**	0.52	0.31	0.28	1.28	1.68	0.72	**0.68**	0.70	0.68	0.76	**0.24**	0.46	0.40
ES	0.32	**0.56**	0.48	0.33	0.33	1.38	1.52	**0.48**	0.74	0.82	0.72	0.90	**0.04**	0.42	0.38
ML	0.53	**0.69**	0.61	0.41	0.51	1.57	1.82	**0.49**	0.51	0.58	0.8	0.96	**0.36**	0.56	0.51
Avg.	0.45	**0.60**	0.56	0.38	0.41	1.33	1.63	**0.54**	0.66	0.71	0.67	0.84	**0.19**	0.43	0.38
SAP	NA					1.18	1.02	0.90	0.68	**0.62**	0.52	0.58	**0.00**	0.30	0.34
CV	NA					0.88	1.08	**0.30**	0.54	0.52	0.60	0.94	**0.00**	0.48	0.18

We then generate feature vectors for each resume for each major skill (Java, Python, Machine Learning, Databases, Embedded Systems). For each skill, we only retain those feature vectors (resumes) which have a non-zero value for x_8, resumes with $x_8 = 0$ have default proficiency level of E_{-1} indicating no exposure to the skill. Table 6 in Appendix shows some basic statistics of this dataset. We set aside two subsets of resumes for each skill – (i) D_s^{train}: training set of 25 resumes used to create 150 resume pairs which are annotated by human experts for *relative* proficiency (0, 1, –1), (ii) D_s^{eval}: evaluation set of 50 resumes which are annotated by human experts for explicit proficiency levels (E_0 to E_4). Relative proficiency labels for each pair of resumes $\langle R_i, R_j \rangle$ created for a skill s, are assigned as follows:

- **1**: R_i has better proficiency in skill s as compared to R_j

- **–1**: R_j has better proficiency in skill s as compared to R_i

- **0**: Both R_i and R_j have comparable proficiency in skill s

We computed inter-annotator agreement for the datasets D_{Java}^{train} (relative labels) and D_{Java}^{eval} (absolute labels) and the Cohen's Kappa coefficients were observed to be 0.54 and 0.35, respectively. This establishes that it is easier for human annotators to assign a relative proficiency label to a pair of resumes as compared to assigning an absolute proficiency label to a resume.

Data Pre-processing: We transform the feature vectors in each skill's dataset so that each feature has a mean of 0 and standard deviation of 1, and the feature values have similar scale across different features.

6.2   Evaluation Metrics and Results

We evaluate the performance of each proficiency estimation technique using the following 3 metrics. **Relative ordering accuracy**: It is computed as the fraction of labelled pairs (in the training datasets D_s^{train}) for which the class label (1/–1/0) was predicted correctly. For supervised techniques, the numbers are reported after 5-fold cross-validation over these datasets. For B_1, M_2 and M_2^{tSNE}, the relative proficiency label is derived from the proficiency level assigned to each resume in a pair. **Average absolute error**: It is the average of absolute differences between the expected and predicted proficiency levels for each resume in the evaluation datasets (D_s^{eval}). **Constraints violation factor**: It is the fraction of resumes in the evaluation datasets (D_s^{eval}) for which the predicted proficiency levels are violating the domain constraints.

Table 3 shows evaluation results of all the techniques on datasets of 5 skills. Although, B_2 is the best technique in terms of relative evaluation, it tends to overestimate the proficiency levels. Hence, it performs poorly in terms of average absolute error with the average error of 2.08 across all datasets and also has the highest constraints violation factor of 0.86. M_1 performs better overall considering all the three evaluation metrics – it gives the best average absolute error and is the second best in terms of the other two metrics. M_2 and M_2^{tSNE} also perform comparably with M_1 in terms of average absolute error and they require lesser annotation efforts as compared to M_1. Unlike B_2 which is trained to learn only the correct relative order and M_2 which does not consider relative ordering explicitly, M_1 is trained to learn relative ordering as well as constraints on the absolute proficiency levels and therefore it is the overall best technique. For the use-cases where only relative ordering of resumes for a particular skill is enough, the techniques B_2 and M_1 can be used. On the other hand, if the use-case requires exact proficiency level to be known, then M_1, M_2 and M_2^{tSNE} would be useful. Table 4 shows evaluation results when the combined models (Sect. 4.5) are used. For the 2 new skills, M_2^{tSNE} performs slightly better than M_1 considering the average absolute error over 2 new skills.

Evaluating Resume Information Extraction: We manually annotated 494 resumes with entity and relation type labels, used 444 resumes for training our entity and relation extraction model, and set aside 50 resumes for evaluating the extraction performance of the model. Table 5 shows the evaluation results.

Table 5. Entity and relation extraction evaluation results

Entity/Relation	TPs	FPs	FNs	Precision	Recall	F1-score
SM (overall)	1636	357	365	0.821	0.818	0.819
Certification (overall)	8	6	24	0.571	0.25	0.348
SM (projects)	474	365	264	0.565	0.642	0.601
Duration (projects)	145	84	33	0.633	0.815	0.713
Role (projects)	65	74	25	0.414	0.644	0.504
Task (projects)	327	565	451	0.367	0.420	0.392
PROJECT (tuple)	132	181	102	0.422	0.564	0.483
PROJECT (sentence-level)	1905	658	939	0.743	0.670	0.705

In accordance with our features, we evaluate SM (skill markers) and Certification entity mentions throughout the resumes and mentions of other entity types mentioned only within projects. For evaluating extracted PROJECT relations, we compare gold-standard project tuples with predicted tuples and consider a match if their project durations match and at least one skill marker is common. Sentence level evaluation measures how accurately sentences within projects are identified by our extraction model by comparing with gold-standard projects. In other words, a sentence is counted as a true positive if it is predicted to be a part of a PROJECT relation (section) and it is also part of a gold-standard project.

7 Conclusions and Further Work

We proposed two techniques for the problem of estimating skill proficiency levels of a person from her resume – (i) supervised neural network based technique, and (ii) weakly supervised clustering based technique. We evaluated our techniques along with two competent baselines on a real-life dataset of IT resumes for 5 major skills. In future, we plan to extend these techniques for several more skills mentioned in resumes across multiple other domains. We also plan to consider dependencies among proficiencies of multiple skills, inter-relationships between skill markers, and recency of the activities.

Appendix

Table 6 shows some basic statistics of the resumes dataset described in Sect. 6.1.

Table 6. Dataset statistics

Feature	Statistic	Java	Python	Database	Emb. system	Mach. learning	SAP	Comp. vision
	#Resumes	6067	1835	7173	2520	4202	1689	458
x_1	Mean	14.69	3.91	23.38	5.10	7.02	18.00	4.27
	StDev	26.32	10.85	32.67	12.75	15.82	40.40	13.37
	Median	6	0	12	0	0	6	0
x_2	Mean	1.75	0.48	2.68	0.66	0.82	1.66	0.47
	StDev	2.97	0.84	3.12	1.09	1.16	3.33	0.66
	Median	1	0	2	0	0	1	0
x_3	Mean	0.867	0.212	1.37	0.20	0.38	0.97	0.18
	StDev	1.48	0.61	1.85	0.60	0.88	1.70	0.51
	Median	0	0	1	0	0	0	0
x_4	Mean	2.95	0.57	2.67	0.65	0.92	0.83	0.46
	StDev	5.60	1.21	3.32	1.01	1.98	0.81	0.63
	Median	1	0	2	0	0	1	0
x_5	Mean	7.14	2.05	10.57	2.31	4.04	6.87	2.03
	StDev	14.42	5.73	14.80	5.40	8.72	11.67	5.72
	Median	1	0	5	0	0	2	0
x_6	Mean	0.18	0.11	0.16	0.18	0.18	0.15	0.12
	StDev	0.22	0.22	0.17	0.29	0.28	0.22	0.22
	Median	0.11	0	0.14	0	0	0.08	0
x_7	Mean	0.05	0.04	0.03	0.005	0.04	0.02	0
	StDev	0.25	0.21	0.21	0.08	0.22	0.18	0
	Median	0	0	0	0	0	0	0
x_8	Mean	5.41	1.80	3.86	2.06	2.23	1.35	1.38
	StDev	6.58	1.80	2.75	1.63	2.96	0.64	0.78
	Median	3	1	3	1	1	1	1

References

1. Corbett, A.T., Anderson, J.R.: Knowledge tracing: modeling the acquisition of procedural knowledge. User Model. User-Adap. Inter. **4**(4), 253–278 (1994)
2. Gugnani, A., Kasireddy, V.K.R., Ponnalagu, K.: Generating unified candidate skill graph for career path recommendation. In: 2018 IEEE International Conference on Data Mining Workshops (ICDMW), pp. 328–333. IEEE (2018)
3. Gugnani, A., Misra, H.: Implicit skills extraction using document embedding and its use in job recommendation. In: Proceedings of the AAAI Conference on Artificial Intelligence, vol. 34, pp. 13286–13293 (2020)
4. Honnibal, M., Montani, I.: spacy 2: natural language understanding with bloom embeddings, convolutional neural networks and incremental parsing (2017). https://spacy.io/
5. Horesh, R., Varshney, K.R., Yi, J.: Information retrieval, fusion, completion, and clustering for employee expertise estimation. In: 2016 IEEE International Conference on Big Data (Big Data), pp. 1385–1393. IEEE (2016)

6. Johns, J., Mahadevan, S., Woolf, B.: Estimating student proficiency using an item response theory model. In: International Conference on Intelligent Tutoring Systems, pp. 473–480. Springer (2006). https://doi.org/10.1007/11774303_47
7. van der Maaten, L., Hinton, G.: Visualizing data using T-SNE. J. Mach. Learn. Res. **9**(86), 2579–2605 (2008)
8. Palshikar, G.K., Srivastava, R., Shah, M., Pawar, S.: Automatic shortlisting of candidates in recruitment. In: ProfS Workshop @ SIGIR (2018)
9. Pawar, S., Thosar, D., Ramrakhiyani, N., Palshikar, G.K., Sinha, A., Srivastava, R.: Extraction of complex semantic relations from resumes. In: ASEA workshop @ IJCAI (2021)
10. Piech, C., et al.: Deep knowledge tracing. In: Proceedings of the 28th International Conference on NeurIPS-Volume 1, pp. 505–513 (2015)
11. Stojkovic, I., Ghalwash, M., Obradovic, Z.: Ranking based multitask learning of scoring functions. In: Joint European Conference on Machine Learning and Knowledge Discovery in Databases. pp. 721–736. Springer (2017). https://doi.org/10.1007/978-3-319-71246-8_44
12. Wauters, K., Desmet, P., Van Noortgate, W.: Monitoring learners' proficiency: weight adaptation in the ELO rating system. In: Educational Data Mining (2010)
13. Xiao, Y., Liu, B., Hao, Z.: A maximum margin approach for semisupervised ordinal regression clustering. IEEE Trans. Neural Networks Learn. Syst. **27**(5), 1003–1019 (2015)
14. Yudelson, M., Rosen, Y., Polyak, S., de la Torre, J.: Leveraging skill hierarchy for multi-level modeling with ELO rating system. In: Proceedings of the Sixth (2019) ACM Conference on Learning@ Scale, pp. 1–4 (2019)

Causal Enhanced Uplift Model

Xiaofeng He[1], Guoqiang Xu[2], Cunxiang Yin[2(✉)], Zhongyu Wei[1(✉)], Yuncong Li[2], Yancheng He[2], and Jing Cai[2]

[1] School of Data Science, Fudan University, Shanghai, China
{hexf20,zywei}@fudan.edu.cn
[2] Tencent Inc., Shenzhen, China
{chybotxu,jasonyin,yuncongli,collinhe,samscai}@tencent.com

Abstract. Uplift modeling refers to approaches to quantify net difference in outcome between applying a treatment and not applying it to an individual. It is a typical causal inference problem which allowing us to design a refined decision rule that only targets those susceptible. The core difficulty of the existing methods is that there is no direct supervision label for uplifts. In this paper, we propose an efficient Causal Enhanced Uplift Model (CEUM) to excavate potential uplift information. Specifically, we first construct contrastive pairs following the properties of partial order relation of uplifts and maintain the structural correlation and consistency during training. Then, we seek for a promising average causal effect in batches to approach the mean of individual estimated uplift. Finally, we benchmark the efficacy of the proposed method by conducting comprehensive experiments and the results show that CEUM achieves the state-of-the-art performance on two real-world datasets.

Keywords: Uplift modeling · Causal inference · Deep learning

1 Introduction

In many applications, researchers and practitioners need to investigate how treatment effects vary across individuals and contexts to personalize treatment regimes and maximize the target profits. For instance, when considering a specific drug, a doctor would prescribe it to patients who have a good medical effect while avoiding those who are at risk of adverse events or harmful side effects; and a company should choose whether to deliver an advertisement to a customer or not in order to maximize gains and reduce costs.

The core difficulty is that we cannot observe both responses of a unit when receiving a certain treatment or not simultaneously, commonly known as *The Fundamental Problem of Causal Inference*. Uplift modeling is a standard predictive method for such tasks and it seeks to forecast the causal effect of an intervention/treatment on a specific outcome at the individual-level. In recent years, various uplift modeling techniques have been developed and applied in a wide

---

X. He and G. Xu contribute equally to this work.

© The Author(s), under exclusive license to Springer Nature Switzerland AG 2022
J. Gama et al. (Eds.): PAKDD 2022, LNAI 13282, pp. 119–131, 2022.
https://doi.org/10.1007/978-3-031-05981-0_10

range of domains [28], including health care [9], digital marketing [14] and public policies [6]. Most of the existing uplift modeling methods are based on potential outcome paradigm [20], including Meta-Learning approaches [13], Direct estimation approaches [18], Transformed outcome approach [1,9] and those based on generalization metric bound [2].

In most cases, Meta-algorithm consists of two phases. Firstly use several base learners to assess the outcomes respectively under control and varied treatments. Secondly use the difference between these estimations to rank candidates. However there is an endogenous drawback that the objectives of these base learners are not perfectly consistent with the objective of estimating uplifts [18]. Each learner is trained independently to promote response classification accuracy without noticing other treatment groups, whereas uplift modeling focuses on the change in response probability. Therefore just maximizing precision of these independent models is no guarantee of reliability in uplift. Furthermore, subtracting predicted scores of two separate models will potentially produce compound errors.

Direct estimation approaches use tree methods to predict individual uplift directly by the uplift measured in terminal nodes. Transformed outcome approach transforms the observed outcome so that the conditional expectation of transformed outcome equals the uplift. Both of direct estimation approaches and transformed outcome approach can model the uplift straightforwardly, but they are optimized based on the response variable and lack of the uplift information. And the transformed outcome approach, which relies substantially on the accurate estimation of propensity score, also suffers from large variation [28]. Bound based approaches concentrate on optimizing the bounds of specific evaluation metrics which are also not directly related to uplifts.

In this work, we proposed a Causal Enhanced Uplift model (CEUM) to alleviate the general problem of missing uplift information. We introduce a new uplift loss function concerning the implicit prior information about uplift on both individual level and group level. Specifically, we utilize the partial order relationship of uplift between positive and negative sample pairs and the uniformity between the mean of individual uplifts and the group treatment effects. To the best of our knowledge, this is the first attempt to utilize implied uplift information in neural network uplift model and the main architecture is simple and universal. Our proposed method is validated on two real-world datasets. The experimental results demonstrate that the proposed model has achieved the state of the art.

The contributions of this work are summarized as follows:

- We introduce a novel Causal Enhanced Uplift Model to capture both the partial order relation of contrastive pairs and average constraint of uplifts.
- The two proposed components enhances the causal effect estimation ability by utilizing and leveraging both individual and global prior information of uplifts.
- We evaluate the proposed model using two diverse real-world public datasets and observe improvements over state-of-the-art baseline methods.

2   Related Work

Uplift is a particular case of causal inference modeling and follows the potential outcomes framework [20]. Uplift refers to modeling behavioral change or business metrics enhancement that results directly from a specific treatment. Instead of modeling the positive response probability, uplift attempts to model the difference between conditional probabilities in the treatment and control groups. Traditional uplift models are developed based on randomized controlled trials settings where both the treatment and outcome are binary random variables.

The intuitive approach to model uplift is to build two classification models [7,13]. This consists of fitting two separated conditional probability models: one for the treated individuals, and another for the controlled. The uplift is estimated as the difference between these two conditional probability models. The advantage of this technique is its simplicity and portability across various existing supervised models such as Logistic Regression and XGBoost. However, it does not perform well in practice [18]. Both models focus on predicting only a one-class probability and information about the other treatment is never provided to the learning algorithm for each model. So the resulting difference is unstable. Using a unique model can solve this drawback. Some traditional uplift estimation methods use the treatment assignment as a feature, adding explicit interaction terms between covariates and the treatment indicator, and then train regression models to estimate the uplift [15]. Transformed outcome Methods [1,9] solve this by creating a new binary target variable of which the expectation conditional mean is equal to uplift. Most off-the-shelf regression algorithms can be applied to estimate this target.

Subsequent research is in the direction of regression trees [16]. These proposed methods view the forests as an adaptive neighborhood metric, and estimate the treatment effect at the leaf node [25,29]. The criteria used for choosing each split during the growth of the uplift trees is based on maximization of the difference in uplifts between the two child nodes. However, in practice, these approaches are prone to overfitting. Another one modeling uplift directly is based on SVM (Support Vector Machines) [27]. Moreover, there are some studies incorporating generalization bounds for CATE metrics, such as PEHE [23], AUUC [2]. They are indicator-oriented methods that have risk in causing inconsistency deviation.

Recently, neural-based representation learning approaches have been proposed for estimating uplift on observational data. An approach emphasizes learning a covariate representation that has a balanced distribution across treatment and outcome [10,23]. Some similar methods try to minimize the distribution difference between treated and control groups in the embedding space [26]. Another approach is based on propensity score matching [19] with neural networks [22]. A new method using neural network representation allows to jointly optimize the difference in conditional means and the transformed outcome losses [17]. Consequently, the model not only estimates the uplift, but also ensures consistency in predicting the outcome. Our method is also a deep learning algorithm, but the difference is that we take advantage of elaborate information directly related to uplift to increase convergence speed, stability and accuracy.

3   Preliminaries

In this part we frame the problem of uplift optimization including the necessary assumptions using the Rubin-Neyman potential outcomes framework [21] and related evaluation metrics.

We denote the $x \in X$ as the features of a unit and $t \in T$ as a potential intervention or action. The correspond potential outcome for a unit x under treatment condition t, is denoted as $Y_t(x) \in Y$. Without sacrificing generality and for simplicity, in this work we exclusively investigate binary treatment set $T = \{0, 1\}$ where 0 indicates the 'control' and 1 represents the "treatment" with outcome variables equal to $Y_0(x)$ or $Y_1(x)$ correspondingly. In this sense, the individualized treatment effect (ITE) can be expressed as $ITE(x) = Y_1(x) - Y_0(x)$. The Fundamental Problem of Causal Inference is that only one outcome variable can be observed for each x.

For uplift modeling, the target quantity to infer is the Conditional Average Treatment Effect (CATE) [28] defined as $CATE = \mathbb{E}[Y_1(x) - Y_0(x)|X = x]$. CATE is the difference between the probability of a unit x who receives a treatment and then becomes active and the probability of the unit x who does not receive a treatment but is active. Another widely used statistic is the Average Treatment Effect (ATE) stated as $ATE = \mathbb{E}[Y_1 - Y_0]$, where Y_1 and Y_0 are the potential treated and control outcomes of the whole population respectively.

In order to make the treatment effect identifiable, we need make the following assumptions: Stable Unit Treatment Value Assumption (SUTVA), Consistency, Unconfoundedness, and Overlap. The interested readers can refer to [28] for details.

With the above notations, we can trivially outline the basic models. Given dataset $D = \{(x_i, t_i, y_i)\}_{i=1}^{n}$ where $y_i = t_i Y_1(x_i) + (1 - t_i)Y_0(x_i)$, we train a classification model $f : X \times T \rightarrow Y$ to fit $\hat{y}_i = f(x_i, t_i)$. Then the unobserved counterfactual outcome $\hat{y}_i^C$ for x_i can be calculated by $f(x_i, 1 - t_i)$. In this way, the uplift value for unit x_i is $\hat{u}_i = (-1)^{1-t_i}(\hat{y}_i - \hat{y}_i^C)$ naturally. And using the ranking of these values, we can detect the partitioning of population with positive treatment return efficiently.

In evaluation, on account of lack of ground truth in individual treatment effects in real world, we refer to the most popular metric in the uplift literature: Area Under the Uplift Curve (AUUC) [2,4]:

Definition 1. *(Area Under the Uplift Curve). Let D_T and D_C specifies the treatment and control subsets of dataset D respectively, with n_T and n_C denoting the corresponding dataset sizes. Let $f\left(D_T, \frac{p}{100}n_T\right)$ and $f\left(D_C, \frac{p}{100}n_C\right)$ be the first p percentages of D_T and D_C respectively when both ordered by prediction of model f. The empirical AUUC of the model f on D_T and D_C is given by:*

$$\widehat{AUUC}(f, D_T, D_C) = \int_0^1 V(f, x)dx \approx \sum_{p=1}^{100} V\left(f, \frac{p}{100}\right), \qquad (1)$$

where

$$V\left(f, \frac{p}{100}\right) = \frac{1}{n_T} \sum_{i \in f\left(D_T, \frac{p}{100} n_T\right)} y_i - \frac{1}{n_C} \sum_{j \in f\left(D_C, \frac{p}{100} n_C\right)} y_j. \tag{2}$$

Intuitively, AUUC measures the ranking ability of the model where the top-ranked should increase more in cumulative uplift.

4 Proposed Method

The overview of CEUM architecture is illustrated in Fig. 1, and it can be regarded as a multi-task model, including one main task and two auxiliary tasks. Overall the causal enhanced uplift model takes a contrastive pair, $\{< x_1, x_2 >\}$, and the actual treatment t of x_1 as external inputs. The main task aims to learn the basic model f to predict the outcome probability of a unit with feature x given treatment t. The first auxiliary task predicts the pair uplifts difference and then utilizes the partial order relationship between the constructed sample pairs to guide prediction of uplift by backpropagation. Another auxiliary task aims at maintaining uplift stability in the average sense. In this process, we implicitly combine the prior information of the uplift in deep learning model, so the main task will also benefit from the auxiliary tasks and obtains better performance. The following subsections explain the framework in detail.

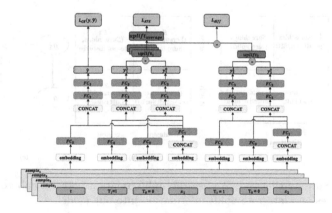

Fig. 1. The architecture of our proposed causal enhanced uplift model.

Pairs Construction. Essentially, we expect to use uplift model to categorize population into four subsets as shown in Fig. 2(a):

- **Sure things** are individuals who always respond, regardless of whether or not they receive treatment. Targeting sure things does not generate additional returns, so this kind of individuals have zero uplift.

- **Persuadables** are individuals who will be more likely to respond because they are exposed to a treatment. These people are exactly the target population we need to distinguish. This kind of individuals have positive uplift.
- **Do-not-disturbs** are individuals who became less likely to respond because of the treatment. Obviously, targeting them has no benefits but brings large additional costs. This kind of individuals have negative uplift.
- **Lost causes** are individuals who never respond, regardless of whether or not they receive the campaign. As treating Lost causes will not generate additional revenues, this kind of individuals have zero uplift.

However, for real-world dataset, it is impractical to find out these four categories. Typically using the data available, we categorize population as illustrated in Fig. 2(b):

- g_1 are individuals who are treated and respond. A unit in g_1 may belong to Persuadables or Sure things.
- g_2 are individuals who are in control group and respond. A unit in g_2 may belong to Sure things or Do-not-disturbs.
- g_3 are individuals who are treated and do not respond. A unit in g_3 may belong to Lost causes or Do-not-disturbs.
- g_4 are individuals who are in control group and do not respond. A unit in g_4 may belong to Persuadables or Lost causes.

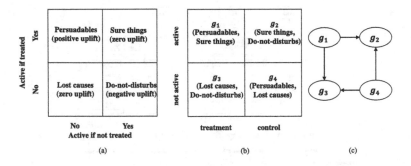

Fig. 2. Four-Subsets Diagram. We use the uplift relation between subgroups to construct instances and integrate uplift information into our model.

We can see that 1) the uplift of a unit from g_1 is bigger than or equal to the uplift of a unit from g_2 and g_3, and 2) the uplift of a unit from g_4 is also bigger than or equal to the uplift of a unit from the uplift of g_2 or g_3. As shown in Fig. 2(c), for each unit (i.e., x_1) from g_1 or g_4, we randomly select one unit from g_2 or g_3 as x_2. Then, we obtain a set of instances, $D^1 = \{(t_1, x_1, x_2, y), y_{diff} = 1\}$. For each unit (i.e., x_1) from g_2 or g_3, we randomly select one unit from g_1 or g_4 as x_2. Then, we obtain a set of instances, $D^2 = \{(t_1, x_1, x_2, y), y_{diff} = -1\}$. Finally, we collect the training pair data as the union of D_k^1 and D_k^2, $D = D^1 \bigcup D^2$.

The external inputs of our causal enhanced uplift model are a pair of units, $< x_1, x_2 >$, and the actual treatment t of x_1. In addition, we denote two auxiliary constant, $T_0 \equiv 0$ and $T_1 \equiv 1$ for convenience.

Treatment Representation. We convert all the treatment inputs (i.e., t, $T_1 \equiv 1$ and $T_0 \equiv 0$) into vectors using a shared embedding layer and then obtain their representations through a linear layer FC_0 (i.e., r_t, $r_{T \equiv 1}$ and $r_{T \equiv 0}$).

User Representation. An individual x include n categorical features, $x = \{x^1, ..., x^i, ..., x^n\}$, where x^i is the index of the i-th feature of the individual. We first embed them into different vectors respectively $v = \{v^1, ..., v^i, ..., v^n\}$. The concatenation of the n vectors, $\{v^1; ...; v^i; ...; v^n\}$ are then fed into a fully connected layer(FC_1) with ReLu as activation function. The output is the encoded representation of feature x. The representation of the user x_1 is r_{x_1} and the representation of the user x_2 is r_{x_2}.

Main Task: Neural-Based S-Learner. We use the concatenated vector of r_t and r_{x_1} as the final fusion representation and feed it into two successive fully connected layers (i.e., FC_2 and FC_3) with ReLu and sigmoid as activation functions. The output of FC_3 is considered to be the probability of outcome. As shown in the Fig. 1, we finally get five probabilities, $\hat{y}$, y_1^1, y_1^0, y_2^1, and y_2^0. $\hat{y}$. y_1^1, and y_1^0 are the response probabilities of the x_1 being persuaded given the actual treatment t, the constructed treatments $T_1 \equiv 1$ and $T_0 \equiv 0$. y_2^1, and y_2^0 are the conversion probabilities of x_2 given $T_1 \equiv 1$ and $T_0 \equiv 0$.

As the main task is simply a classification task predicting the outcome probability $\hat{y}$ of x_1 given the actual treatment t, a natural choice is the cross-entropy loss:

$$L_{CE}(y, \hat{y}) = -\sum_{i=1}^{n} (y_i \log \hat{y}_i + (1 - y_i)\log(1 - \hat{y}_i)), \tag{3}$$

where y is the ground truth label.

Auxiliary Task1: Uplifts Difference in Pairs. Each pair implies a contrastive characteristics in uplifts. During training, we estimate the uplift of x_1 as $uplift_{x1} = y_1^1 - y_1^0$, and the uplift of x_2 as $uplift_{x2} = y_2^1 - y_2^0$. So the difference between $uplift_{x_1}$ and $uplift_{x_2}$ is $diff_{uplift} = uplift_{x_1} - uplift_{x_2}$. We use the hinge loss to optimize the uplift order:

$$L_{diff} = \sum_{i=1}^{n} [margin_1 - y_{diff} diff_{uplift}]_+. \tag{4}$$

where $margin_1$ is a hyperparameter quantifying discrimination threshold and y_{diff} is a indicative label. If x_1 has bigger uplift than x_2, $y_{diff} = 1$, otherwise, $y_{diff} = -1$.

Auxiliary Task2: Consistency in Batch. A precise uplift model should also preserve the accuracy and consistency of the overall causal effect:

$$ATE = \mathbb{E}[Y_1 - Y_0] = \mathbb{E}_X[\mathbb{E}[Y_1(x) - Y_0(x)|X = x]] = \mathbb{E}_X[CATE]. \quad (5)$$

Practically, We try to reduce the ATE prediction gap in each batch:

$$L_{ATE} = \sum_{i=1}^{B}[-margin_2 + \|\hat{\tau}_i - \tilde{\tau}_i\|^2]_+ \quad (6)$$

where $\hat{\tau}_i$ is the batch ATE calculated by $\hat{\tau}_i = \sum_{j \in b_i} \frac{1}{n_{i_1}} t_j y_j - \frac{1}{n_{i_0}}(1 - t_j) y_j$, $\tilde{\tau}_i$ is the average estimated uplifts of x_1 in batch b_i: $\tilde{\tau}_i = \frac{1}{n_i} \sum_{j \in b_i} uplift_j$, and B is the number of batches.

Training. In Causal enhanced uplift model, we optimize three loss functions jointly, and the overall loss is:

$$L = L_{CE} + \lambda L_{diff} + \beta L_{ATE}, \quad (7)$$

where λ and β are the corresponding regularization coefficients of the losses. In practice, we train our model by minimizing (7) using the stochastic gradient descent (SGD) optimizer.

5   Experiment

5.1   Datasets

We verify the effect of the model on two open, real-world unbiased datasets. Hillstrom's Email Advertisement dataset [8] includes data tracked in a randomized trial following an email advertising campaign. The customers were evenly distributed into two treatment groups and a control group at random. We choose receiving a "Women's merchandise Email" as treatment and the outcome is customers' visit status. Criteo Uplift modelling dataset (Criteo-v2) [5] is another recent dataset collected from a randomized control experiment in digital advertising business. And the treatment indicator represents whether the customer receives a promotional advertisement and the outcome indicates the visit status of the customer. See Table 1 for details about datasets.

5.2   Experiment Settings

Implementation Details. In the experiment, each dataset is split into train (70%) and test (30%) sets. And following [4] we conduct 10 repeated experiments for each model and take their average as the final results for reducing the impact of fluctuations. We implement our model through the Keras framework based on Tensorflow. We choose Adam as the optimizer [12], and the initial learning rate

Table 1. Summary statistics of datasets.

Metric	Hillstrom	Criteo-v2
Total size	42,693	13,979,592
Treatment/control ratio	1.0038	5.6667
# Features variables	8	12
Treatment positive rate	0.1514	0.04854
Control positive rate	0.1062	0.03820
Group positive rate	0.1288	0.04699
Uplift initial campaign	0.0452	0.01034

is $\alpha = 0.001$. The two margins are 0.4 and 0.001 and the numbers of neurons in the dense layer FC_0, FC_1, FC_2, FC_3 are 8, 64, 128, 1 respectively. The batch sizes are 256 and 2048 respectively. In the evaluation phase, we follow previous works [2, 4] and use AUUC as the metrics. We also define a new metric: $\epsilon_{ATE} = \frac{|\overline{ATE} - \widetilde{ATE}|}{\overline{ATE}}$, where $\overline{ATE}$ means sample ATE calculated in test-set, and $\widetilde{ATE}$ is the average uplifts of test-set. So ϵ_{ATE} measures the error rate of ATE prediction.

Compared Methods. We compare our proposed method with both basic and advanced methods as follows: (1) **Meta-learner algorithms** [13] where we choose the top ranked methods including S-learner, T-learner, and Doubly Robust (DR) learner with the powerful LGBM [11] as base regressor; (2) **Transformed algorithm** CVT [1] also with LGBM as base estimator; (3) **Tree-based algorithms** [18] based on uplift tree/random forests on KL divergence, Euclidean Distance, and Chi-Square; (4) **Neural-network-based algorithms** including TARNet [23] and DragonNet [24]; (5) **Bound-based algorithm** AUUC-max [2] specially optimizing the bound of AUUC.

Most of these baselines are implemented using an open source Python package (CausalML [3]) except AUUC-max and PCG where we just quote the results of original papers. Note that PCG just selects a proportion of Criteo-v2 dataset randomly.

5.3 Experimental Result Analysis

Main Results. Table 2 shows the performance achieved by all the tested methods. The results indicate that CEUM achieves a statistically superior performance (tied with other competitor in Criteo-v2). On Hillstrom dataset, CEUM wins by a relatively large margin across all competing techniques in both AUUC and ϵ_{ATE}. All methods have lower AUUC on Criteo-v2. This may be due to the low positive ratio and average uplift of the "visit" outcome. Our method is still comparatively more robust. The superiority is expected as we leverage the uplift gap in contrastive pairs and the overall average property of uplifts. All methods have lower AUUC on Criteo.

Figure 3(a) shows the Uplift Curves on the Hillstrom dataset (higher is better). Likewise, we can observe that our approach achieves a higher and more stable AUUC, which demonstrates the effectiveness of our method.

Ablation Studies. We perform ablation studies to ascertain each component's contribution. Summarized results are presented in the penultimate line of Table 2. We can observe that contrastive pairs promote the AUUC significantly, and the average causal effect balance consistently exhibits advantages over ϵ_{ATE}, though the advantageous margins vary over datasets. The full model leverages both aspects and achieves high comprehensive performance.

Table 2. Test results on Hillstrom (left) and Criteo-v2 (right). Higher is better for AUUC and lower is better for ϵ_{ATE}.

Models	Hillstrom		Criteo-v2	
	AUUC±S.E.	ϵ_{ATE}±S.E.	AUUC±S.E.	ϵ_{ATE}±S.E.
t-learner	.02369 ± .00050	.1372 ± .0068	.01485 ± .00002	.3985 ± .0009
s-learner	.02927 ± .00052	.1048 ± .0055	.01522 ± .00007	.4577 ± .0048
dr-learner	.03066 ± .00021	.1187 ± .0263	.01473 ± .00002	.4100 ± .0017
CVT	.02836 ± .00047	.1453 ± .0027	.01530 ± .00002	.4473 ± .0008
Rand.F.(K.L.)	.03037 ± .00058	.1427 ± .0272	.01466 ± .00012	.4124 ± .0051
Rand.F.(E.D.)	.03022 ± .00068	.1584 ± .0210	.01482 ± .00015	.4360 ± .0054
Rand.F.(Chi.)	.03005 ± .00051	.1411 ± .0258	.01474 ± .00010	.4306 ± .0047
AUUC-max(deep) [2]	.02999 ± .00326	–	.00924 ± .00001	–
PCG [4]	.03055 ± N/A	–	**.01601 ± N/A**	–
TARNet	.02820 ± .00153	.3787 ± .1266	.01497 ± .00014	.3843 ± .0800
DragonNet	.03016 ± .00042	.1669 ± .0605	.01505 ± .00017	**.2636 ± .0435**
w/o L_{diff} and L_{ATE}	.02977 ± .00104	.2037 ± .1018	.01507 ± .00017	.4696 ± .0995
w/o L_{diff}	.03049 ± .00055	**.0320 ± .0250**	.01524 ± .00015	**.1995 ± .0366**
w/o L_{ATE}	**.03099 ± .00067**	.1344 ± .0571	**.01534 ± .00015**	.6202 ± .0294
CEUM	**.03124 ± .00032**	.0570 ± .0399	**.01534 ± .00016**	.2560 ± .0431

Hyper-parameter Sensitivity. We further test the sensitivity of CEUM w.r.t. the hyper-parameters $margin_1, margin_2, \lambda, \beta$ and batch size b. We report the results on the extracted validation set of Hillstrom in Fig. 3.

Figure 3(b) shows that both an extremely small or large $margin_1$ will destroy performance, and Fig. 3(c) demonstrates that the trend of ϵ_{ATE} is consistent with $margin_2$ while AUUC fluctuates. From Fig. 3(d) and 3(e) we conclude that it is beneficial to extend the uplift distance of the pairs appropriately but if the penalty is too large, the model will shrink extremely. And L_{ATE} can help constrain uplift in a reasonable scope on average. However when β are large, it faces the risk of homogenizing uplifts and weaken the ability to rank. Furthermore, our method benefits from larger batch sizes and more training

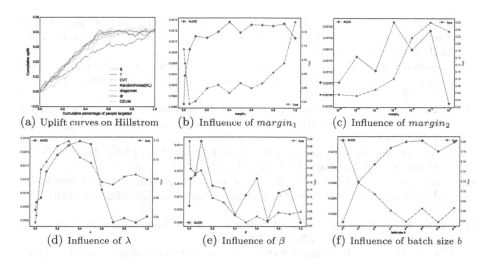

(a) Uplift curves on Hillstrom   (b) Influence of $margin_1$ (c) Influence of $margin_2$

(d) Influence of λ (e) Influence of β (f) Influence of batch size b

Fig. 3. AUUC curves and effect of hyper-parameter

steps and small batch size will undermine the accuracy of batch ATE estimation leading to training instability in Fig. 3(f).

6 Conclusion

We introduced a novel multitasks framework for uplift model. The proposed model alleviates the lack of direct supervision or explicit label for uplift, which the preceding supervised classifiers often suffer. CEUM constructs a sequence of pairs by performing random sampling from opposite subsets, exploiting the partial order relation for instances between subgroups. Furthermore, CEUM performs ATE correction simultaneously with the aim of progressively reducing prediction deviation. Experiments on two real-world datasets demonstrate its promising effectiveness and outperform competitive baselines.

References

1. Athey, S., Imbens, G.W.: Machine learning methods for estimating heterogeneous causal effects. Stat. **1050**(5), 1–26 (2015)
2. Betlei, A., Diemert, E., Amini, M.R.: Uplift modeling with generalization guarantees. In: Proceedings of the 27th ACM SIGKDD Conference on Knowledge Discovery and Data Mining, pp. 55–65 (2021)
3. Chen, H., Harinen, T., Lee, J.Y., Yung, M., Zhao, Z.: Causalml: python package for causal machine learning. arXiv preprint arXiv:2002.11631 (2020)
4. Devriendt, F., Van Belle, J., Guns, T., Verbeke, W.: Learning to rank for uplift modeling. IEEE Trans. Knowl. Data Eng. (2020)
5. Eustache, D., Artem, B., Renaudin, C., Massih-Reza, A.: A large scale benchmark for uplift modeling. In: Proceedings of the AdKDD and TargetAd Workshop, KDD, London, United Kingdom, 20 August 2018, ACM (2018)

6. Grimmer, J., Messing, S., Westwood, S.J.: Estimating heterogeneous treatment effects and the effects of heterogeneous treatments with ensemble methods. Polit. Anal. **25**(4), 413–434 (2017)
7. Hansotia, B.J., Rukstales, B.: Direct marketing for multichannel retailers: issues, challenges and solutions. J. Database Market. Customer Strat. Manage. **9**(3), 259–266 (2002)
8. Hillstrom, K.: The minethatdata e-mail analytics and data mining challenge (2018). https://blog.minethatdata.com/2008/03/minethatdata-e-mail-analytics-and-data.html
9. Jaskowski, M., Jaroszewicz, S.: Uplift modeling for clinical trial data. In: ICML Workshop on Clinical Data Analysis, vol. 46 (2012)
10. Johansson, F., Shalit, U., Sontag, D.: Learning representations for counterfactual inference. In: International Conference on Machine Learning, pp. 3020–3029. PMLR (2016)
11. Ke, G., et al.: LightGBM: a highly efficient gradient boosting decision tree. Adv. Neural Inf. Process. Syst. **30**, 3146–3154 (2017)
12. Kingma, D.P., Ba, J.: Adam: a method for stochastic optimization. arXiv preprint arXiv:1412.6980 (2014)
13. Künzel, S.R., Sekhon, J.S., Bickel, P.J., Yu, B.: Metalearners for estimating heterogeneous treatment effects using machine learning. Proc. Natl. Acad. Sci. **116**(10), 4156–4165 (2019)
14. Li, S., Vlassis, N., Kawale, J., Fu, Y.: Matching via dimensionality reduction for estimation of treatment effects in digital marketing campaigns. In: IJCAI, pp. 3768–3774 (2016)
15. Lo, V.S.: The true lift model: a novel data mining approach to response modeling in database marketing. ACM SIGKDD Explor. Newsl. **4**(2), 78–86 (2002)
16. Loh, W.Y.: Classification and regression trees. Wiley Interdiscipl. Rev. Data Mining Knowl. Disc. **1**(1), 14–23 (2011)
17. Mouloud, B., Olivier, G., Ghaith, K.: Adapting neural networks for uplift models. arXiv preprint arXiv:2011.00041 (2020)
18. Radcliffe, N.J., Surry, P.D.: Real-world uplift modelling with significance-based uplift trees. White Paper TR-2011-1, Stochastic Solutions, pp. 1–33 (2011)
19. Rosenbaum, P.R., Rubin, D.B.: The central role of the propensity score in observational studies for causal effects. Biometrika **70**(1), 41–55 (1983)
20. Rubin, D.B.: Estimating causal effects of treatments in randomized and nonrandomized studies. J. Educ. Psychol. **66**(5), 688 (1974)
21. Rubin, D.B.: Causal inference using potential outcomes: design, modeling, decisions. J. Am. Statist. Assoc. **100**(469), 322–331 (2005)
22. Schwab, P., Linhardt, L., Karlen, W.: Perfect match: a simple method for learning representations for counterfactual inference with neural networks. arXiv preprint arXiv:1810.00656 (2018)
23. Shalit, U., Johansson, F.D., Sontag, D.: Estimating individual treatment effect: generalization bounds and algorithms. In: International Conference on Machine Learning, pp. 3076–3085. PMLR (2017)
24. Shi, C., Blei, D.M., Veitch, V.: Adapting neural networks for the estimation of treatment effects. arXiv preprint arXiv:1906.02120 (2019)
25. Wager, S., Athey, S.: Estimation and inference of heterogeneous treatment effects using random forests. J. Am. Statist. Assoc. **113**(523), 1228–1242 (2018)
26. Yao, L., Li, S., Li, Y., Huai, M., Gao, J., Zhang, A.: Representation learning for treatment effect estimation from observational data. Adv. Neural Inf. Process. Syst. **31**, 1–10 (2018)

27. Zaniewicz, L., Jaroszewicz, S.: Support vector machines for uplift modeling. In: 2013 IEEE 13th International Conference on Data Mining Workshops, pp. 131–138. IEEE (2013)

28. Zhang, W., Li, J., Liu, L.: A unified survey of treatment effect heterogeneity modelling and uplift modelling. ACM Comput. Surv. (CSUR) 54(8), 1–36 (2021)

29. Zhao, Y., Fang, X., Simchi-Levi, D.: A practically competitive and provably consistent algorithm for uplift modeling. In: 2017 IEEE International Conference on Data Mining (ICDM), pp. 1171–1176. IEEE (2017)

An Incentive Dispatch Algorithm for Utilization-Perfect EV Charging Management

Lo Pang-Yun Ting, Po-Hui Wu, Hsiu-Ying Chung, and Kun-Ta Chuang$^{(\boxtimes)}$

Department of Computer Science and Information Engineering, National Cheng Kung University, Tainan City, Taiwan
{tinglo,phwu}@netdb.csie.ncku.edu.tw, ktchuang@mail.ncku.edu.tw

Abstract. Due to the rapid growth of electric vehicles (EVs), the charging scheduling of EVs has become highly important. In order to reduce the total operating cost, how to arrange the charging of each EV becomes the main issue. However, existing scheduling methods usually obtain schedules without considering EVs users' charging willingness, which will let EVs users be reluctant to follow the arranged charging schedule, thereby incurring low charging utilization and high operational overhead. To solve this problem, we devise an online charging registration mechanism, an incentive-based framework called *POSIT*, to provide a feasible schedule for different EVs to enhance the quality of user experience. In the proposed mechanism, the charging scheduler will provide a relevant reward (as an incentive) for users to properly enhance users' willingness to accept the arranged schedule. In addition, the interactive learning is adopted to improve the next recommendation based on the user's feedback. The *POSIT* framework is able to satisfy the energy demand of EVs charging and the commercial building. The implemented experiments indicate that the proposed framework can not only increase the charging utilization, but also can significantly reduce the electrical operating costs and increase the revenue at the cost of small incentives.

1 Introduction

With the increase in environmental concerns related to carbon emission, electric vehicles (EVs) have received significant attention in recent years. According to the statistics report of the International Energy Agency (IEA) [3], EV sales will overtake hybrid vehicles sales in 2021 and may reach 7 million in 2022. Due to the growing need of EV charging infrastructure, the related techniques and cooperation have been thriving in many aspects. For example, retailers are deemed to be ideal operators to serve as site hosts for EV charging stations since retailers not only make a profit from vehicle charging fees but also may get extra in-store retail sales when the customers stay in the stores during the charging time [11].

J. Gama et al. (Eds.): PAKDD 2022, LNAI 13282, pp. 132–146, 2022.
https://doi.org/10.1007/978-3-031-05981-0_11

However, uncoordinated charging activities may cause adverse impacts. Generally, the retail building operator declares a demand named contracted capacity and signs an electricity contract with a power company. If the customers charge their EVs at the peak-electricity-demand time period of a retail building, a higher electricity penalty will be introduced due to the excesses of contracted capacity of peak electricity usage, which represents the maximum power requirement of the customer at any given time. To handle this issue, various approaches have been used to investigate the most profitable EV charging schedule [8,13,15]. The most common method is to predict the charging demand for the next day and then rearrange the charging schedule of EVs in order to achieve the highest revenue [14,15]. The challenge is that not every EV user is willing to accept the prearranged charging schedule. EV users may have different levels of willingness to follow different schedules. Therefore, it is highly necessary to develop an effective strategy which stimulates EV users to obey the prearranged charging schedule to reach the optimal usage of charging station with maximum profit.

In this paper, we propose a new charging registration mechanism to investigate an acceptable charging schedule for individual EV user with a relevant incentive. The goal is to develop a win-win strategy that each user can charge an EV at an acceptable time period with incentives and the operator of charging station can obtain the highest revenue by simultaneously considering the electricity system of the retail building and the EV charging station. In the proposed mechanism, each EV user can online reserve their preferred length of charging time (e.g., 2 h) in the registration system. Subsequently, the system would display the recommended charging time periods with various relevant incentives to promote a user's willingness to follow the arranged schedule. However, it is a challenging issue to decide a feasible arrangement between the charging time periods and value of incentives because each EV user's acceptance level for the recommended arrangements is unknown. Note that to some extent, most users would like to charge their EVs during weekends, causing congestion during the rush time. To resolve this issue, a novel framework *POSIT* (**P**ersonalized **O**nline **S**cheduling with Adaptive **I**ncentives in a **T**hreshold Control) is proposed to explore the acceptable arrangements for all EVs users under a fixed incentive budget.

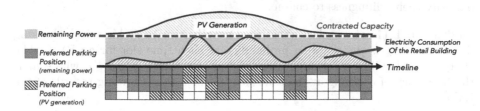

Fig. 1. The example of the preferred parking position estimation.

Two main factors, recommended charging periods and provided incentives, which may affect a user's willingness to follow the arrangement are considered in the *POSIT* framework. To investigate the best combination of these two factors, we design an interactive recommendation process to explore the most feasible scheduling arrangement.

Firstly, the number of available EVs charging positions for each time period in the next day, which is named *preferred parking positions* in this paper, is estimated by predicting the solar photovoltaic (PV) power generation and electricity consumption of a retail building. As shown in Fig. 1, the remaining power of each time slot can be estimated by subtracting the predicted electricity consumption of the retail building from contracted capacity. Hence, the remaining power and PV generation can be transferred to the preferred parking positions for each time period. After the user completes the charging registration, the recommended charging time periods (selected from available charging positions) with incentives are determined based on the user's historical charging records. In addition, for better scheduling, the user's selection results will be used as an online user feedback manner, affecting the next recommended charging arrangement. The objective of *POSIT* is to maximize the utilization of EV charging stations to obtain the highest revenue and minimize the operating cost by considering users' willingness.

The structure of this paper is presented as follows. We present the problem statement in Sect. 2, followed by describing the proposed methods in Sect. 3. The experimental results are exhibited in Sect. 4. Section 5 reviews the relevant studies, and Sect. 6 concludes this work.

2   Problem Definition

Before introducing our framework, we give the description of the necessary symbols and definitions. Firstly, let V be a set of EVs, and each EV $v \in V$ is associated with a set of historical charging records R^v, where each $x_t^v \in X^v$ represents the electrical charging value of an EV v at a discrete time slot $t \in \{1, 2, ..., T\}$. Secondly, let E^p and E^h be the sets of PV generation and the electrical consumption of a retail building, respectively, where each $e_t^p \in E^p$ and $e_t^h \in E^h$ denote the power generated and the power used at each time slot, respectively.

Here we formulate the charging probability of EV users in each time slot to quantify their willingness to charge.

Definition 1 (Historical Charging Distribution): A historical charging distribution P_h^v is the charging probability at each time slot for an EV v. The charging probability is estimated by $P_h^v(X = t) = n(t)/\sum_{t'} n(t')$, where n_t is the frequency of charging at the time slot t in historical charging records of the EV v.

Definition 2 (Incentive Charging Distribution): An incentive charging distribution P_i^v is the possible charging probability at each time slot for an EV v after providing incentives. Based on the law of the diminishing marginal utility [2], we assume that the increased charging willingness will be gradually diminished when the provided incentive rises. Accordingly, the increased charging willingness $f(\theta_t)$ at time slot t with the provided incentive θ_t is designed as:

$$f(\theta_t) = f(\theta_t - \Delta c) + x\theta_t^{-y}, \tag{1}$$

where Δc is the value difference between provided incentives, and x and y are constant values.

Let $I(t)$ be the changed willingness, the incentive charging distribution P^v is defined as:

$$P_i^v(X = t) = I(t)/\sum_{t'} I(t'), \ I(t) = P_h^v(X = t) + f(\theta_t). \tag{2}$$

Definition 3 (Trend Charging Distribution): A trend charging distribution P_g is the charging probability at each time slot for *all EVs*. The probability is estimated by $P_g(X = t) = \bar{n}(t)/\sum_{t'} \bar{n}(t')$, where $\bar{n}(t)$ is the average number of charging at the time slot t in historical charging records of all EVs.

Problem Statement: Given the set of EVs, namely V, historical charging distributions $\mathbf{P_h} = \{P_h^v | v \in V\}$, incentive charging distributions $\mathbf{P_i} = \{P_i^v | v \in V\}$, trend charging distribution P_g, PV generation data F^p, electrical consumption data E^h, and the fixed incentive budget B, our goal is to investigate the optimal EVs charging schedule which can minimize the operating cost, taking into account the electricity demand of EVs and the retail building simultaneously. A recommended schedule for an EV v is denoted by $M^v = \{(\tau_1, \theta_1), (\tau_2, \theta_2), ...\}$, where τ and θ represent the recommended charging time slot and the corresponding incentive, respectively.

By recommending a feasible charging time slot with incentive, the willingness of an EV user to comply with our scheduling suggestion can be increased and the total electricity consumption can be prevented from exceeding the contract capacity. In addition, the utilization of EV charging stations can be maximized to achieve the highest revenue.

3   The *POSIT* Framework

In this section, we introduce the structure of our charging strategy and present the *POSIT* (**P**ersonalized **O**nline **S**cheduling with Adaptive **I**ncentives in a **T**hreshold Control) framework to recommend a feasible arrangement for each EV user. Firstly, we estimate the optimal charging capacity of each time slot for the next day by predicting the future PV generation and the electricity consumption of the retail building. We subsequently propose an online charging scheduling system to provide profitable charging arrangements. Finally, the recommended charging arrangements for the next EV user according to the previous user's feedback can be determined, which helps to investigate the users' willingness of accepting different kinds of recommended arrangements.

3.1   Optimal Charging Capacity Estimation

In order to reduce the operating cost of the integrated charging station providing the power for EV charging and the retail building, it is important to forecast the electricity charging capacity for the next day. We use a collection of the historical PV generation data $E^p = \{e_t^p | t \in \{1, 2, ..., T\}\}$ and the electricity consumption of the retail building $E^h = \{e_t^h | t \in \{1, 2, ..., T\}\}$ as the input data. To derive an effective multi-step time series forecasting model, the well known LSTM model is applied since it has been shown to outperform other traditional models [9]. Hence, the predicted PV generation data $\tilde{E}^p = \{\tilde{e}_t^p | t \in \{T + 1, T + 2, ..., T + h\}\}$ and the predicted electricity consumption of the retail building $\tilde{E}^h = \{\tilde{e}_t^h | t \in \{T + 1, T + 2, ..., T + h\}\}$ can be approximately acquired by such prediction manners, and the optimal charging capacity of the next day can also be estimated. According to the estimated charging capacity, we could infer the optimal number of parking positions for each time slot, which are defined as *preferred parking positions* as follows.

Definition 4 (Preferred Parking Position): Let $L = \{l_t | t \in \{T + 1, T + 2, ..., T + h\}\}$ be the preferred parking position set, where each l_t represents the maximum number of charging positions for EVs at the time slot t (as illustrated in Fig. 1). Each l_t is estimated based on the predicted PV generation $\tilde{e}_t^p$ and the predicted electricity consumption of a retail building $\tilde{e}_t^h$. Assuming that the charging speed of all EVs is equivalent[1], the charging speed of EVs is defined as ω, and the electricity load of the contracted capacity in a time slot is denoted by e^c. Hence, the number of preferred parking positions for a time slot is designed as:

$$l_t = (e^c + \tilde{e}_t^p - \tilde{e}_t^h)/\omega. \qquad (3)$$

By obtaining the preferred parking positions L for the next day, we are able to arrange the EVs charging schedule without exceeding the contracted electrical capacity, and it is easier to handle the unexpected arrivals of EVs.

3.2   Online Charging Scheduling

We aim to discover the best charging arrangements to meet the requests of all EV users. However, it is not a practical assumption to know charging requests of all EVs in advance. Thus, we formulate an online charging mechanism that optimizes the charging schedule only with the charging requests and arrangement results of EV users who have completed the charging registration so far. Accordingly, we develop the charging scheduling algorithm that emerges from an analogy to the well known online knapsack problem (OKP for short) [18].

Online Charging Mechanism: In our scenario, the incentive budget is regarded as the knapsack capacity, and the goal is to investigate the optimal

---

[1] For ease of explanation, we assume the equal charging speed in our discussion. It is believed that the implementation issue of variant charging speeds can be seamlessly extended in POSIT.

charging arrangement consisting of multiple options (charging time periods and incentives). An EV user can choose and follow one of the options as the example illustrated in Fig. 2. Assuming that the benefit score and the total cost of the recommended arrangement are the stone value and weight in OKP [18], our goal is to find a set of profitable charging arrangements where each arrangement with a value-to-weight ratio higher than the threshold.

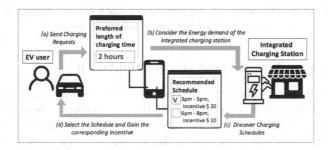

Fig. 2. The overview of the proposed online charging mechanism.

Benefit Scores of Recommended Time Slots: A recommended arrangement for an EV v is denoted by $M^v = \{(\tau_1, \theta_1), (\tau_2, \theta_2), ...\}$, where each tuple (τ, θ) is a charging option, and τ and θ represent the recommended charging time slot and the corresponding incentive, respectively.

Two attributes are mainly considered to judge the benefit level of a recommended charging time slot τ: (a) the personal charging willingness after providing the incentive, and (b) the group charging willingness in the historical records. Given the incentive charging distribution P_i^v for an EV v and the trend charging distribution P_g as defined in Sect. 2, the benefit score ε_τ of the time slot τ can be designed as follows:

$$\varepsilon_\tau = \begin{cases} 0 & , \text{ if } \min\{\tilde{l}_\tau, ..., \tilde{l}_{\tau+h_v-1}\} = 0 \\ \alpha P_i^v(X = \tau) + (1-\alpha)(1 - P_g(X = \tau)) & , \text{ otherwise} \end{cases},$$
(4)

in which $\tilde{l}_t$ represents the remaining preferred parking positions of the time slot t, h_v denotes the requested length of charging time given by the EV v, and α is a constant value to balance the personal charging willingness and the trend charging willingness. Specifically, a zero score will be obtained if the remaining parking positions are not sufficient during the recommended charging time period. A user with high charging willingness and other users with low charging willingness at the time slot τ will obtain a higher benefit score.

Profitable Charging Arrangement: According to the benefit score, the value-to-weight ratio of a charging arrangement can be estimated to discover the most profitable arrangements. The profitable arrangement set $\mathbf{M}_p^v$ for an EV v should follow the condition:

$$\mathbf{M}_p^v = \{M^v | v \in V, \left(\frac{\mathcal{E}}{C}\right)^{M^v} \geq \varphi\}, \tag{5}$$

in which φ is a threshold to guarantee the quality of a charging arrangement, and $(\mathcal{E}/C)^{M^v}$ denotes the value-to-weight ratio of an arrangement M^v, where $\mathcal{E}$ is the sum of all benefit scores ε_τ (τ represents each the recommended time slot in M^v), and C is the sum of all incentives θ provided in M^v.

3.3   Interactive Arrangement Recommendation

According to the profitable charging arrangements determined in the online charging scheduling, it is expected that, to some extent, these arrangements can lead to the good result. However, the real charging willingness of users after providing incentives is still unknown. To increase the probability that a user will select an options from recommended profitable arrangement, an interactive planning process is devised in this paper. Specifically, the reinforcement learning [6] technology is utilized to enable the selection of the most appropriate action by receiving user feedback. We model our interactive arrangement scheduling as a MDP (Markov Decision Processes) problem, which includes a sequence of states, actions, and rewards defined as follows:

States S: A state set $S = \{s_1, ..., s_n\}$ is defined as the online scheduling history that our algorithm recommends to the EV user. Each s_i contains the information of recommended charging options (time slots and provided incentives) and the historical charging distribution of the recommended EV user at the i-th iteration.

Actions A: An action $a_i \in A$ is to recommend a charging arrangement (with multiple charging options) to an EV user at each iteration.

Rewards R: A reward $r_i \in R$ is given from EV users' feedback. If an EV user accepts the recommended arrangement, then the reward $r(s_i, a_i) = 1$; otherwise, $r(s_i, a_i) = 0$ as default.

Q-Learning for Online Charging Recommendation: Q-learning aims to select an action a_t that can maximize the Q value $Q^*(s_t, a_t)$ at each state s_t. Different from the standard scenario of Q-learning, our scenario is to decide whether to recommend multiple charging options while a user can only choose one of them. Considering that a user's decision may be affected when providing different charging options at the same time, we further formulate the Q value estimation for our online recommendation mechanism to recommended the optimal charging arrangement $M^v \in \mathbf{M}_p^v$, which consists of multiple charging options (τ, θ).

According to the *Nearest Neighbor Q-Learning* (NNQL) algorithm [12], the Q value of a specific (s, a) affected by not only rewards it gains, but also the Q values of all observations that lie in its neighborhood. Inspired by NNQL, we redesign the Bellman operator to consider the effects of other options for a specific one that in the same charging arrangement. Given estimated Q values $Q = \{q(s_i, a_i)|s_i \in S, a_i \in A\}$ with the finite states $S = \{s_i\}_{i=1}^n$, for each state-action pair $(s, a) \in S \times A$, the Bellman optimality operator Ψ_B for our online charging mechanism is denoted by:

$$\Psi_B(s, a) = r(s, a) + \gamma \mathbb{E}[\max_{a'} \Psi_{OCR}(s', a')|s, a], \tag{6}$$

and the operator Ψ_{OCR} is designed as follows:

$$\Psi_{OCR}(s, a) = \sum_{i=1}^{n} \sum_{j=1}^{|M_i|} \frac{q(s_{ij}, a)}{|M_i|}, \tag{7}$$

where M_i represents the recommended charging arrangement at the i-th iteration, and $q(s_{ij}, a)$ is the Q value of selecting the action a (recommend or not recommend) for the j-th charging option in M_i.

We now introduce the Q-learning policy for the online charging recommendation. In each iteration, the policy runs a number of time steps, and the iteration count is update form k to $k + 1$ with all state-action (s, a) pairs visited at least once. Let $O_k(s, a)$ be the counter which records how many times (s, a) has been learned from the beginning of iteration k till the current time stamp, and denote $1/(O_k(s, a) + 1)$ by η, the Q value of the online charging recommendation at each time stamp within iteration k is updated with current state s_t as:

$$\Psi_{OCR}^k q^k(s_t, a_t) = (1 - \eta)(\Psi_B^k q^k)(s_t, a_t) + \eta(r_t + \gamma \max_{a_{t+1}} \Psi_{OCR}^k(s_{t+1}, a_{t+1})). \tag{8}$$

At the end of the k-th iteration, a new q^{k+1} is generated for each (s, a) as follows:

$$q^{k+1}(s, a) = (1 - \beta_k)q^k(s, a) + \beta_k(\Psi_B^k q^k)(s, a), \tag{9}$$

where $0 \leq \beta_k \leq 1$ is a given weight.

Here we use the ϵ-greedy method for balancing exploration and exploitation. The charging option selection policy is described as follows:

$$a_i = \begin{cases} random(\mathcal{A}), & if\ \xi \leq \epsilon. \\ \arg\max_{a_i \in \mathcal{A}} Q^*(s_t, a_t), & otherwise. \end{cases} \tag{10}$$

Finally, the optimal policy for actions that maximizing rewards can be discovered. For a EV user v, we are able to estimate which charging arrangements in $\mathbf{M}_p^v$ (defined in Eq. 5) is the most effective by averaging the new Q values of all charging options in each charging arrangement.

4 Experiments

We conduct experiments to answer three questions. First, we wonder whether the proposed *POSIT* framework can outperform other baselines. Second, by varying the values the weights of the personal charging willingness and the trend charging willingness which affect the benefit score ε_τ, how does the performance of methods evolve? Third, can Q-Learning technique really discover the real charging willingness of users and enhance the acceptance rates of recommended charging arrangements?

4.1 Experimental Setup

Dataset Description: We employ three real-world datasets for the experiments. For the electricity demand of a retail building, we apply a real-world data collected at a supermarket in Romania [10]. The data contains hourly energy consumption retrieved from the smart metering devices for the whole year 2016. For the PV generation data, the solar power dataset provided by the electricity distribution company Ausgrid is utilized [1]. The data includes the power generated from PV (accumulated every 30 min) of each solar home in Australia from 1 July 2012 to 30 June 2013. The third dataset we used is ACN-data [7], which contains the EV charging information of 54 EV chargers during 25 April 2018 to 20 January 2021 from a university in Pasadena, California. The charging information includes the user ID, the time an EV plugged in and unplugged, and the amount of energy delivered during the session.

For the purpose of this study, we combine three datasets and select one month data as the training dataset for the electricity forecasting model, and the next 7-day data is regarded as the testing dataset to evaluate our online charging scheduling. In order to simulate whether an EV user will accept the recommended charging arrangement or not, we assume that the acceptance probability of users follows a normal distribution $\mathcal{N}(0,1)$, where the y-axis represents the probability of accepting the charging arrangement. In addition, the x-axis represents the value of (*the recommended time slot - the time slot a user used to charge EV*)/*the provided incentive*, which means that a user will have higher accepting probability when the recommended time slot is closer to the one that a user used to charge and the provided incentive is higher.

Comparative Methods: To demonstrate the effectiveness of the proposed method, we compare *POSIT* with several baselines. OR is the method that users' charging behaviors follow the 7-day testing dataset without registering the charging mechanism. PRE is the method which the charging mechanism recommends the most popular charging time slots in the training dataset for each EV (the distribution of the acceptance rate follows each user's historical charging distribution). EL and EL-Q are the methods which provide the same value of incentives each time, while EL-Q considers the accepting results of users to improve the next recommendation. OSIT is similar to *POSIT* but without the electricity forecasting model to estimate the preferred parking positions.

Table 1. The structure of Time-of-Use rates (unit price: US$).

Classification				Charge (US$)	
				Summer	Non-summer
Capacity rate	per kW			8.50	
Energy rate	Peak time	Weekdays 07:30–22:30	per kWh	0.12	0.11
	Semi-peak time	Saturday 07:30–22:30		0.08	0.07
	Off-peak time	Other time periods		0.06	0.05

Evaluation Settings: We evaluate the methods by considering two attributes: (i) the effectiveness of the integrated charging station and (ii) the feasibility of time slots at which users charge. For the first attribute, we examine the revenue and the operating cost of each method. The revenue is the total earning (paid by users during charging) of the charging station after deducting the operating cost. Based on the charging price of Tesla superchargers[2], we set that a user should pay US$ 0.25 per kWh when charging an EV. The operating cost includes the incentive budget, the electricity costs of the energy delivered from the chargers and the electricity consumption of the retail building.

The estimation of the electricity costs is based on designed Time-of-Use rates which is shown in Table 1. In our scenario, the contracted electrical capacity is set as 30 kW. If the peak electricity demand in 7 days does not exceed the contracted electrical capacity, a fixed basic charge is levied ($30 * 8.50 =$ US$ 255). Otherwise, a penalty charge estimated based on two times of the basic rate is levied. For example, if the peak electricity demand is 50 kw, then the penalty charge is $(50 - 30) * 2 * 8.50 =$ US$ 340. Also, an energy charge should be paid based on how many kWh used during peak periods, semi-peak periods, and off-peak periods. Hence, the electrical operating cost consists of a basic charge, an energy charge, and a penalty charge. The fixed incentive budget of each day is set as US$ 7.2, which is 20% of the average penalty charge per day in the training dataset.

For the second attribute, the feasibility of time slots at which users charge, we estimate the overload percentage, which represents the average percentage of the times of exceeding the contracted electrical capacity in each time slot for each day. Also, the divergence percentage is estimated to evaluate the difference of users' charging behaviors before and after our recommendation. Let $\Delta\tau_v$ is the difference value of the time slots a user v chooses from our recommended arrangement and a user v used to charge at in the past, the divergence percentage of a day is defined as $\sum_v^V \Delta\tau_v / |T||V|$, where $|T|$ is the number of time slots within a day. Note that we assume that if a user accepts the recommended charging arrangement, he or she will indeed go to charge at the selected charging time slot.

---

[2] Please refer to https://www.solarreviews.com/blog/tesla-supercharger-guide.

4.2   Experimental Results

POSIT **Performance:** To answer the first question, we compare *POSIT* with aforementioned baselines. The comparison is summarized in Table 2. Convincingly, our *POSIT* framework, which discovers profitable charging arrangement in the online charging mechanism, utilizing users' selecting feedback to improve the next charging recommendations can achieve the smallest operating cost and the largest revenue.

Table 2. Evaluation of the feasibility and the effectiveness with the consideration of the PV generation. (Numbers inside parentheses denote the performance difference compared to *POSIT*.)

Effectiveness evaluation						
Metric	OR	PRE	EL	EL-Q	OSIT	**POSIT**
Operating cost (US$)	870.4(+361.5)	1,771.3(+1,262.4)	640.9(+132.0)	636.4(+127.5)	618.9(+110.0)	**508.9**
Revenue (US$)	−193.6(−361.5)	−1,094.5(−1,262.4)	35.8 (−132.0)	40.3(−127.5)	57.8(−110.0)	**167.9**
Feasibility evaluation						
Metric	OR	PRE	EL	EL-Q	OSIT	**POSIT**
Overload percentage (%)	12.5(+11.8)	6.4(+5.7)	6.0(+5.3)	4.8(+4.1)	3.6(+3.0)	**0.6**
Divergence percentage (%)	−	10.0(+8.3)	2.0(+0.3)	2.1(+0.3)	2.6(+0.9)	**1.7**

In addition, *POSIT* can also acquire the smaller overload percentage and the smaller divergence percentage than PRE, EL, EL-Q, and OSIT, which represents that our method hardly exceeds the contracted electrical capacity during the online scheduling, and our recommended charging time periods are close to the time periods at which users used to charge. Note that OR does not apply the recommendation mechanism, so the divergence percentage cannot be estimated. In Table 2, we can observe that it can not earn any revenue without the charging scheduling (OR). PRE causes higher operating cost because it aims to recommend the most popular charging time periods, which let the electricity consumption exceed the contracted electrical capacity during the recommended time periods. The results of EL, EL-Q, and OSIT, which remove the techniques of electricity forecasting and the consideration of incentive costs in the discovery of the profitable charging arrangements, also are not as good as our method.

Parameter Sensitivity Analysis: To answer the second question, we analyze the sensitivity of the weight α which affect the value of the benefit score ε_τ in Eq. 4. The larger α means that the personal charging willingness has a higher impact on discovering profitable charging arrangements. Otherwise, the group charging willingness has higher impact. By varying α as $\{0.2, 0.4, 0.6, 0.8\}$, we estimate the operating costs and the revenues of our method, EL, EL-Q, and OSIT, and the result is shown in Fig. 3. We can obviously see that most of methods have the highest operating cost and the lowest revenue when α is set as 0.8, which means that if the estimation of the benefit score is determined merely based on the group charging willingness would make the least profit. On the other hand, we can observe that our method *POSIT* can obtain the largest revenue when α is set as 0.4. That is to say, balancing the impact of the personal charging willingness and the group charging willingness can help to enhance the online charging scheduling.

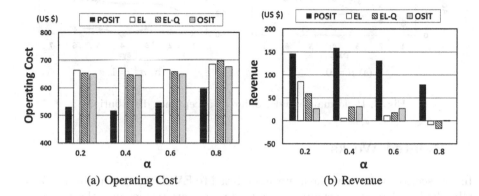

(a) Operating Cost   (b) Revenue

Fig. 3. Performance of varying α.

Effectiveness of the Interactive Recommendation: To answer the third question, we simulate that the accepting probability of users follows different normal distributions. To reveal the effectiveness of Q-learning for the online charging scheduling, we show the users' average acceptance rate of the recommended charging arrangement in 7-day testing data by applying *POSIT*. In Fig. 4(a), we simulate four normal distributions $\mathcal{N}(\mu, \sigma^2)$ by setting the mean μ as 0 and varying the variance σ^2 as $\{0.2, 0.5, 1, 5\}$. We can see that the acceptance rate of each day is higher when the value of σ^2 is smaller. In Fig. 4(b), we also simulate four normal distributions by setting the variance σ^2 as 1 and varying the variance σ^2 as $\{-3, -2, -1, 0\}$. When the value of μ is closer to 0, the acceptance rate of each day becomes higher.

From the above results, we can know that the larger revenue can be achieved when the users' acceptance probability follows the normal distribution $\mathcal{N}(0, 0.2)$. Although the performances are quite different with each distribution, we can still observe that the acceptance rate rises with the increase of the testing days, which reveals the effectiveness of Q-learning for the online charging recommendation.

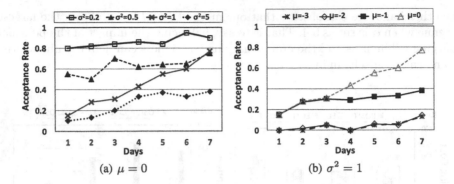

(a) $\mu = 0$ (b) $\sigma^2 = 1$

Fig. 4. Performance under different acceptance distributions.

5   Related Works

In this section, we review the literature related to EV charging scheduling. Generally, the EV charging mechanisms can be divided into two groups: the centralized charging control and the distributed charging control. The centralized charging control method is able to calculate the optimal schedule as all the information is available to it. In the study like [4], which derives a mixed integer linear programming (MILP) to maximize the aggregator revenue. Yan *et al.* [15] propose a four-stage optimization and control algorithm to improve the operating performance by utilizing the whole EVs related information. In contrast, in the distributed charging control, each EV is equipped with some computing capability, and the decision to charge or not is taken by each EV in the collaboration with the aggregator. Yang *et al.* [16] propose a model predictive control to discover the optimal charging solution by using the locally generated wind power. Kabir *et al.* [5] utilize the game theory to conduct the distributed scheduling with the goal of minimizing the EV charging cost. Zhou *et al.* [17] propose an incentive-based distributed control scheme of EV charging to facilitate the load management of utility company under constrained charging requests. Li *et al.* [8] formulate the real-time EV charging scheduling problem as a constrained Markov Decision Process to minimize the user's electricity cost.

However, most EV charging mechanisms put emphasis only on maximizing the revenue and ignore the users' charging willingness which has a key impact on the success of the EV charging scheduling. In our work, we aim to discover the profitable charging arrangements which to enhance EV users' charging willingness and benefit the integrated charging station with the online charging manner. Therefore, these studies are orthogonal to our work.

6   Conclusions

In this paper, we investigate how to recommend the profitable charging arrangements for each EV user in the online charging manner. In the proposed *POSIT* framework, we first estimate the optimal charging capacity of each time slot for the next day and propose an online charging scheduling system to provide profitable charging arrangements. In addition, we conduct the interactive learning to improve the next charging recommendation according to the previous user's feedback. Finally, we get the fact that our method *POSIT* can outperform other baselines and enhance the users' acceptance rate under different acceptance distributions.

Acknowledgement. This work was supported in part by Ministry of Science and Technology, R.O.C., under Contract 109-2221-E-006-187-MY3, 110-2221-E-006-001 and 111AT16B.

References

1. Ausgrid. Solar home electricity data (2012)
2. Gossen, H.H.: Gossen, hermann heinrich: Entwickelung der gesetze des menschlichen verkehrs, und der daraus fließenden regeln für menschliches handeln. Die 100 wichtigsten Werke der Ökonomie (2019)
3. International Energy Agency (IEA). IEA global annual hybrid, plug-in hybrid and battery electric vehicle sales 2021 (2020)
4. Jin, C., Tang, J., Ghosh, P.K.: Optimizing electric vehicle charging with energy storage in the electricity market. IEEE Trans. Smart Grid **4**, 311–320 (2013)
5. Kabir, M.E., Assi, C.M., Tushar, M.H.K., Yan, J.: Optimal scheduling of EV charging at a solar power-based charging station. IEEE Syst. J. **14**, 4221–4231 (2020)
6. Kuan, C.-P., Young, K.Y.: Reinforcement learning and robust control for robot compliance tasks. J. Intell. Rob. Syst. **23**, 165–182 (1998)
7. Lee, Z.J., Li, T., Low, S.H.: ACN-data: analysis and applications of an open EV charging dataset. In: Proceedings of the Tenth ACM International Conference on Future Energy Systems (2019)
8. Li, H., Wan, Z., He, H.: Constrained EV charging scheduling based on safe deep reinforcement learning. IEEE Trans. Smart Grid **11**, 2427–2439 (2020)
9. Liu, Y., Hou, D., Bao, J., Qi, Y.: Multi-step ahead time series forecasting for different data patterns based on LSTM recurrent neural network. In: WISA (2017)
10. Kuo, C.-C., Piedad, E.J.: A 12-month data of hourly energy consumption levels from a commercial-type consumer. Mendeley Data (2018)
11. Satterfiled, C., Nigro, N.: Public EV charging business models for retail site hosts (2020)
12. Shah, D., Xie, Q.: Q-learning with nearest neighbors. In: NeurIPS (2018)
13. Sun, X., Qiu, J.: A customized voltage control strategy for electric vehicles in distribution networks with reinforcement learning method. IEEE Trans. Industr. Inf. **17**, 6852–6863 (2021)
14. Wang, R., Wang, P., Xiao, G.: Two-stage mechanism for massive electric vehicle charging involving renewable energy. IEEE Trans. Veh. Technol. **65**, 4159–4171 (2016)
15. Yan, Q., Zhang, B., Kezunovic, M.: Optimized operational cost reduction for an EV charging station integrated with battery energy storage and PV generation. IEEE Trans. Smart Grid **10**, 2096–2106 (2019)

16. Yang, Yu., Jia, Q.-S., Guan, X., Zhang, X., Qiu, Z., Deconinck, G.: Decentralized EV-based charging optimization with building integrated wind energy. IEEE Trans. Autom. Sci. Eng. **16**, 1002–1017 (2019)
17. Zhou, Y., Kumar, R.E., Tang, S.: Incentive-based distributed scheduling of electric vehicle charging under uncertainty. IEEE Trans. Power Syst. **34**, 3–11 (2019)
18. Zhou, Y., Chakrabarty, D., Lukose, R.: Budget constrained bidding in keyword auctions and online knapsack problems. In: Papadimitriou, C., Zhang, S. (eds.) WINE 2008. LNCS, vol. 5385, pp. 566–576. Springer, Heidelberg (2008). https://doi.org/10.1007/978-3-540-92185-1_63

A Two-Stage Self-adaptive Model for Passenger Flow Prediction on Schedule-Based Railway System

Boyu Li[1], Ting Guo[1(✉)], Ruimin Li[2], Yang Wang[1], Amir H. Gandomi[1], and Fang Chen[1]

[1] Data Science Institute, University of Technology Sydney, Sydney, Australia
boyu.li@student.uts.edu.au,
{ting.guo,yang.wang,amirhossein.gandomi,fang.chen}@uts.edu.au
[2] Bureau of Statistics and Analytics, Transport for New South Wales, New South Wales, Australia
ruimin.li2@transport.nsw.gov.au

Abstract. Platform-level passenger flow prediction is crucial for addressing the overcrowding problem on platforms that endangered the passengers' safety and experience in railway systems. Although some studies exist on this topic, it remains difficult to apply these methods in the real world *e.g.*, the data deficiency in older railway systems, potential impacts of dynamic interchange passenger flows, real-time predictive ability. Thus, we propose a two-stage self-adaptive model for accurately and timely predicting platform-level flow. In the first stage, a self-attention-based prediction model is introduced to predict the next-day passenger flow based on the historical boarding record. The proposed decomposing components transferring the discrete boarding records into continuous patterns make the module able to deliver a robust minute-level prediction. In the second stage, a real-time fine-tuning model is developed to adjust the predicted flow based on the real-time emergencies in passenger flows. The combination of offline deep learning mechanism and real-time reallocation algorithm ensures the real-time response without loss of accuracy. The experiments show that our model can offer accurate predictions to trip planners for timetable design and provide timely decision support for controllers when emergencies happen, and our end-to-end framework has been applied to the railway system in Sydney, Australia.

Keywords: Passenger flow prediction · Railway system · Attention mechanism · Deep learning · ITSs

1 Introduction

The performance of the railway system is suffering from an inescapable challenge of the overcrowding problem on the platforms with high denied boarding rate

J. Gama et al. (Eds.): PAKDD 2022, LNAI 13282, pp. 147–160, 2022.
https://doi.org/10.1007/978-3-031-05981-0_12

[4,8]. Therefore, accurate estimation of the degree of crowdedness on platform level is badly needed for both train timetable planning and railway emergency management [13]. Passenger flow prediction task can be regarded as a typical time series prediction problem as the passenger flow normally shows strongly repeatable patterns at the same location (station/platform) and similar time periods in the real world.

Earlier studies mainly focus on applying the traditional mathematical or statistical methods to model and predict the passenger flow, e.g., auto-regressive moving average (ARIMA) [18] and regression modeling methods [6,16]. However, those methods' performance cannot satisfy industrial application standards. With the development of information technology, more and more railway systems gradually adopt electronic equipment, like smart card ticketing systems and Advanced Train Management System (ATMS), which generates massive traffic data every day [3,7]. Thus, data-driven methods like the emerging deep learning methods have proved to perform well on the time series prediction tasks. Some studies have achieved reasonably good results by applying the recurrent neural networks (RNNs) to learn the temporal dependencies of passenger patterns [10,11]. Meanwhile, the attention mechanism gains increasing interest due to its ability to identify the information in a historical input most pertinent [17], which some studies have proven it can achieve better prediction performance in large urban railway systems [5,21]. As the coarse-grained data acquisition from passenger inbound or train boarding records, the mainstream studies mainly stay on forecasting the station-level passenger flow (each station contains multiple platforms), while platform-level passenger flows are more concerned by the railway managers and controllers. Some researchers have tried to predict the station-level passenger flow firstly and then adopted the origin-destination matrix to estimate passengers on different platforms [19], but the passenger interchange between platforms is hardly estimated, especially when emergencies happen. Thus, it still remains some enormous difficulties for applying existing theoretical models to the real world:

- Lack of **fine-grained data:** Existing models always rely on fine-grained training data to achieve fine-grained prediction. However, many railway systems around the world have operated for decades and only coarse-grained data can be obtained [1,2]. For small stations, the departure time interval is large and existing models applied on coarse-grained data hardly capture the continuous flow change according to timetable adjustment and emergencies.
- Lack of **considering delay and interchange:** For a complex railway network, delays caused by unpredictable special events and sudden incidents can significantly influence the crowdedness of the platforms. It can cause not only the delays of other trips but also the deferred interchange passenger flow, which is a big part of passenger flow on platforms. To the best of our knowledge, no existing passenger prediction model considers both delays and interchange passengers.
- Lack of **real-time predictive ability:** Considering that the ultimate goal of passenger flow prediction is to provide timely decision support to managers

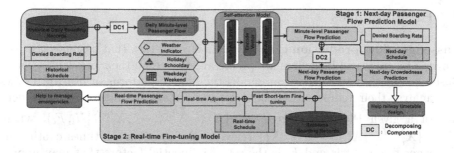

Fig. 1. The overall framework of our proposed two-stage self-adaptive model.

and controllers in daily operation, the prediction methods must be able to work in real-time without lags. Even though the deep learning models show stunning performance on time series prediction tasks, the hard requirements for long running time and sufficient data are still the main obstacles.

In this paper, we propose a two-stage self-adaptive model to address these challenges. In Stage 1, we use a behavior-based data decomposing component to reallocate the data from discrete boarding records to continuous estimation. Denied boarding rate and passenger arrival pattern are considered in this stage. The reallocated data with relevant external features are then fed into a self-attention model for minute-level next-day passenger flow prediction. To better respond to emergencies and short-term variation on passenger flow, we propose a real-time fine-tuning model in Stage 2. The delay propagation and interchange flow are considered. The overall framework of the proposed model is as shown in Fig. 1. To sum up, the main contributions of this paper are as follows:

- We propose a two-stage self-adaptive model that can provide the schedule-based daily passenger flow estimation based on deep learning with high accuracy, and the real-time response and adjustment for emergencies.
- We introduce a novel modeling method that migrates coarse-grained historical data, influential factors and real-time boarding records for fine-grained daily and real-time prediction. Besides, different from existing methods, we develop a reallocation model to take delay propagation and interchange into consideration, which can help achieve high performance on railway networks.
- Besides the passenger flow prediction, we also provide the crowdedness prediction for each platform by taking scheduled timetable, denied boarding rate, interchange and real-time emergencies into consideration, which has more practical significance in train control and emergency response.
- The performance of the proposed model is evaluated with the comparison of other benchmark methods, and its effectiveness and superiority withstand the deployment on a real-world railway system of Greater Sydney Area. The daily operation shows our data-driven model can offer the accurate prediction that benefits the trip planners for timetable design and also provide timely decision support for controllers when emergencies happen.

2 Preliminaries

Passenger flow prediction is a typical time series task that assumes the passenger's future travel behaviors are predictable based on their previous records. Based on the assumption, we can model the hidden patterns of historical flows and analyse their temporal dependencies to forecast the future N times density of a given Platform i as $\widehat{J_i} = \{\widehat{X_i^{t+1}}, \widehat{X_i^{t+2}}, \cdots, \widehat{X_i^{t+N}} | H_i, F_i\}$, where $H_i = \{X_i^{t-M+1}, X_i^{t-M+2}, \cdots, X_i^t\}$ refers to the previous M times continuous passenger flow records and F_i is the set of influential factors that may impact the passenger flow volume. In this paper, we focus on two time dimensions: *Next-day prediction* (long-term) - the minute-level prediction for the entire next day; *Real-time prediction* (short-term) - the following minute-level prediction.

Coarse-grained data *v.s.* fine-grained data: 'Coarse-grained' means the data acquisition is infrequent and irregular, while fine-grained data is obtained in a systematic and regular way. Specifically, the coarse-grained data in this paper is the trip boarding data which records how many passengers boarding a trip at a given platform. As the service gap[1] is changing with time (*e.g.* peak hours/off-peak hours) and locations, the trip boarding data is a kind of coarse-grained data. Applying this kind of data hardly captures the continuous flow change according to timetable adjustment and emergencies.

Passenger crowdedness is the number of passengers on Platform i at a given time point t. It is determined by inbound passenger flow, interchange passenger flow, last train arrival time, next train arrival time, and denied boarding rate.

Denied boarding rate is the proportion of passengers that cannot get on the current arrived train. The reason for denied boarding could be (1) Overloaded train; (2) Missing train; (3) The train cannot reach those passengers' destination.

Passenger flow relative independence : The inbound passenger flow is insensitive to emergencies and temporary changes on the scheduled timetable. Note that, the interchange passenger flow is not under this hypothesis.

3 Stage 1: Next-Day Passenger Flow Prediction Model

In the first stage, we proposed the prediction model based on discovering the long-term passenger flow patterns from historical data as shown in Fig. 1.

3.1 Model Inputs Decomposing Component (DC₁)

To fully utilize the coarse-grained trip boarding data in time-series-based prediction learning, we have to transfer it into continuous data as model inputs. We proposed the DC_1 is to generate smooth minute-level historical passenger flow.

---

[1] **Service gap** is the time period between the arrival time points of two consecutive trips at a given platform. It means the maximum waiting time for the passengers.

The trip boarding data keep a record of the number of passengers get on board at a given platform, *i.e.* $\mathcal{B}_i = \{B_i^{T_1}, B_i^{T_2}, B_i^{T_3}, \cdots\}$. We use B_i^T to represent the number of boarding passengers from Platform i at time point T. Considering that not all the passengers can get on the arrival trip due to the denied boarding rate. Thus, the relationship between the accumulated passenger flow during service gaps and the trip boarding records can be formulated as:

$$B_i^{T_k} = (1 - \rho_i^{T_k})(I_i^{[T_{k-1}, T_k]} + \rho_i^{T_{k-1}}(I_i^{[T_{k-2}, T_{k-1}]} + \rho_i^{T_{k-2}}(I_i^{[T_{k-3}, T_{k-2}]} + \ldots)))$$
(1)

where $I_i^{[T_1, T_2]}$ represents the accumulated passenger flow between T_1 and T_2 (Service gap). ρ_i^T is the denied boarding rate at the train arrival time T. As the maximal missing trips by most of the passengers is 2, we only consider the previous two trips for simplicity. Then Eq. 1 can be simplified as:

$$B_i^{T_k} = (1 - \rho_i^{T_k})(I_i^{[T_{k-1}, T_k]} + \rho_i^{T_{k-1}}(I_i^{[T_{k-2}, T_{k-1}]} + \rho_i^{T_{k-2}} I_i^{[T_{k-3}, T_{k-2}]}))$$
(2)

By the given trip boarding records and the denied boarding rates, we can calculate the accumulated passenger flow between the service gaps one by one using the chain rule. Then the minute-level passenger flow can be derived based on the accumulated passenger flow. As a smooth input is preferable to the prediction model learning, we assume the passenger flow is a uniform distribution in any service gap. The minute-level passenger flow estimation for model input can be formulated as: $P_i^t = I_i^{[T_1, T_2]}/(T_2 - T_1)$, where P_i^t is a kind of fine-grained data we defined in Sect. 2. The minute-level passenger flow estimation provides regular and smooth data samples without losing any information compared with the infrequent and irregular coarse-grained data. Based on our experiments, the fine-grained inputs can help the model significantly improve its predictive ability.

By applying DC_1, we can transfer a set of coarse-grained trip boarding records $\mathcal{B}_i = \{B_i^{T_1}, B_i^{T_2}, B_i^{T_3}, \cdots, B_i^{T_m}\}$ into a series of fine-grained minute-level passenger flow records $\mathcal{P}_i = \{P_i^{t_1}, P_i^{t_2}, P_i^{t_3}, \cdots, P_i^{t_n}\}$, where $n \gg m$.

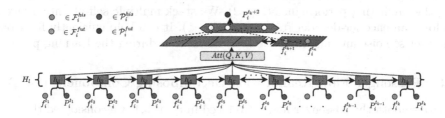

Fig. 2. An illustration of the proposed self-attention layer.

3.2 Self-attention-Based Prediction Component

In this paper, we adopt a self-attention-based encoder-decoder framework for passenger flow prediction as shown in Fig. 2. The advantage of self-attention is its ability to identify the information in an input most pertinent to accomplishing a task, increasing the prediction performance, especially in time-series forecasting [17].

In our prediction task, given a series of historical minute-level passenger flow $\mathcal{P}_i^{his} = \{P_i^{t_1}, P_i^{t_2}, P_i^{t_3}, \cdots, P_i^{t_k}\}$. Meanwhile, based on our studies, the passenger flow pattern is highly related with **external influential factors**[2] $\mathcal{F}_i = \mathcal{F}_i^{his} + \mathcal{F}_i^{fut} = \{f_i^{t_1}, f_i^{t_2}, f_i^{t_3}, \cdots, f_i^{t_k}, \cdots, f_i^{t_n}\}$, the aim is to predict the future passenger flow $\mathcal{P}_i^{fut} = \{P_i^{t_{k+1}}, P_i^{t_{k+2}}, P_i^{t_{k+3}}, \cdots, P_i^{t_n}\}$.

The core idea of the self-attention mechanism is to calculate relevance with Query (Q), Key (K) and Value (V), matrices which can be obtained as:

$$Q = H_l \cdot W^Q; \quad K = H_l \cdot W^K; \quad V = H_l \cdot W^V \tag{3}$$

where H_l is the input of Layer l of self-attention module and the initial input is $H_0 = \mathcal{P}_i^{in} \oplus \mathcal{F}_i^{his}$. W is the learnable weighted matrix. To illustrate, to calculate the relevance among different input passenger flow and other influential factors, we can calculate the similarity between the Query of the input vector and the Key of other input vectors as weights. Then the weighted average of Values (including itself) is used to obtain the self-attention output vector as:

$$\text{Att}(Q, K, V) = \text{softmax}\left(\frac{Q \cdot K^T}{\sqrt{d_k}}\right) \cdot V \tag{4}$$

where $\sqrt{d_k}$ refers to the scaling factor that is used to reduce the dimension of $Q \cdot K^T$. By combining the future external factors, the final output of the self-attention module is:

$$O = \mathbf{FC}(\text{Att}(Q, K, V) \oplus \mathcal{F}_i^{fut}) \tag{5}$$

where $\mathbf{FC}(\cdot)$ is a fully connected layer. It means the historical input passenger flow is a weighted average with historical influential factors, and the predicted passenger flow is fused with future influential factors to generate a new vector. Benefiting from this, the relevant influential factors can be captured for improving the predicting performance (Fig. 2). We stack multiple self-attention layers to form an encoder-decoder framework (Fig. 1). It can help efficiently filter out the useless noise and maintain the key information during the learning process.

3.3   Passenger Flow Reallocation Decomposing Component (DC_2)

As we smooth the passenger flow introduced in Sect. 3.1, we considered the real-world passenger behavior in DC_2, including both the bimodal inbound passenger flow and interchange passenger flow into consideration.

---

[2] The influential factors adopted in this work are: (1) Weekday/weekend; (2) Weather (rainfall and temperature, which the future weather information can be collected from weather forecasting website.); (3) School day/public holiday.

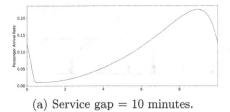

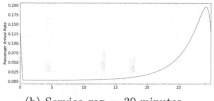

(a) Service gap = 10 minutes.   (b) Service gap = 30 minutes.

Fig. 3. Passenger arrival rate in different service gaps: the x-axis refers to the time and the y-axis refers to the arrival rate. The left peak in two subgraphs (a) and (b) is caused by the passengers missing the passed trains while the right peak is the dense stream of passengers reached the platform near the arrival of the next trains.

Bimodal Inbound Flow Estimation. Passengers tend to arrive at a period close to the scheduled train arrival time and the increase of inbound passenger flow can be observed near the scheduled train arrival time accordingly. Lack of waiting buffer time can also cause denied boarding due to passengers' late or early arrival of the trains. [14] have explored this phenomenon and found that the inbound passenger flows always form a bimodal distribution between two consecutive trips (as illustrated in Fig. 3) and follow the Johnson S_R distribution during any service gap which can be formulated as:

$$f_{sd(a,b,\alpha_1,\alpha_2)}(t) = \begin{cases} \frac{\alpha_2(b-a)}{(t+b-\delta_{sd}-a)(\delta_{sd}-t)\sqrt{2\pi}}e^{-0.5\left\{\alpha_1+\alpha_2\ln\left(\frac{t+b-\delta_{sd}-a}{\delta_{sd}-t}\right)\right\}} & \text{if } a < t < \delta_{sd} \\ \frac{\alpha_2(b-a)}{(t-\delta_{sd}-a)(b+\delta_{sd}-t)\sqrt{2\pi}}e^{-0.5\left\{\alpha_1+\alpha_2\ln\left(\frac{t-\delta_{sd}-a}{b+\delta_{ss}-t}\right)\right\}} & \text{if } \delta_{sd} < t < b \\ 0 & \text{otherwise} \end{cases}$$

(6)

where α_1 and α_2 are used to describe the shape of the distribution that can be fitted with the hysterical observations from the platform. Moreover, due to the familiarities of passengers with the train schedule (including special events like delay) being various, δ_{sd} is used to control the shift of schedule-dependent distribution with the headway. Overall, the passenger arrival distribution can be composed of the two types of passengers as:

$$f_a(t, \alpha_1, \alpha_2) = r_{sd} \cdot f_{si} + r_{si} \cdot f_{sd(a,b,\alpha_1,\alpha_2)}$$

(7)

where a and b refer to the headway of two consecutive trains and t refers to a given time point between them, r_{sd} and r_{si} refers to the rate of the schedule-dependent and schedule-independent passengers. It is noteworthy that the rate between the two types of passengers depends on the unique attributes of different platforms, which the feature can be learned with historical statistics.

Interchange Passenger Reallocation. The interchange platforms offer passengers transfer trips as there is no through train to their destinations in complex railway systems. To estimate the interchange passengers, we adopt the interchange rate between different interchange platform pairs. Here we use r_{ij} as the

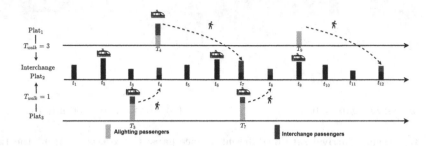

Fig. 4. Interchange passenger flow reallocation.

dynamic passenger interchange rate from Platform i to Platform j in a period which can be learned from historical data.

As the outputs of the prediction model contain both the inbound passengers and the interchange passengers, we need to split the predicted passenger flow into two corresponding parts. Specifically, given the predicted passenger flow for Platform i's service gap $[t_m, t_n]$ as $\mathcal{P}_i^{fut} = \{P_i^{t_{m+1}}, P_i^{t_{m+2}}, P_i^{t_{m+3}}, \cdots, P_i^{t_n}\}$, the accumulated passenger flow of this service gap is as:

$$I_i^{[t_m, t_n]} = \sum_{k=1}^{n-m+1} P_i^{m+k} = I_{i(\text{Inbound})}^{[t_m, t_n]} + I_{i(\text{Interchange})}^{[t_m, t_n]} \qquad (8)$$

If the interchange rates of all transfer trips to Platform i are $\{r_{1 \cdot i}, r_{2 \cdot i}, r_{3 \cdot i}, \cdots, r_{u \cdot i}\}$, the two passenger flow can be calculated as:

$$I_{i(\text{Inbound})}^{[t_m, t_n]} = (1 - \sum_{j=1}^{u} r_{j \cdot i}) I_i^{[t_m, t_n]}; I_{j \cdot i(\text{Interchange})}^{[t_m, t_n]} = r_{j \cdot i} I_i^{[t_m, t_n]}; I_{i(\text{Interchange})}^{[t_m, t_n]} = \sum_{j=1}^{u} I_{j \cdot i(\text{Interchange})}^{[t_m, t_n]}$$
$$(9)$$

$I_{j \cdot i(\text{Interchange})}^{[t_m, t_n]}$ is used in the bimodal inbound flow estimation in Sect. 3.3, while the interchange passenger flow from Platform j to Platform i, $I_{j \cdot i(\text{Interchange})}^{[t_m, t_n]}$, is predicted to enter Platform i at time point $T_j + T_{\text{walk}}$, where T_j is the planned train arrival time at Platform j and T_{walk} is the walking time from Platform j to Platform i. An illustration is as shown in Fig. 4.

Based on the scheduled timetable and the denied boarding rate, the passenger crowdedness for a given platform i at time point T_k is:

$$C_i^T = I_i^{[T_{k-1}, T_k]} + \rho_i^{T_{k-1}} (I_i^{[T_{k-2}, T_{k-1}]} + \rho_i^{T_{k-2}} I_i^{[T_{k-3}, T_{k-2}]}) \qquad (10)$$

4   Stage 2: Real-Time Fine-Tuning Model

The next-day minute-level passenger flow prediction can provide visionary customer-oriented information for timetable planning, but it cannot respond to emergencies and temporary timetable changes. Thus, we will adjust the two parts based on the real-time changes in Stage 2 and develop a fast short-term fine-tuning method to adjust the prediction results quickly.

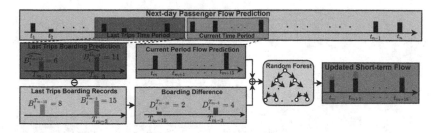

Fig. 5. The workflow of fast short-term fine-tuning.

4.1 Fast Short-Term Fine-Tuning

To better respond to the short-term passenger flow variation and further improve the real-time performance, we develop a fast short-term fine-tuning method as shown in Fig. 5. In this stage, we extract the short-term prediction passenger flow from the next-day prediction outputs of Stage 1 as our fine-tuning base. If the target time period is $[t, t+M]$, the short-term predicted passenger flow on Platform i is $\mathcal{P}_i^{[t,t+M]} = \{\widehat{P_i^t}, \widehat{P_i^{t+1}}, \widehat{P_i^{t+2}}, \cdots, \widehat{P_i^{t+M}}\}$ and the predicted passenger flow for "just-passed" time period is $\mathcal{P}_i^{[t-M,t-1]} = \{\widehat{P_i^{t-M}}, \widehat{P_i^{t-M+1}}, \widehat{P_i^{t-M+2}}, \cdots, \widehat{P_i^{t-1}}\}$ extracted from next-day prediction $\widehat{\mathcal{P}_i^{fut}}$. As the real-time operating records are still the trip boarding data, we have to transfer the minute-level prediction results into trip boarding format for the comparison purpose:

$$\widehat{B_i^{T_k}} = (1 - \rho_i^{T_k})(I_i^{\widehat{[T_{k-1},T_k]}} + \rho_i^{T_{k-1}}(I_i^{\widehat{[T_{k-2},T_{k-1}]}} + \rho_i^{T_{k-2}} I_i^{\widehat{[T_{k-3},T_{k-2}]}})) \quad (11)$$

where $\widehat{I_i^{[T_1,T_2]}} = \sum_{a=T_1}^{T_2} \widehat{P_i^a}$ is the predicted accumulated passenger flow. Therefore, the difference between the actual and the predicted boarding volume of the passed trip at time point T_k is:

$$D_i^{T_k} = B_i^{T_k} - \widehat{B_i^{T_k}}, \quad t - M \leq T_k < t \quad (12)$$

$D_i^{T_k}$ and $\mathcal{P}_i^{[t,t+M]}$ are then concatenated and fed into the fine-tuning model for short-term prediction adjustment. Here we choose Random Forest (RF) model as our fast fine-tuning model because of its high computing speed and multi-stream precise which are suitable for online tasks [9]:

$$\mathcal{P}_i'^{[t,t+M]} = \mathbf{RF}([D_i^{T_{k-n+1}}, D_i^{T_{k-n}}, \cdots, D_i^{T_k}] \oplus \mathcal{P}_i^{[t,t+M]}) \quad (13)$$

where n means the boarding volumes of passed n trips are compared for the difference exploration.

4.2 Real-Time Adjustment for Interchange and Crowdedness

For the interchange and passenger crowdedness adjustment when emergencies or timetable changes occur, we will recalculate the numbers by applying the same

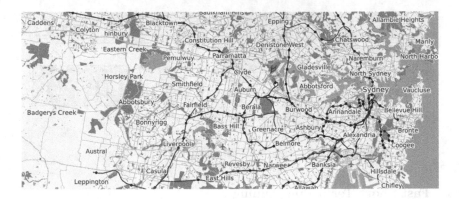

Fig. 6. The railway networks in the Greater Sydney Areas.

methods introduced in Sect. 3 based on the updated timetable. For example, if the arrival time of one trip at Platform j will delay M minutes, the interchange passenger flow from Platforms j to i is recalculated from Eq. 9 as:

$$I_{j \cdot i(\text{Interchange})}^{[t_m, t_n + M]} = r_{j \cdot i} \times I_i^{[t_m, t_n + M]} \tag{14}$$

and the corresponding passenger crowdedness on Platform j with the rescheduled arrival time of the trip is recalculated from Eq. 10 as,

$$C_i^T = I_i^{[T_{k-1}, T_k + M]} + \rho_i^{T_{k-1}} I_i^{[T_{k-2}, T_{k-1}]} + \rho_i^{T_{k-2}} I_i^{[T_{k-3}, T_{k-2}]} \tag{15}$$

5   Experiments

Our proposed model is deployed on a real-world urban railway system (Sydney Trains) operating in the Greater Sydney Area as shown in Fig. 6.

5.1   Experimental Dataset and Setting

The experiments are conducted on the pre-COVID time period of 7 months (May, 2019–November, 2019) historical trip boarding records of 713 platforms located over the whole railway network collected by Sydney Trains. The dataset is split into 85% (first 6 months) for training and 15% (last month) for testing. And we adopt two standard evaluations metrics in the flow prediction tasks: *Mean Absolute Error* (MAE) and *Root mean squared error* (RMSE).

5.2   Benchmark Methods

As the flow prediction is a typical time-series task, we choose advanced methods as comparisons: (1) **ARIMA** is a traditional statistical model that uses past observations of the time series data to predict the future [15]. (2) **SVR** (Support

Table 1. Performance of next-day prediction model and benchmark methods.

Model	Weekdays		Weekends	
	MAE	RMSE	MAE	RMSE
ARIMA	55.63	58.38	52.87	53.75
SVR	48.34	50.74	44.78	42.87
Random Forest	42.57	45.68	36.88	38.84
LSTM	37.58	42.87	31.87	34.24
NDP Model -Without IF	29.78	37.45	20.87	21.12
NDP Model	**28.13**	**35.52**	**18.45**	**20.87**

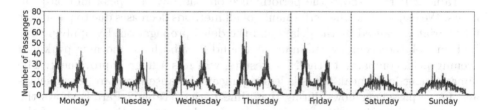

Fig. 7. Comparison between next-day prediction trip boarding and the ground truth in a week (Nov 11–Nov 17, 2019) on Platform 24 of Sydney Central Station: the red curve refers to the prediction results while the blue curve refers to the ground truth. (Color figure online)

Vector Regression) is a machine learning model that can construct hyperplanes in high dimensional space to fit the data [6]. (3) **Random forest** is an ensemble technique capable of performing in regression tasks with multi decision trees [12]. (4) **LSTM** (long-short-term memory) is a variant of the RNN-based model which can maintain the information with a memory mechanism [20].

5.3 The Evaluation of Next-Day Passenger Flow Prediction

The next-day passenger flow prediction experiments are conducted to evaluate the accuracy of our prediction based on the actual trip boarding records. As the trip boarding records are only recorded when the trains arrive at the platforms, we compare the actual passenger number with the aggregated predicted passenger flows by considering the denied boarding rate at the arrival time points as Table. 1 shows. We can observe that our next-day passenger flow prediction (NDP) model performs much better than other benchmarks methods because our model well defines and processes the passenger flow records well and introduces the self-attention mechanism to capture the different impacts of historical input series on the prediction. Figure 7 further intuitively show the high degree of fitting between next-day prediction *v.s.* the ground truth for a week period. Meanwhile, to evaluate the importance of influential factors, we also evaluate the performance of our model without introducing influential factors (**With**

& **Without IF**). The experiments show that the influential factors can help improve the performance of more than 3% in MAE.

5.4   The Evaluation of Real-Time Fine-Tuning

To evaluate the effectiveness of our real-time fine-tuning (RTF) model, we select the platforms with major delays (OTR is smaller than 85%) to demonstrate the performance comparison. Here we define the major delay as the actual arrival time at the platform exceeds the scheduled arrival time more than 5 min. Therefore, 11 eligible days (8 weekdays and 3 weekends) are selected from the testing time period (November, 2019).

Table 2 Part A shows the performance on bad days for peak and off-peak hours. We can see that the performance of all methods decreases due to passenger flow variation caused by the delays and the delay propagation. By applying our RTF model, the accuracy can improve around 25% both the morning peak and evening peak compared to the NDP results, which is a huge improvement. The improvement in off-peak hours is less significant because there are fewer delays and fewer passenger flows during that time period. Table 2 Part B shows the performance by using different target time windows as the time period of the short-term prediction. We can see that the performance drops with the increased size of the target time window which suggests that the size of the target time window is an influential factor for real-time passenger flow prediction.

Table 2. Real-time prediction model evaluation: Part A refers to prediction evaluation on bad days for peak and off-peak hours, the morning peak refers to 7:00–10:00, the evening peak refers to 17:00–22:00, and off-peak hours refers to the rest. Part B refers to prediction evaluation on bad days for different target time windows.

Experiments	A						B					
Model	Morning Peak		Evening Peak		Off-peak hours		15 Min		20 Min		30 Min	
	MAE	RMSE	MAE	RMSE	MAE	RMSE	MAE	RMSE	MAE	RMSE	MAE	RMSE
ARIMA	75.45	77.94	70.54	74.81	62.84	66.14	67.30	70.14	67.68	70.45	68.80	71.08
SVR	70.84	72.05	65.78	68.14	57.57	59.67	58.84	62.87	59.05	63.85	59.78	64.13
Random Forest	63.15	65.17	58.85	62.10	45.87	48.18	51.37	55.57	52.27	55.90	52.74	54.01
LSTM	60.08	64.57	53.12	57.19	41.57	44.11	47.01	50.24	47.22	51.87	47.98	52.93
NDP Model	55.74	62.27	46.87	48.75	35.78	39.87	40.87	43.87	40.90	45.57	41.78	45.12
RTF Model	**38.12**	**45.01**	**34.68**	**36.14**	**32.00**	**33.37**	**34.58**	**34.58**	**34.84**	**37.74**	**35.57**	**38.01**

To better illustrate the effectiveness of the RTF compared to the NDP model, we conduct the case study under three different situations as shown in Fig. 8: delayed trips, delayed interchange passengers and canceled trips. When an emergency occurs, our RTF will adjust the predicted passenger flow and passenger crowdedness based on the real-time updated timetable. Specifically, for the case of delayed trips in (a), the passengers waiting on the platform will accumulate until the delayed trip arrived. If a delay happened on an interchange trip that

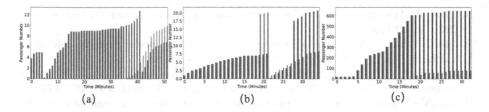

Fig. 8. Real-time fine-tuning results under timetable changes. The blue bars refer to the updated passenger crowdedness by applying the RTF model while the yellow bars represent the next-day prediction results. Sub-graph refers to different situations: (a) Delay trips. (b) Delayed interchange passengers. (c) Canceled trips. (Color figure online)

caused the deferred arrival of interchange passengers, our RTF model can still capture the change and adjust the interchange passenger flow accordingly as shown in (b). Similar real-time adjustment is like (c) the canceled trip.

6   Conclusion

In this paper, we proposed a two-stage self-adaptive model for passenger flow prediction on the large-scale railway network. The two-stage model can provide significant and accurate passenger flow prediction to railway managers and controllers for better timetable planning and emergency response in time. Benefits from the two-stage model, the timely response-ability gives the controllers more flexibility of on-site regulation. The deployment of the real-world urban railway system has proven the efficiency and effectiveness of our model.

References

1. Anderson, M., Murray, M.H., Ferreira, L., Lake, N.J.: Collection and use of railway track performance and maintenance data. In: Conference on Railway Engineering (2004)
2. Bendfeldt, J., Mohr, U., Muller, L.: RailSys, a system to plan future railway needs. WIT Trans. Built Environ. **50** (2000)
3. Bergquist, B., Söderholm, P.: Data analysis for condition-based railway infrastructure maintenance. Qual. Reliab. Eng. Int. **31**(5), 773–781 (2015)
4. Chodrow, P.S., Al-Awwad, Z., Jiang, S., González, M.C.: Demand and congestion in multiplex transportation networks. PLoS ONE **11**(9), e0161738 (2016)
5. Cui, Y., Jin, B., Zhang, F., Sun, X.: A deep spatio-temporal attention-based neural network for passenger flow prediction. In: 16th MobiQuitous (2019)
6. Drucker, H., Burges, C.J., Kaufman, et al.: Support vector regression machines. In: NIPS, vol. 9, pp. 155–161 (1997)
7. Flammini, F., et al.: Towards wireless sensor networks for railway infrastructure monitoring. In: Electrical Systems for Aircraft, Railway and Ship Propulsion. pp. 1–6. IEEE, Bologna (2010)
8. Lam, W.H., Cheung, C., Poon, Y.: A study of train dwelling time at the Hong Kong mass transit railway system. J. Adv. Transport. **32**(3) (1998)

9. Liaw, A., Wiener, M., et al.: Classification and regression by randomforest. R News **2**(3), 18–22 (2002)
10. Liu, L., Chen, R.C.: A novel passenger flow prediction model using deep learning methods. Transp. Res. Part C Emerg. Technol. **84**, 74–91 (2017)
11. Liu, Y., Liu, Z., Jia, R.: DeepPF: a deep learning based architecture for metro passenger flow prediction. Transp. Res. Part C Emerg. Technol. **101**, 18–34 (2019)
12. Liu, Y., Wu, H.: Prediction of road traffic congestion based on random forest. In: 2017 10th International Symposium on Computational Intelligence and Design. vol. 2, pp. 361–364. IEEE (2017)
13. Lu, K., Han, B.: Congestion risk evaluation and precaution of passenger flow in metro stations. Open Civil Eng. J. **10**(1) (2016)
14. Luethi, M., Weidmann, U., Nash, A.: Passenger arrival rates at public transport stations. In: 86th TRB. pp. 07–0635. Transportation Research Board (2007)
15. Moayedi, H.Z., Masnadi-Shirazi, M.: ARIMA model for network traffic prediction and anomaly detection. In: 2008 International Symposium on Information Technology, vol. 4, pp. 1–6. IEEE (2008)
16. Sun, H., Liu, H.X., Xiao, H., Ran, B.: Short term traffic forecasting using the local linear regression model. In: RB 2003 Annual Meeting CD-ROM (2002)
17. Vaswani, A., et al.: Attention is all you need. In: NIPS (2017)
18. Williams, B.M., et al.: Urban freeway traffic flow prediction: application of seasonal autoregressive integrated moving average and exponential smoothing models. Transp. Res. Record. **1644**(1), 132–141 (1998)
19. Yang, D., et al.: Urban rail transit passenger flow forecast based on LSTM with enhanced long-term features. IET Intell. Transp. Syst. **13**(10) (2019)
20. Zhao, Z., et al.: LSTM network: a deep learning approach for short-term traffic forecast. IET Intell. Transp. Syst. **11**(2), 68–75 (2017)
21. Zhou, X., et al.: Predicting multi-step citywide passenger demands using attention-based neural networks. In: Proceedings of the 11th WSDM, pp. 736–744 (2018)

ToothCR: A Two-Stage Completion and Reconstruction Approach on 3D Dental Model

Haoyu Zhu[1], Xiuyi Jia[1(✉)], Changdong Zhang[2], and Tingting Liu[2]

[1] School of Computer Science and Engineering, Nanjing University of Science and Technology, Nanjing, China
jiaxy@njust.edu.cn
[2] School of Mechanical Engineering, Nanjing University of Science and Technology, Nanjing, China

Abstract. Deep neural networks have made a number of achievements both in tooth segmentation and arrangement on complete 3D dental models. But few studies have used deep learning methods on the tooth completion and reconstruction on the incomplete dental models. To rebuild the missing tooth from incomplete dental model, we propose a two-stage approach ToothCR which takes advantage of the powerful learning capabilities of deep neural networks. In the first stage, ToothCR introduces a geometry-aware transformer encoder into the 3D dental model completion task. Self-attention mechanism in transformers could better model long-range dependencies in point cloud and ensure the predicted missing parts to have precise geometric structures. In the second stage, ToothCR uses a novel surface reconstruction algorithm to recover the surface of the predicted missing tooth. The reconstruction algorithm guarantees the generated surface to be watertight and avoids holes or redundant meshes which traditional methods may produce. Extensive experiments conducted on 3D dental datasets show that our approach outperforms state-of-the-art methods both in qualitative and quantitative results.

Keywords: 3D dental model · Point cloud completion · Surface reconstruction

1 Introduction

High-precision 3D dental models obtained by intraoral scanners play an important role in CAD dental diagnosis, and using deep neural networks to process dental models and solve dental problems becomes more and more popular. Neural networks have shown amazing performance in both tooth segmentation and arrangement areas [11,23,30]. As for the tooth completion and reconstruction

This work was supported by National Key Research and Development Program of China (2019YFB1706900), and the Fundamental Research Funds for the Central Universities (30920021131).

J. Gama et al. (Eds.): PAKDD 2022, LNAI 13282, pp. 161–172, 2022.
https://doi.org/10.1007/978-3-031-05981-0_13

problems, most work generally use non-learning approaches to rebuild the missing tooth based on geometric priors [3,14,28]. Ping *et al.* [17] provides a learning-based tooth completion method based on voxel grids. But the time and memory consumption increases squarely with the voxelization resolution, thus it is not the optimal solution for tooth completion. Tian *et al.* [19] proposes a two-stage generative adversarial network to reconstruct the occlusal surface from depth images of dental crown. However, depth images provide less geometric information than the other 3D representation. And their method could only recover the occlusal surface, not the whole surface of a single tooth. From a 3D dental model with missing tooth, completing the missing parts and recovering the surface remain challenging. Therefore, we propose a two-stage approach ToothCR to solve the problems above.

ToothCR consists of two stages of work: tooth completion and surface reconstruction. In the first stage, we take point cloud as representation of 3D dental model, and predict the point cloud of the missing tooth. We introduce a geometry-aware transformer encoder into the completion task and operate directly on the raw 3D coordinates. ToothCR takes advantage of the self-attention mechanism to better leverage the inductive bias about 3D geometric relationships of the point cloud, and predicts the missing tooth in a coarse-to-fine manner through a multi-scale generator. Our completion network is efficient for learning the dependencies among points, and guarantees accurate recovery of local details on the predicted point cloud. To the best of our knowledge, this is the first work conducted on the 3D coordinates of the original dental model using a transformer structure.

In the second stage, ToothCR adopts a novel surface reconstruction algorithm to acquire the final triangle mesh of the predicted tooth. When facing point cloud with noise, traditional surface reconstruction methods [1,5,9] often produce holes and redundant triangles, which cannot recover smooth and complete surface. To handle with this problem, our reconstruction algorithm uses a KNN algorithm to form an initial triangle mesh without holes from the point cloud. Then an octree represents it to build geometric connections in neighboring points. Finally ToothCR constructs the final surface by isosurface extraction. Our approach eliminates some degree of noise and ensures that the generated surface has no holes. The final surface recovered by ToothCR is a 2-manifold topology, which means the surface is watertight.

Our contributions are summarized as follows:

- We propose a two-stage approach ToothCR to complete dental model and reconstruct the surface of the missing tooth. As far as we know, this is the first work conducting dental model completion on the raw 3D coordinates of point clouds.
- We introduce a geometry-aware transformer encoder and a multi-scale generator in to the 3D dental completion task. Self-attention mechanism helps to better model the long-range dependencies in point cloud, and the multi-scale generator ensures the geometric details on the predicted tooth.
- We propose a simple and efficient surface reconstruction algorithm that eliminates the negative impact of noisy points in the point clouds and guarantees the reconstructed surface to be watertight.

2   Related Work

Traditional methods adopt voxel grids or distance fields to represent 3D objects to solve the completion problems. However, these methods suffer from high time and memory consumption, because they need higher resolution to get better completion performance while the computational burden increases sharply. To deal with such problems, researchers use point cloud as the representation of 3D objects for the small memory consumption and powerful characterization ability. PointNet [4] and PointNet++ [18] firstly operate on 3D coordinates and inspires PCN [29], the pioneer work in point cloud completion. PCN is based on learned global features from the partial input, and uses a FoldingNet [26] to recover complete point cloud. After PCN, many other methods come up and provide better results with richer local geometry details. MSN predicts a complete but coarse-grained point cloud with a collection of parametric surface elements and then merges it with input to get the final output [12]. GRNet uses cubic feature sampling layers to extract features of neighboring points [25]. SpareNet presents the channel-attentive Edge-Conv to fully exploit local structures [24]. PF-Net estimates the missing point cloud hierarchically by utilizing a feature-points-based multi-scale generating network [8]. VRCNet designs a self-attention kernel and a point selective kernel module to exploit relational point features which refines local shape details conditioned on the coarse completion [15]. PoinTr introduces point cloud transformers into completion task to model the local geometric relationships explicitly [27]. All these methods use different technologies to preserve and recover local details which make the completion results better. Similar to our work, Ping et al. [17] introduces self-attention to the implicit function network for completion of gum and the missing tooth. Their approach converts every point cloud into voxel grids, which takes a heavy burden of time and memory consumption. On contrast, our approach directly operates on 3D coordinates to make the time and memory cost more acceptable.

As for surface reconstruction, traditional reconstruction methods mainly include two paradigms: explicit reconstruction and implicit reconstruction. Explicit reconstruction methods such as Alpha Shapes [5], Ball Pivoting [1], Zippering [20], Delaunay Triangulation [2], resort to the local surface connectivity estimation and connect the sampled points directly by triangles. Implicit reconstruction methods approximate the point cloud with a field function and employ the marching cube algorithm [13] to extract the isosurface of the field function to reconstruct surfaces [9]. However, both explicit reconstruction and implicit reconstruction methods fail in dealing with 3D dental models with missing tooth, which encourages us to design another suitable algorithm as proposed in our paper.

3   Our Approach

The pipeline of our ToothCR is summarized in Fig. 1. ToothCR consists of two stages, incomplete dental model completion and surface reconstruction. More details about ToothCR will be described in the following subsections.

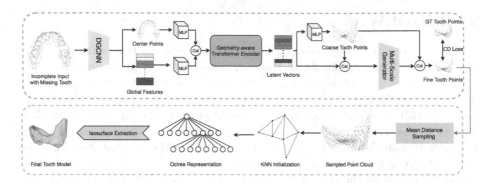

Fig. 1. The pipeline of ToothCR. In the first stage shown in the blue frame, ToothCR takes incomplete point cloud after FPS as input, and DGCNN module helps to group center points and extract local features. A geometry-aware transformer encoder is adopted to furtherly extract latent vectors which describe the missing tooth. Multi-scale generator concatenates the latent vectors and the coarsely predicted point cloud to generate final precise point cloud of the missing tooth. The surface reconstruction process is shown in the red frame. KNN initialization forms an initial mesh using KNN algorithm from point cloud after mean distance sampling. ToothCR uses an octree to represent the initial mesh and recovers the final surface by isosurface extraction through searching connections among nodes in the octree. (Color figure online)

3.1   Incomplete Dental Model Completion

Given an incomplete dental model $X \in \mathbb{R}^{N \times 3}$, where X indicates the incomplete point cloud with N points. Our goal is to predict the missing tooth. We indicate the ground truth missing tooth with $Y_{gt} \in \mathbb{R}^{M \times 3}$, and the predicted result with $Y_{pred} \in \mathbb{R}^{M \times 3}$. We constrain the ground truth tooth and the predicted tooth to have the same number of M points to facilitate the calculation of the loss.

DGCNN Grouper. High-precision 3D dental models usually contain tens or even hundreds thousands of points which result in a huge time and memory consuming when operating directly on the original data. Our completion network firstly conducts furthest point sample (FPS [4]) on the dataset to make the inputs to have the same number of points N. ToothCR adopts a widely known Dynamic Graph CNN (DGCNN [22]) to group the input into a smaller set of point centers $\mathcal{P} \in \mathbb{R}^{n \times 3}$ with n key coordinates. DGCNN extracts the features of each point described as $\mathcal{F} \in \mathbb{R}^{n \times 3}$ which are provided as local features. Then two different MLPs are adopted on the point centers and local features to expand them to a same higher dimension to benefit our geometry-aware transformer encoder, which are computed as followed:

$$\mathcal{F}' = \mathcal{M}_1(\mathcal{P}) + \mathcal{M}_2(\mathcal{F}), \tag{1}$$

where $\mathcal{M}_1$ and $\mathcal{M}_2$ indicte two similar MLPs which expand the dimension of point centers and the corresponding features, respectively, $\mathcal{F}'$ indicates the final features after DGCNN.

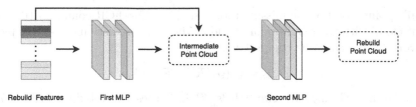

Fig. 2. Our multi-scale generator. TooCR feeds the rebuilt features obtained by encoder into the first 3-layer perceptron to get an intermediate point cloud, and concatenates it with the input to feed into the second 3-layer perceptron to rebuilt a finer point cloud.

Geometry-Aware Transformer Encoder. Thanks to the self-attention mechanism, transformers in both NLP and CV fields have proven to be efficient to tasks containing long sequence input [16,21,27], which is perfectly suitable for our tooth completion problem. Inspired by [27], ToothCR uses a geometry-aware transformer encoder to model the geometric dependencies between points. The geometry-aware encoder contains six continuous blocks with self-attention modules. After DGCNN grouper obtaining the final features $\mathcal{F}' \in \mathbb{R}^{n \times 3}$, the encoder queries the features according to the key coordinates, then learns the local geometric structures by feature aggregation with a linear layer followed by max pooling. ToothCR combines the geometric features learned by the encoder and the semantic features provided by center points, and maps them to the original dimensions. Through operations above, our encoder extracts the latent vectors which describes the missing part coarsely. An MLP is adopted to roughly rebuild the missing tooth with a small number of points. ToothCR concatenates the coarsely rebuilt tooth with the latent vectors as the input of the multi-scale generator. Operations above could be computed as:

$$\mathcal{R} = \varphi(\mathcal{F}') \oplus \mathcal{M}_3(\varphi(\mathcal{F}')), \tag{2}$$

where $\mathcal{F}'$ is the results of DGCNN grouper, φ indicates the encoder, $\varphi(\mathcal{F}')$ indicates the latent vectors, $\mathcal{M}_3$ is an MLP to coarsely predict the missing tooth, $\mathcal{R}$ is the rebuilt features concatenated by $\varphi(\mathcal{F}')$ and $\mathcal{M}_3(\varphi(\mathcal{F}'))$, and $\oplus$ indicates the concatenation operation.

Multi-scale Generator. The multi-scale generator consists of two 3-layer perceptrons as shown in Fig. 2. The goal of this module is to predict the missing tooth $Y_{pred} \in \mathbb{R}^{M \times 3}$ from the rebuilt features $\mathcal{R}$ which describe the general shape and the numerical range of the missing tooth. Our multi-scale generator builds the final point cloud in a coarse-to-fine manner using different scales of points and features. The results of the geometry-aware encoder provide a coarse point cloud $Y_c \in \mathbb{R}^{n \times 3}$ with a small number of points. The generator concatenates the coarse point cloud and global features as the rebuilt features $\mathcal{R}$, passes them through the first MLP. After obtaining the intermediate point cloud, the second MLP continuously combines it with the rebuilt features, and outputs the rebuilt point cloud of the missing tooth. At last, ToothCR combines the coarse point cloud with the rebuilt point cloud as the final prediction of our completion network, which could be computed as follow:

$$Y_{pred} = \mathcal{G}_2(\mathcal{G}_1(\mathcal{R}) + \mathcal{R}) \oplus Y_c, \tag{3}$$

where $\mathcal{G}_1$ and $\mathcal{G}_2$ indicate two MLPs, $\mathcal{R}$ indicates the rebuilt features, $Y_c = \mathcal{M}_3(\varphi(\mathcal{F}'))$ indicates the coarse point cloud, and $\oplus$ indicates the concatenation operation.

Loss Function. Two permutation-invariant metrics are introduced in [6] to compare two unordered point sets which are Chamfer Distance (CD) and Earth Mover's Distance (EMD):

$$CD(S_1, S_2) = \frac{1}{|S_1|} \sum_{x \in S_1} \min_{y \in S_2} ||x - y||_2 + \frac{1}{|S_2|} \sum_{y \in S_2} \min_{x \in S_1} ||y - x||_2, \tag{4}$$

$$EMD(S_1, S_2) = \min_{\phi: S_1 \to S_2} \frac{1}{|S_1|} \sum_{x \in S_1} ||x - \phi(x)||_2, \tag{5}$$

where S_1 and S_2 indicte two point clouds. We choose CD with $\mathcal{O}(N \log N)$ complexity as our loss function during training. As mentioned before, we rebuild the missing tooth in a coarse-to-fine manner, our loss function consists of two terms:

$$\mathcal{L}(Y_c, Y_{pred}, Y_{gt}) = CD(Y_c, Y_{gt}) + \alpha CD(Y_{pred}, Y_{gt}), \tag{6}$$

where α is a weight hyperparameter. The first term calculates the squared distance between the coarse point cloud and the ground truth, and the second term calculates the squared distance between the fine predicted point cloud and the ground truth. This design increases the proportion of the feature points, leading the training process to focus on the feature points, resulting in that the completed missing tooth has more geometric details.

3.2 Surface Reconstruction

After the completion, we get a point cloud about the missing tooth with noise in some degree. The goal of ToothCR in the second stage is to reconstruct the surface of the missing tooth from this rough point cloud through the following operations as shown in the red frame in Fig. 1.

Mean Distance Sampling. The rough point cloud obtained by the first stage may contain points which are too close to others. These points will lead to too many redundant faces when conducting surface reconstruction. ToothCR uses a mean distance sampling firstly by computing the mean distance d between each pair of points in the point cloud, and filters the points with a threshold τ, which is set to $\tau = d/3$ in our experiments. After filtering, point clouds with a relatively uniform distribution, but without losing the original geometric details so much are acquired.

KNN Initialization. Triangle meshes represent a surface by combining three close points into a triangle. ToothCR queries a point cloud to itself with KNN algorithm, and gets K neighbors of each point. Aligning them three by three, our approach could form $K - 1$ triangles which describe the surface around each point. Then ToothCR removes the duplicated and degenerate triangles to get the initial surface of the input rough point cloud. It is worth noting that, small K value may lead to holes in the initial mesh while large K value may produce redundant triangles which describe the neighboring geometric relationships wrongly. K is set to 32 in our experiments.

Octree Representation. In order to conduct isosurface extraction, we need to convert initial surface into the signed distance filed, which means using a uniform grid is the simplest way. ToothCR sets the length of the unit cube as 0.1 in the uniform grid to obtain a precise mesh, and the signs are recorded in these cubes which include positive, zero or negative. Our algorithm extracts the surface between positive cubes and zero cubes to represent the isosurface. Inspired by [7], ToothCR adopts an octree to store cubes to make the representation more efficient. Starting with a single root node, the octree stores the effective bounding box and the set of triangles that the bounding box contains or intersects. ToothCR keeps subdividing the occupied nodes (we define the nodes whose triangle set is none-empty as occupied nodes) into eight children unless they reach the final resolution (0.1) and resigns the triangle set of the node to its children. Then our algorithm builds connections between the occupied nodes with others in a recursive way. Starting from the root, ToothCR builds connections inside each child node, the recursion ends when both children nodes are not occupied nodes.

Isourface Extraction. We define the isosurface as the face between positive nodes and occupied nodes. Firstly, if there is a path from a certain node to the boundary which is not occupied by any triangle, we define this node is positive. Positive and negative nodes indicate whether the space is inside the surface or not, and space between positive and negative nodes is the isosurface to reconstruct surface. After initializing all boundary nodes to be positive, our algorithm applies a BFS algorithm to expand the positive nodes. To search for all these faces, ToothCR recursively visits all occupied nodes and their neighbors to detect occupied-positive connections. Then the relations between occupied cubes and triangles they contain are built, it is convenient to compute the nearest triangle to a certain vertex. After traversing all vertices of the initial mesh, we could get all triangles that form the final surface.

4 Experiments

Datasets. We conduct experiments on a real 3D dental dataset which consists of 906 patients' complete oral scan. Since what we need is incomplete dental model and single missing tooth to form an input-output pair, we extract every

Table 1. Qualitative comparison results of our methods and state-of-art methods. We use CD-l_1, CD-l_2, EMD and F-score metrics, and bold the best result under each metric. Our methods outperform others in all metrics.

Methods	CD-l_1	CD-l_2	EMD	F-score
PCN	0.006272	0.000120	0.019825	0.811700
PF-Net	0.006991	0.000362	0.030689	0.867447
VRCNet	0.004598	0.000059	0.020524	0.942281
PoinTr	0.005279	0.000078	0.026651	0.900436
Ours	**0.004473**	**0.000055**	**0.019302**	**0.947553**

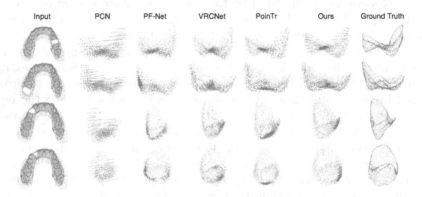

Fig. 3. The comparison of results from different point cloud completion methods. From left to right: input incomplete tooth, PCN [29], PF-Net [8], VRCNet [15], PoinTr [27], our method and the ground truth missing tooth. Our method recovers not only the external shape of the missing tooth, but also avoids clustered points and noisy points.

tooth from the complete 3D model as the ground truth, together with the model with a hole which indicates the missing tooth to form a pair. After filtering single tooth point cloud who has less than 2048 points, we construct our dataset with 6109 pairs of incomplete tooth and ground truth missing tooth. We randomly split our dataset into two parts: 5469 for training and 640 for testing.

Evaluation Metrics. In line with other state-of-the-art methods, we evaluate our completion network by computing Chamfer Distance (Eq. 4) between the predicted point cloud Y_{pred} and the ground truth point cloud Y_{gt}. Following the previous work, we use two versions of Chamfer Distance: CD-l_1 which uses l_1-norm to calculate the distance between two points and CD-l_2 which uses l_2-norm instead. We also adopt another two metrics: Earth Mover Distance (Eq. 5) and F-score [10] which is defined as the harmonic mean between precision and recall. Through these four metrics, we could better evaluate how well our method performs.

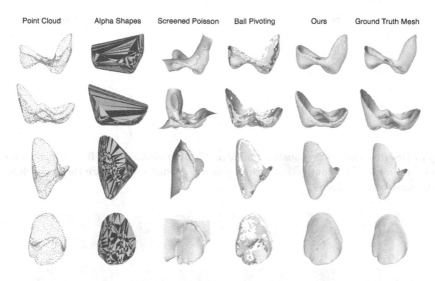

Fig. 4. The comparison of results from different surface reconstruction methods. We evaluate teeth in different locations. Alpha Shapes and Ball Pivoting fail with holes on the surface, while Screen Poisson Reconstruction produce redundant planes. Our method correctly recovers the complete surface of the input point cloud.

Implementation Details. We implement our networks on PyTorch. All modules in our completion network are trained with an ADAM optimizer, and the initial learning rate is set to 0.0001 with a decay rate 0.7 every 40 epoches. We set our batch size to 64 and numbers of workers to 10. α in the loss function (Eq. 6) is 0.1. All training work is done with two NVIDIA RTX3090 GPUs.

4.1 Incomplete Dental Model Completion Results

As mentioned above, there are 5469 pairs for training and 640 pairs for testing in the dataset. We train our network and other comparative methods with the same epochs (500 epochs). It is worth noting that, we change the loss function of comparison networks to the same as ours. And the final predicted numbers of points to 2048. We compare our network with four SOTA point cloud methods: PCN [29], PF-Net [8], VRCNet [15], and PoinTr [27], the quantitative results are shown in Table 1. It could be clearly observed that our completion network outperforms all the other methods. The qualitative comparison results are shown in Fig. 3. Our method is able to generate more precise shapes with richer details than others. Our results have less clustered points and noisy points, which make them to be more like to the ground truth missing teeth.

4.2 Surface Reconstruction Results

To evaluate the performance of our surface reconstruction algorithm, we choose three well known algorithms for comparison: Alpha Shapes [5], Ball Pivoting [1]

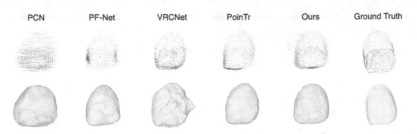

Fig. 5. Results using our reconstruction algorithm on point clouds from different completion networks. Our completion results generate surfaces which are the most close to the ground truth surfaces.

Fig. 6. Results after combining reconstructed surface with the input incomplete tooth model. The predicted teeth are marked with circles upper, and the corresponding ground truth dental models are lain below.

and Screened Poisson Reconstruction [9]. In detail, we set the tradeoff parameter α to 0.5 in Alpha Shapes, and the radii in Ball Pivoting to be twice the mean distance between points in the source point cloud. In Screen Poisson Reconstruction, we set the depth of the octree to be 10. We only take a small number of the ground truth samples to conduct experiment with implements by [31]. The qualitative results are shown in Fig. 4. We can see that both Alpha Shapes and Ball Pivoting generate holes, while the Screen Poisson Reconstruction produces redundant surfaces and cannot cover the point cloud correctly. Our surface reconstruction algorithm could recover the correct shape of point cloud, and the geometry details like bumps on the tooth are well preserved. As shown in Fig. 5, we test our algorithm on the results of different completion networks mentioned in Sect. 3.1. Point clouds generated by our completion network benefit from our reconstruction algorithm better than other methods, which implicitly suggests that our completion network recovers point cloud the most close to the ground truth. Combine our reconstruction results with the incomplete dental models, we can get the final repaired models as shown in Fig. 6, which indicates that our ToothCR obtains satisfactory qualitative results.

5   Conclusion

We propose a two-stage approach ToothCR for 3D dental model completion and reconstruction. Our completion network uses a geometry-aware encoder to model local dependencies in point cloud and a multi-scale generator to predict the missing tooth with precise geometric details. Our surface reconstruction algorithm uses octree to store point information and recovers surface by isosurface extraction. Our approach avoids clustered points and holes traditional methods may produce. Experiments show that our approach outperforms state-of-the-art methods both in qualitative and quantitative comparison.

References

1. Bernardini, F., Mittleman, J., Rushmeier, H., Silva, C., Taubin, G.: The ball-pivoting algorithm for surface reconstruction. IEEE Trans. Visual. Comput. Graph. 5(4), 349–359 (1999)
2. Boissonnat, J.D., Geiger, B.: Three-dimensional reconstruction of complex shapes based on the Delaunay triangulation. In: Biomedical Image Processing and Biomedical Visualization, pp. 964–975 (1993)
3. Buchaillard, S.I., Ong, S.H., Payan, Y., Foong, K.: 3D statistical models for tooth surface reconstruction. Comput. Biol. Medi. 37(10), 1461–1471 (2007)
4. Charles, R., Su, H., Kaichun, M., Guibas, L.J.: PointNet: deep learning on point sets for 3D classification and segmentation. In: Proceedings of the IEEE/CVF Conference on Computer Vision and Pattern Recognition, pp. 77–85 (2017)
5. Edelsbrunner, H., Kirkpatrick, D., Seidel, R.: On the shape of a set of points in the plane. IEEE Trans. Inf. Theory 29(4), 551–559 (1983)
6. Fan, H., Su, H., Guibas, L.J.: A point set generation network for 3D object reconstruction from a single image. In: Proceedings of the IEEE/CVF Conference on Computer Vision and Pattern Recognition, pp. 605–613 (2017)
7. Huang, J., Su, H., Guibas, L.: Robust watertight manifold surface generation method for shapenet models. arXiv preprint arXiv:1802.01698 (2018)
8. Huang, Z., Yu, Y., Xu, J., Ni, F., Le, X.: PF-Net: point fractal network for 3D point cloud completion. In: Proceedings of the IEEE/CVF Conference on Computer Vision and Pattern Recognition, pp. 7662–7670 (2020)
9. Kazhdan, M., Hoppe, H.: Screened Poisson surface reconstruction. ACM Trans. Graph. 32(3), 1–13 (2013)
10. Knapitsch, A., Park, J., Zhou, Q.Y., Koltun, V.: Tanks and temples: benchmarking large-scale scene reconstruction. ACM Trans. Graph. 36(4), 1–13 (2017)
11. Lian, C., Wang, L., Wu, T.H., Wang, F., Yap, P.T., Ko, C.C., Shen, D.: Deep multi-scale mesh feature learning for automated labeling of raw dental surfaces from 3D intraoral scanners. IEEE Trans. Med. Imaging 39(7), 2440–2450 (2020)
12. Liu, M., Sheng, L., Yang, S., Shao, J., Hu, S.M.: Morphing and sampling network for dense point cloud completion. In: Proceedings of the AAAI Conference on Artificial Intelligence, pp. 11596–11603 (2020)
13. Lorensen, W.E., Cline, H.E.: Marching cubes: a high resolution 3D surface construction algorithm. ACM Siggraph Comput. Graph. 21(4), 163–169 (1987)
14. Martorelli, M., Ausiello, P.: A novel approach for a complete 3D tooth reconstruction using only 3D crown data. Int. J. Interact. Des. Manuf. 7(2), 125–133 (2013)
15. Pan, L., Chen, X., Cai, Z., Zhang, J., Zhao, H., Yi, S., Liu, Z.: Variational relational point completion network. In: Proceedings of the IEEE/CVF Conference on Computer Vision and Pattern Recognition, pp. 8524–8533 (2021)
16. Parmar, N., Vaswani, A., Uszkoreit, J., Kaiser, L., Shazeer, N., Ku, A., Tran, D.: Image transformer. In: International Conference on Machine Learning, pp. 4055–4064 (2018)

17. Ping, Y., Wei, G., Yang, L., Cui, Z., Wang, W.: Self-attention implicit function networks for 3D dental data completion. Comput. Aided Geomet. Des. **90**, 102026 (2021)
18. Qi, C.R., Yi, L., Su, H., Guibas, L.J.: Pointnet++: Deep hierarchical feature learning on point sets in a metric space. In: Advances in Neural Information Processing Systems, pp. 5105–5114 (2017)
19. Tian, S., Wang, M., Dai, N., Ma, H., Li, L., Fiorenza, L., Sun, Y., Li, Y.: DCPR-GAN: dental crown prosthesis restoration using two-stage generative adversarial networks. IEEE J. Biomed. Health Inform. **26**(1), 151–160 (2021)
20. Turk, G., Levoy, M.: Zippered polygon meshes from range images. In: Proceedings of the 21st Annual Conference on Computer Graphics and Interactive Techniques, pp. 311–318 (1994)
21. Vaswani, A., Shazeer, N., Parmar, N., Uszkoreit, J., Jones, L., Gomez, A.N., Kaiser, Ł., Polosukhin, I.: Attention is all you need. In: Advances in Neural Information Processing Systems, pp. 5998–6008 (2017)
22. Wang, Y., Sun, Y., Liu, Z., Sarma, S.E., Bronstein, M.M., Solomon, J.M.: Dynamic graph CNN for learning on point clouds. ACM Trans. Graph. **38**(5), 1–12 (2019)
23. Wei, G., Cui, Z., Liu, Y., Chen, N., Chen, R., Li, G., Wang, W.: TANet: Towards fully automatic tooth arrangement. In: European Conference on Computer Vision, pp. 481–497 (2020)
24. Xie, C., Wang, C., Zhang, B., Yang, H., Chen, D., Wen, F.: Style-based point generator with adversarial rendering for point cloud completion. In: Proceedings of the IEEE/CVF Conference on Computer Vision and Pattern Recognition, pp. 4619–4628 (2021)
25. Xie, H., Yao, H., Zhou, S., Mao, J., Zhang, S., Sun, W.: GRNet: gridding residual network for dense point cloud completion. In: European Conference on Computer Vision, pp. 365–381 (2020)
26. Yang, Y., Feng, C., Shen, Y., Tian, D.: FoldingNet: point cloud auto-encoder via deep grid deformation. In: Proceedings of the IEEE/CVF Conference on Computer Vision and Pattern Recognition, pp. 206–215 (2018)
27. Yu, X., Rao, Y., Wang, Z., Liu, Z., Lu, J., Zhou, J.: PoinTr: diverse point cloud completion with geometry-aware transformers. In: Proceedings of the IEEE/CVF International Conference on Computer Vision, pp. 12498–12507 (2021)
28. Yuan, T., Liao, W., Dai, N., Cheng, X., Yu, Q.: Single-tooth modeling for 3D dental model. Int. J. Biomed. Imaging **2010** (2010)
29. Yuan, W., Khot, T., Held, D., Mertz, C., Hebert, M.: PCN: Point completion network. In: International Conference on 3D Vision, pp. 728–737 (2018)
30. Zhang, L., et al.: TSGCNet: discriminative geometric feature learning with two-stream graph convolutional network for 3D dental model segmentation. In: Proceedings of the IEEE/CVF Conference on Computer Vision and Pattern Recognition, pp. 6699–6708 (2021)
31. Zhou, Q.Y., Park, J., Koltun, V.: Open3D: a modern library for 3D data processing. arXiv preprint arXiv:1801.09847 (2018)

BaDumTss: Multi-task Learning
for Beatbox Transcription

Priya Mehta$^{(\boxtimes)}$ ⓘ, Meet Maheshwari ⓘ, Brihi Joshi ⓘ,
and Tanmoy Chakraborty ⓘ

Department of CSE, IIIT-Delhi, Delhi, India
{priya20033,meet20012,brihi16142,tanmoy}@iiitd.ac.in

Abstract. The challenge of transcribing audio into symbolic notations is a well-known problem in music information retrieval. In this work, we explore a novel task – *automatic music transcription for Beatbox sounds*, also known as *Vocal Percussions*. As Beatbox sounds cannot be created in a synthetic manner, they inherently vary within the same speaker as well as across different speakers. To address this, we propose BaDumTss, which makes use of a pretraining strategy over a novel sequence traversal method, thereby ensuring robustness and efficiency against new Beatbox sequences. Furthermore, BaDumTss is agnostic to time-based stretches and warps, as well as amplitude changes in the Beatbox sequence. It predicts both onsets and frame-set in a multi-task manner while gaining a whopping 56% and 326% relative improvement frame-set and onset-level F1 scores over the best performing baseline respectively. We also release an annotated dataset of monophonic Beatbox sequences along with their corresponding MIDI labels, the first of its kind comprising Beatbox samples with different variations such as time-stretches, pitch shifts, and added noise.

Keywords: Automatic music transcription · Multi-task learning · Sequence modelling

1 Introduction

Vocal percussion, *aka* Beatboxing, is perhaps the most intuitive way to express a rhythm without the use of an actual instrument. A Beatbox makes use of several techniques to replicate different instruments in an actual drum. For example, both Beatboxing and Drumming have instruments like kicks, Hi-hats, etc. that have distinct sounds. However, like any other musical piece, transcribing it is not only tedious but also an immensely challenging task. It is difficult even for humans to identify which beat was played when and for how long. Thus, the need for automatic transcription methods is dire. Along with these challenges, Beatboxing brings forth additional constraints like biometric factors, since each person has a different voice, style and volume. One way to transcribe a Beatbox

---

P. Mehta and M. Maheshwari—Equal contribution.

© The Author(s), under exclusive license to Springer Nature Switzerland AG 2022
J. Gama et al. (Eds.): PAKDD 2022, LNAI 13282, pp. 173–186, 2022.
https://doi.org/10.1007/978-3-031-05981-0_14

sequence into valid symbolic representation is to convert it into MIDI (Musical Instrument Digital Interface) representations. A MIDI sequence is a lossless way of encoding information of the audio as well as the metadata.

In our work, we focus our attention on the problem of **Beatbox transcription** and address the following two problems:

1. How can we create Beatbox transcription datasets that adhere to natural constraints like biometric factors that Beatboxing suffers from, as well as maintain the general percussive patterns of the Beatbox sequences?
2. Can we also design an efficient yet robust system that improves previous transcription results but with lesser additional resources?

In this work, we study the problem of Beatbox transcription in detail – from developing a novel, fully annotated dataset to designing a new method for the task. Our primary contributions in this work can be summarised as follows:

1. We present a novel dataset[1] consisting of Beatbox sequences, accompanied by corresponding MIDI annotations. We also conduct a user-study that validates that our generated Beatbox sequences are *musically cohesive*, and distinct from random sounding Beatbox instruments that are stitched together.
2. Inspired by the recent success of pretraining in other domains, we propose BaDumTss, a U-Net based unsupervised pretraining unit, followed by a novel Bottled Sequence Traversal Unit and a multi-task autoencoder to strengthen BaDumTss towards high-level changes like new sequences, and low-level changes like time and amplitude shifts, as well as jointly predicting onsets and frame-sets.

2   Related Work

2.1   Automatic Music Transcription

Automatic Music transcription (AMT) has long been studied, and various techniques have been proposed for the ultimate goal of representing any kind of instrument in a human/machine-readable file format. Early machine learning techniques for AMT relied on hand-crafted feature extraction. One such work on polyphonic piano transcription [20] proposes a discriminative model based on frame-level SVM for pitch classification and note-level Hidden Markov Models [1]. However, such methods have become outdated as machine learning based techniques generalize poorly on a large corpus of multiple instruments.

Drum Transcription. Vocal percussion sounds are an imitation of drum sound patterns, and parallels can be drawn between these two transcription tasks. Most datasets for Automatic Drum Transcription (ADT) are still of limited size, which has led to DrummerNet [3], an unsupervised method and the state-of-the-art system for ADT. It contains a series of U-net [13] variants, which

---

[1] The source code and dataset are available at https://github.com/LCS2-IIITD/BaDumTss-PAKDD22.

in turn consist of 1-D convolution layers, recurrent layers, and gated Sparse-max activation. Pedersoli and Yi [14] explored the idea of pre-stacking a U-net, originally proposed to improve biomedical image segmentation. For drum-based sequences containing rhytmic patterns, RNNs showed competitive results [21]. This has also led to similar variants like the usage of GRU in a solo drum context as well as in polyphonic drum context. Furthermore, the application of a bidirectional-RNN with soft attention mechanism [4] has also been explored.

Beatbox Transcription. As compared to other instruments like piano, Beatbox transcription has been sparsely explored. A few techniques have been proposed for vocal percussion sound recognition; however, they do not intro-duce Beatbox sounds. One of the early studies [9] presented an Autonomous Classifier Engine (ACE) based on C4.5 decision trees for classifying Beatbox-ing sounds. It experimented on five type of instruments where each clip has only one label/instrument. Other studies [10] included a Hidden Markov Model (HMM) based model for the Beatbox instrument classification, in addition to pitch analysis and onset detection for feature extraction. A variant of this work [11] trained a fusion model based on HMM and Gaussian mixture model on a larger vocabulary.

All the aforementioned models work well with (and are applied only on) isolated samples. There is no evidence for good performance when applied to full-length rhythmic sequences of vocal percussion. Moreover, none of these models deal with onset detection of a beat in the music, which is necessary for a complete transcription pipeline in order to generate the corresponding MIDI sequences.

2.2 Input Audio Representation

Earlier studies focused on feature extraction from raw audio signals, but the same is now obsolete with the advent of deep learning models. Cheuk et al. [16] showed that a significant improvement in transcription accuracy is possible by choosing the right input representation. They concluded that the Mel spectrogram ensures high transcription results even though it is a compact representation of the original audio itself. Log-Mel Spectrogram is a variant of a mel spectrogram with log-scaled amplitude values and is a popular choice used by current transcription models for multiple instruments like piano [5] and drum [18].

2.3 Datasets for AMT

Standard datasets are available for popular instruments such as piano performance and their corresponding MIDI files [8], and for drum performances and aligned MIDI files [7,24]. However, Beatboxing itself does not include one specific instrument; rather it is a human imitating different percussion instrument sounds. The most related work released a dataset [19], which was proposed for vocal percussion analysis with the aim of improving algorithms for drum pattern query by vocal imitation. Samples in these dataset are recorded by amateurs with little or no experience in the art of Beatboxing and contain four different labels,

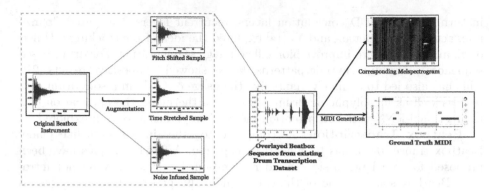

Fig. 1. Data curation pipeline: on each individually extracted Beatbox sample, we arbitrarily apply augmentations of three different types, which are then sewn together into a sequence. This sequence is constructed by overlaying the samples over existing drum-based transcription datasets. These Beatbox sequences are then used to generate the ground-truth MIDI samples, and their corresponding Mel Spectrograms are used as input to BaDumTss.

namely *kick*, *snare*, *hithat open*, and *hihat close*. This work concluded that onset detection for Beatboxing is not a trivial task and invites a novel approach.

3   Proposed Dataset

Beatboxing is the art of generating percussive sounds from the human body. As it is with any other human-centered dataset, the collection and annotation of Beatbox samples and sequences is a very tedious task. Limited data is a common problem for any automatic music transcription task, and Beatbox datasets impose several additional constraints. Firstly, we cannot prepare a completely synthetic dataset using software and other tools as human voice is required for the music to sound natural, which distinguishes a Beatbox sequence from a regular Drum sequence. Secondly, since Beatbox samples need to be recorded using some recording equipment, it adds other ungovernable factors affecting sound quality while recording such as environmental noise. Lastly, for utilizing any supervised learning algorithm, annotations in the form of a regular textual symbolic representation like the MIDI file of the corresponding musical file is essential; yet the MIDI-level annotation for Beatbox sequences is not only tedious, but also prone to hearing errors. In this section, we detail our data curation strategy – individual sample collection, data augmentation, Beatbox sequence generation, and ground-truth MIDI generation. Figure 1 demonstrates the architecture of the data collection process.

3.1   Beatbox Sample Collection

Given that Beatbox sequences are difficult to annotate, we begin our data collection at each sample level. We first begin by collecting individual Beatbox audio samples for four instruments - *kick*, *snare*, *clap*, *hihat*. These are samples that are clipped from online tutorials for Beatboxing and our own samples. The ratio of male to female Beatboxers in our dataset is 6:4, ensuring enough diversity with respect to the fundamental frequencies ($F0$). With this, we end up with a total of 176 unique Beatbox samples. The breakdown of samples at the instrument level is provided in Table 1.

3.2   Data Augmentation

Using 176 collected samples, we create a semi-synthetic dataset by leveraging data augmentation methods as described in [15] and implemented in [12]. Augmented samples are generated by applying a random series of transformations for heterogeneity. The augmentations that are applied are provided in detail below:

1. **Pitch Shift:** Keeping in mind the biometric factors responsible for each human voice to be different from the other, pitch shifts are introduced in the original audio samples. We generate a random fractional number, which is then used to shift the pitch of an audio by a factor that ranges between 10 half-steps in both the directions – increasing or decreasing the pitch.
2. **Time Stretching:** In order to produce a factor of temporal variation, we also introduce a random factor for time-stretch based augmentations to our Beatbox samples. In the final dataset, the length of individual Beatbox samples ranges between 85%–115% of the original duration.
3. **Noise Injection:** A tolerable amount of noise is injected in varying proportions to the audio clips for different sound qualities. In order to inject realistic levels of noise to mimic real-world recordings that do not overshadow the actual content of the sample, we add random noise whose amplitude reaches a peak of 0.5% of the maximum amplitude of the sample.
4. **Amplitude Shift:** Keeping in mind the rhythmic sequence as well as the distance of a recorder from the microphone, we shift the amplitude by 75%–125% of the original sample.

A total of 5,457 augmented sample from original Beatbox samples are created. Table 1 shows the breakdown of samples at the instrument level.

3.3   Audio Sequence Generation

One of the ways to create a Beatbox sequence comprising the above generated samples is to randomly stitch the samples together. However, such sequences cannot be termed as Beatbox as they do not contain any percussive nature or musicality, and end up sounding like noise.

In order to combine Beatbox samples, we make generate Beatbox sequences from already existing drum-based transcription datasets. *Extended Groove* [7] is one such large dataset with varying number of drum styles (like Jazz, Rock, etc.). It is a polyphonic dataset wherein there are multiple overlapping beats at the same time. For this work, we focus only on monophonic sequences; thus,

Table 1. For each of the four classes, Audio Samples (Original) denotes the total number of raw vocal percussion utterances, and Audio Samples (Augmented) denotes total number of vocal percussion utterances generated with controlled synthetic augmentation.

Instrument	#Original samples	#Augmented samples
Kick	54	1681
Snare	48	1467
HiHat	53	1653
Clap	21	656
Total	176	5457

we convert the drum sequences from polyphonic to monophonic by selecting the instrument with the highest intensity at any given time-frame. We then overlay the Beatbox samples, that are selected randomly within each instrument class, as per the drum sequences present in the dataset.

3.4   User Study on the Generated Audio Sequences

We conduct a user study to determine the qualitative musicality of our dataset. We curate a set of 10 randomly picked generated Beatbox sequences. Furthermore, to this set, we add 10 Beatbox sequences with randomly stitched samples, i.e., the generation process includes randomly picking the Beatbox samples and lining them one after another. We then randomly order these 20 sequences, and ask our annotators to distinguish which of the sequences are randomly generated and which of the sequences sound distinctly percussive.

For our study, we employ 27 annotators, out of which 17 have professional music training. The inter-annotator agreement was seen to be substantially high, for categorising the sequences into random and musically cohesive (Cohen's Kappa (κ) is 0.87). This validated that our generated dataset in fact was musically cohesive, and hence, appropriate to be used for a transcription task.

4   Proposed Methodology

Our proposed model architecture, BaDumTss, comprises of three primary components - **U-Net based pretraining** followed by **Bottled Sequence Traversal** and lastly, **Multi-task Autoencoders**. In this section, we explain our efforts in designing these three components as well as their relevance in the overall task of Beatbox transcription.

4.1   Pretraining Unit

A standard U-Net [13] architecture, designed originally for the task of semantic segmentation in images, consists of three components – (i) *a contracting*

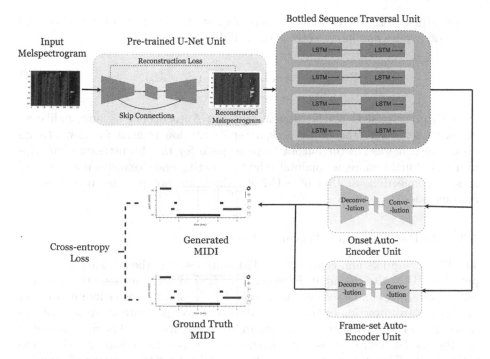

Fig. 2. A schematic diagram of our proposed model, BaDumTss.

path, which is a vanilla convolutional neural network, morphing the input to a bottle-necked representation of a certain dimension; (ii) *an expansive path*, which performs deconvolutions to generate a feature map of the same size as the input image, from the bottle-necked representations; and (iii) *skip connections*, which is the stacking of corresponding layers of the convolutional stack with the deconvolutional stack, used to prevent the catastrophic forgetting of information learnt during the expansive path. One difference between the final feature map and the input image is that the depth of the final feature map is same as the number of classes that would be used for the final pixel classification.

To design our model, we make use of a U-Net for pretraining our entire network structure. We begin by extracting the Mel spectrogram (M) of the audio sequence, that is of the dimension $N \times T$, where N is a pre-specified number of Mels and T is the number of time frames in our Mel spectrogram. We treat M as an input image which is passed through a standard U-Net architecture, denoted by $f_\Theta(\cdot)$. However, we attempt to use U-Net to reconstruct the input Mel spectrogram in the output feature map, $Z = f_\Theta(M)$. Therefore, the number of channels in our output feature space is the same as that of the input. We make two major changes to the vanilla U-Net architecture – firstly, the number of channels in the output feature map is the same as the input, thus deviating from its intended usage for semantic segmentation; and secondly, we make use of a reconstruction loss between the output feature map and the input image, as given in Eq. 1. Therefore, our pretraining step is entirely unsupervised and does not require any labels such as the MIDI sequence.

$$\text{Reconstruction Loss} = \sum_{i=1}^{n} (Z_i - M_i)^2 \qquad (1)$$

This pretrained U-Net is different from an Autoencoder because unlike an Autoencoder where the bottle-necked representation is used for downstream tasks, we make use of the output feature space for the downstream transcription task. Furthermore, we maintain the contracting and expansive paths as well as skip connections as that of a U-Net, which are not present in a standard Autoencoder.

4.2   Bottled Sequence Traversal Unit

The BST comprises four composite LSTM units - each of them taking in the output of the U-Net as the input. However, each LSTM unit processes the sequence differently. As it can be seen in Fig. 2, each LSTM unit processes its input either in the forward direction or the reverse direction. We permute all such combinations of two LSTMs together, resulting in four permutations. The final output of the BST is the average of the output representations of these four LSTM permutation units. Thus inherently making the model robust to a variety of samples, without actually requiring these samples.

4.3   Multi-task Autoencoder Unit

The final unit of our model is a **Multi-task Autoencoder Unit**. The architecture of an Autoencoder is exactly the same as that of the U-Net, however, an Autoencoder does not have skip connections. However, we branch out two Autoencoders, both which take in the output provided by the BST as input. One of the Autoencoders, called the **Frame-set Auto-Encoder Unit** is used to predict the frame-sets, i.e., the Beatbox instrument corresponding to every time-frame. The output of this Autoencoder is then compared with the ground-truth frame-set matrix, with the help of a Cross-entropy Loss.

The second autoencoder, the *onset autoencoder unit*, is used to predict the onsets, i.e., the start times of every note. The onset prediction is carried out as follows - for each frame, a binary prediction is made, which characterises whether an onset occurs in the given time-frame, which is then used to compare with the ground truth onset matrix using the binary cross-entropy loss.

Thus, our model learns both the onsets and the frame-sets in a multi-task manner, by sharing the Pretrained U-Net and the BST to learn generic sequence-level information. The Multi-task autoencoder unit helps in learning frame-set and onset level information separately, which are then trained in an end-to-end fashion.

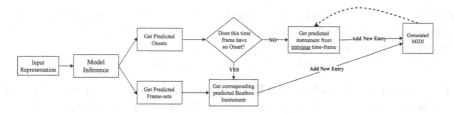

Fig. 3. The flowchart detailing the process of generating the MIDI transcription for an input representation using BaDumTss. Based on the predicted onsets, the inference pipeline decides whether a new Beatbox instrument is played, or the previous instrument is held, for the given time-frame.

Table 2. Performance of BaDumTss and two baselines – **AB** and **NE2E**.

Frame-level metrics	AB						NE2E		
	XGB	ADB	RF	SVM	LR	MLP	PRESTK	RNNDRUM	BaDumTss
Frame-set F1	0.21	0.11	0.14	0.27	0.41	0.38	0.63	0.62	**0.97**
Frame-set accuracy	0.43	0.31	0.37	0.40	0.57	0.40	0.59	0.57	**0.98**
Onset F1	-	-	-	-	-	-	0.08	0.23	**0.98**

4.4   Inference-Time MIDI Generation

After BaDumTss has been trained end-to-end, we can generate the transcribed MIDI in the following manner:

1. We convert the test audio into a melspectrogram and pass it through the model. We receive two outputs from the model - a frame-set matrix and an onset matrix.
2. Given the onset matrix, for every onset in a time frame, we search for the corresponding time-frame in the frame-set matrix and extract the predicted Beatbox instrument. For any frame in the onset matrix where there are no onsets found, we assume that the instrument is 'held', and thus, the Beatbox instrument in the previous time-frame is used.
3. Thus, we now have time-dependent Beatbox instrument predicted, that can be easily converted into the MIDI format.

Here, we note that the onsets prove to be very important, as they help us determine when an instrument is being played newly, or carried over from the previous time-frame. This technique is often used in polyphonic transcription tasks, following [5]. Figure 3 details the process with the help of a flowchart.

5   Experimental Results

In our study, we focus only on frame-level and onset-level results, and not note-level results. This is primarily because note-level results in a monophonic audio is equivalent to developing an instrument classification pipeline for all the samples present in the sequence; whereas, frame-level and onset-level predictions directly address the transcription task and is more like a localisation task - we need to predict *which sample* occurs at *what time*.

We compare our method with two distinct categories of baselines –

1. **Algorithmic Baselines (AB):** Certain algorithms do not explicitly predict onsets; rather, it requires an algorithmic onset detection, followed by frame-set prediction that is learnt by a model. Therefore, we tweak such algorithms for a monophonic usecase - we begin by identifying onsets (the time at which a sample starts playing) of each sample by first calculating the onset strength envelope (a technique which determines where possible onsets occur) and then further, selecting the peaks from this envelope. The audio is then chunked on the basis of these detected onsets, which is then fed into a classification pipeline, used for instrument classification. These two-component baselines are then assembled together to create a transcription pipeline for monophonic Beatbox sequences. For these baselines, since onset detection is done algorithmically and not learnt, onset-level F1 scores are not available.

   For these baselines, we make use of the Librosa [12] implementation for onset detection, followed by 7 state-of-the-art machine learning algorithms - XGBoost (**XGB**), AdaBoost (**ADB**), Random Forests (**RF**), Support Vector Machines (**SVM**), Logistic Regression (**LR**), Multi-layer perceptron (**MLP**) and 2-Nearest Neighbours (**KNN**).

2. **Neural End-to-End Baseline (NE2E):** For these baselines, we focus on those methods which are end-to-end – they do the onset as well as frame-set prediction, together, and produce the final MIDI matrix during inference. The baselines that we use are:

   (a) Pre-stacked U-Net for Transcription (**PRESTK**) [14] - This model makes use of a pre-stacked U-Net followed by CNNs for the task of polyphonic transcription.

   (b) RNN-based drum transcription (**RNNDRUM**) [21] - This model makes use of an RNN-based architecture for the task of drum transcription.

Table 2 shows the performance of the competing methods in terms of Frame-set accuracy and F1 score, as well as Onset F1 scores, for those cases where it can be calculated. As we can observe, the algorithmic baselines fail to perform competitively, when placed in comparison to neural end-to-end methods. This is because for the latter, the onset prediction and instrument classification go hand-in-hand with the end-to-end transcription process, as it has been shown earlier in [5] for polyphonic transcription. BaDumTss beats the best baseline (PRESTK), adding large relative improvements of 56% and 326% in Frame-set and onset-level F1 scores.

Ablation Study. Components of our architecture and conduct an ablation study with different components separately. The architectures used in the ablation study are shown below:

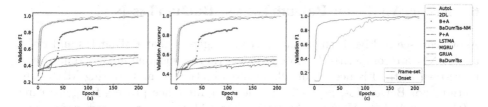

Fig. 4. Ablation study and analysis: we observe that our proposed model not only performs better, but also converges faster, when compared to other models used in the ablation study. (a) and (b) demonstrate the Validation F1 and Accuracy scores respectively with respect to training epochs. (c) demonstrates the Validation F1 scores of Frame-set and Onset prediction for BaDumTss.

1. Autoencoder + Linear (**AutoL**): This makes use of the Autoencoder unit followed by a linear layer.
2. BST + Autoencoder Unit (**B+A**): This architecture is the same as our proposed model, minus the pretraining unit.
3. Pretraining Unit + Autoencoder Unit (**P+A**): This architecture is the same as our proposed model, minus the BST.
4. BaDumTss with GRU (**MGRU**): This is the same as our proposed model, but we replace LSTMs in the BST with GRUs.
5. Single forward LSTM + Autoencoder Unit (**LSTMA**: Here, instead of using our LSTM permutations as described in the BST, we make use of a single LSTM, followed by an autoencoder Unit.
6. Single forward GRU + Autoencoder Unit (**GRUA**): Same as above, but with a single forward GRU rather than a single forward LSTM
7. 2D Convolutions + Linear Layers (**2DL**): We tweak the CNNs being used so far, and instead make use of a series of 2D CNNs, followed by a linear layer.
8. BaDumTsswith only one autoencoder Unit (**BaDumTss-NM**): This is the non-multitask version of BaDumTss, wherein, we only make use of frame-set level predictions for final MIDI generation. This does not contain an onset autoencoder Unit.

Figure 4 shows the validation F1 and accuracy scores with respect to the training epochs. We can clearly notice that BaDumTss not only converges faster as compared to other configurations of its units, but also performs better by a large margin. Another model that closely reaches the same F1 and accuracy is MGRU and BaDumTss-NM, which is the same architecture as that of BaDumTss; however, the LSTM units are replaced by that of GRU in the BST and BaDumTss-NM only predicts the frame-set. However, BaDumTss converges faster than MGRU. Even though BaDumTss-NM seems to perform competitively with BaDumTss, it encounters severe issues during inference time, which are discussed in Sect. 6.

6   Error Analysis: **BaDumTss** vs. **BaDumTss-NM**

As we can see in Fig. 4, the ablation study suggests that BaDumTss and BaD-umTss-NM are competitively close to each other. Thus, one might argue why a multi-task method is needed at all. This distinction is made during MIDI generation. For BaDumTss, Sect. 4.4 detailed our inference process for generating MIDI transcriptions. However, for BaDumTss-NM, there are no predicted onsets. Therefore, the predicted frame-sets are directly converted to MIDI. This process has a caveat. It is observed that for sequences longer than 5 seconds, even though BaDumTss-NM is able to correctly predict frames, their placement throughout the sequence duration is haphazard. Most samples are off by a couple of frames, thereby rendering the generated MIDIs useless, as it is not a direct transcription of the original audio. BaDumTss on the other hand, is able to accurately predict onsets and frames at the same time, thus, ensuring that the frames are *correctly* placed throughout the sequence duration. This observation validates the usage of a multi-task model for jointly predicting onsets and frames at the same time.

Another important observation is found in the Validation curves for Onset and Frame-set prediction in Fig. 4. One can notice that Onset prediction takes a *significant* amount of time to reach the same level of performance as frame-set prediction. This can sometimes hamper the transcription performance, if the training is stopped midway. Thus, even though BaDumTss is better at the overall transcription task than BaDumTss-NM, it loses out on the training speed, as onset predictions take time to reach the ideal performance. This can be improved by employing pretraining strategies for learning onset behaviour, that can save time while training the actual transcription model. We conclude that this is out of scope for our current work.

7   Conclusion

In this work, we firstly developed a novel dataset consisting of Beatbox sequences, along with their corresponding transcribed MIDI sequences as ground-truth. We attempted to capture biometric differences and natural noise by modifying recorded Beatbox samples with semi-synthetic augmentations. On this collected dataset, we also proposed BaDumTss, a new method that is demonstratively robust to new Beatbox sequences, as well as agnostic to time and amplitude-based variations. It shows a 56% and 326% relative improvement on frame-set and onset-level F1 scores over the best performing baseline. For future work, we aim to further explore transformer-based pretraining strategies that would also help us learn coherent musical structure present in musical sequences.

Acknowledgement. The authors would like to acknowledge the support of the Ramanujan Fellowship (SERB, India), Infosys Centre for AI (CAI) at IIIT-Delhi, and ihub-Anubhuti-iiitd Foundation set up under the NM-ICPS scheme of the Department of Science and Technology, India.

References

1. Cazau, D., Wang, Y., Adam, O., Wang, Q., Nuel, G.: Improving note segmentation in automatic piano music transcription systems with a two-state pitch-wise HMM method. In: ISMIR (2017)
2. Ishizuka, R., Nishikimi, R., Nakamura, E., Yoshii, K.: Tatum-level drum transcription based on a convolutional recurrent neural network with language model-based regularized training (2020)
3. Choi, K., Cho, K.: Deep unsupervised drum transcription (2019)
4. Southall, C., Stables, R., Hockman, S.: Automatic drum transcription for polyphonic recordings using soft attention mechanisms and convolutional neural networks. In: Proceedings of the 18th International Society for Music Information Retrieval Conference (2017). https://doi.org/10.5281/zenodo.1415616
5. Hawthorne, C., et al.: Onsets and frames: dual-objective piano transcription (2018)
6. Wang, Y., Salamon, J., Cartwright, M., Bryan, N.J., Pablo Bello, J.: Few-shot drum transcription in polyphonic music. In: Proceedings of the 21th International Society for Music Information Retrieval Conference, ISMIR 2020, Montreal, Canada, 11–16 October 2020
7. Callender, L., Hawthorne, C., Engel, J.: Improving perceptual quality of drum transcription with the expanded groove MIDI Dataset (2020)
8. Hawthorne, C., et al.: Enabling factorized piano music modeling and generation with the MAESTRO dataset. In: International Conference on Learning Representations (2019)
9. Sinyor, E., McKay, C., Fiebrink, R., McEnnis, D., Fujinaga, F.: Beatbox classification using ACE. In: ISMIR (2005)
10. Picart, B., Brognaux, B., Dupont, S.: Analysis and automatic recognition of Human BeatBox sounds: a comparative study. In: 2015 IEEE International Conference on Acoustics, Speech and Signal Processing (ICASSP)
11. Evain, S., et al.: Beatbox sounds recognition using a speech-dedicated HMM-GMM based system. In: Models and Analisys of Vocal Emission for Biomedical Applications Firenze, Italy (2019)
12. librosa/librosa: 0.8.0. https://doi.org/10.5281/zenodo.3955228
13. Weng, W., et al.: U-Net: convolutional networks for biomedical image segmentation. IEEE Access 9, 16591–16603 (2015)
14. Pedersoli, F., Tzanetakis, G., Yi, K.M.: Improving music transcription by pre-stacking AU-Net. In: ICASSP 2020–2020 IEEE International Conference on Acoustics, Speech and Signal Processing (ICASSP)
15. Cartwright, M., Bello, J.P.: Increasing drum transcription vocabulary using data synthesis. In: Proceedings of the International Conference on Digital Audio Effects (DAFx) (2018)
16. Cheuk, K.W., Agres, K., Herremans, D.: The impact of Audio input representations on neural network based music transcription. In: 2020 International Joint Conference on Neural Networks (IJCNN) (2020)
17. Kong, Q., Li, B., Song, X., Wan, Y., Wang, Y.: High-resolution Piano transcription with pedals by regressing onsets and offsets times (2020)
18. Jacques, C., Roebel, A.: Data augmentation for drum transcription with convolutional neural networks. In: 2019 27th European Signal Processing Conference (EUSIPCO) (2019)
19. Delgado, A., McDonald, S., Xu, N., Sandler, M.: A new dataset for amateur vocal percussion analysis. In: Proceedings of the 14th International Audio Mostly Conference: A Journey in Sound (2019)

20. Poliner, G.E., Ellis, D.P.W.: A Discriminative Model for Polyphonic Piano Transcription. EURASIP J. Adv. Signal Process. **2007**(1), 1–9 (2007). https://doi.org/10.1155/2007/48317
21. Poliner, G.E., Ellis, D.P.W.: A Discriminative Model for Polyphonic Piano Transcription. EURASIP Journal on Advances in Signal Processing **2007**(1), 1–9 (2007). https://doi.org/10.1155/2007/48317
22. Vogl, R., Dorfer, M., Knees,P.: Drum transcription from polyphonic music with recurrent neural networks. In: 2017 IEEE International Conference on Acoustics, Speech and Signal Processing (ICASSP) (2017). https://doi.org/10.1109/ICASSP.2017.7952146
23. Vogl, R., Widmer, G., Knees, P.: Towards multi-instrument rum transcription. In: Proceedings of the 21st International Conference on Digital Audio Effects (DAFx 2018), 4–8 September 2018, Aveiro, Portugal (2018)
24. Gillet, O., Richard, G.: ENST-drums: an extensive audio-visual database for drum signals processing. In: Proceedings of the 7th International Conference on Music Information Retrieval (2006). https://doi.org/10.5281/zenodo.1415902
25. Emiya, V., Bertin, N., David, B., Badeau, R.: MAPS - a piano database for multipitch estimation and automatic transcription of music. Res. Rep. **11.**, 00544155 (2010)

Detail Perception Network for Semantic Segmentation in Water Scenes

Cuixiao Liang[1], Wenjie Cai[2], Shaowu Peng[1], and Qiong Liu[1(✉)]

[1] South China University of Technology, Guangzhou, China
liuqiong@scut.edu.cn
[2] Shunfeng Technology (Shenzhen) Co., Ltd., Shenzhen, China

Abstract. Semantic segmentation in the water scene is significant for water environment monitoring. Recent water scene segmentation methods usually regard all floating objects as only one foreground category that limits the understanding of the water scene. Considering various floating objects, we propose a Detail Perception Network (DPNet) to address two challenges for water scene segmentation with multi-categories floating objects. One is the sample imbalance among objects of different scales, which leads to low accuracy on small objects covering a few pixel samples. Another is the weak discriminability of features among the categories that are close to the blurred edge, which leads to the miss-segmentation in the blurred edge region. For sample imbalance, we design Distance Field Loss (DF Loss) to strengthen the learning of small objects by a pixel-wise weight calculated from a distance field during training. To address the weak discriminability of features among categories that are close to the blurred edge, we propose a Category Edge Perception Pyramid (CEPP) module that learns the edge feature of each category as prior knowledge to enhance edge features. For training and evaluating relative models, we also establish a dataset named ColorWater, which contains 1279 images with 9 semantic labels over various water scenes. Extensive experiments demonstrate that our model performs favorably against the state-of-the-art models on our ColorWater dataset and the public Aeroscapes dataset.

Keywords: Semantic segmentation · Data imbalance · Feature enhancement

1 Introduction

Water environment monitoring is very significant to the sustainable development of the water ecological environment. Compared with image classification and object detection, semantic segmentation in the water scene is important for water environment monitoring as it can finely describe scenes.

However, recent water scene segmentation methods usually regard all floating objects as only one foreground category, which is too rough for comprehensively

© The Author(s), under exclusive license to Springer Nature Switzerland AG 2022
J. Gama et al. (Eds.): PAKDD 2022, LNAI 13282, pp. 187–199, 2022.
https://doi.org/10.1007/978-3-031-05981-0_15

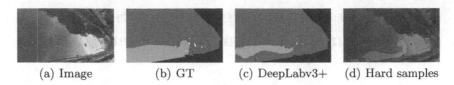

(a) Image   (b) GT   (c) DeepLabv3+   (d) Hard samples

Fig. 1. Examples in ColorWater, including image, the ground truth, the predicted mask of DeepLabv3+ [2], and the hard sample distribution. Gradient norm [13] is used to measure the difficulty of sample recognition by the model. The larger the value is, the more difficult the model is to identify the sample. The red area in the last column is the sample with the gradient norm larger than 0.8, which is considered as the hard sample.

understanding the water scene. These methods mainly focus on specific applications, such as salient foreground segmentation [23], water body segmentation [6], oil spill monitoring [4], and obstacles segmentation [1,3]. This is due to the lack of a water scene dataset with annotation of various floating objects to provide data for training and benchmark evaluating of water scene segmentation models with multi-categories floating objects. To address this, we establish a dataset named ColorWater, which contains 1279 images with pixel-wise semantic annotation for 9 categories. Compared with existing datasets for water scenes analysis, ColorWater is the first to provide semantic labels for multiple floating objects, which can accurately describe quantitative information of objects such as area, shape, and perimeter of objects, providing richer information for water scenes analysis.

As shown in Fig. 1, water scene segmentation with multi-categories of floating objects has its own challenges. The first challenge is the sample imbalance among objects of different scales. As the pixel number of small objects is much smaller than that of large objects, the influence of small objects on loss is too weak for model to learn. The second challenge is the weak discriminability of features among categories that are close to the blurred edge, which leads to the miss-segmentation in the blurred edge region. Salient edge features is beneficial to improve the discriminability of features. However, it is difficult to extract salient edge features from the blurred edge region without prior knowledge. Meanwhile, downsampling operation in network further increases the difficulty.

To address these challenges, we propose Detail Perception Network (DPNet). (1) For the imbalance problem, we consider that the distance field [5] (without direction), which represents the nearest distance for each pixel to the object's boundary, could appropriately represent the small objects. As pixels in small objects are much closer to the object's boundary than that in large objects, the smaller the value of the distance field is, the greater the probability that the pixel is located in the small objects. Based on the distance field we propose Distance Field Loss (DF Loss) to assign weight to each pixel and avoid learning invalid noise samples. The pixel-wise weight increases with the decrease of the distance field value and vice versa. (2) To address the weak discriminability of features among categories that are close to the blurred edge, we propose a Category Edge Perception Pyramid (CEPP) module that learns the edge feature of each category as prior knowledge to enhance edge features. In CEPP, an auxiliary task is used to learn multi-level category edge features by multi-categories edge detection. A Category Edge Perception Attention (CEPA) module is used to fuse the category edge features and the semantic features.

The contributions of this paper are as follows. (1) We firstly build a ColorWater dataset, which contains 1279 images with pixel-wise semantic annotation for 9 categories. (2) We propose DPNet, in which DF loss is used to solve the imbalance problem and CEPP is used to improve the discriminability of features among categories that are close to the blurred edge. (3) Extensive experiments show that our DPNet can effectively improve the segmentation performance and achieve state-of-the-art results on our ColorWater dataset and a public Aeroscapes [17] dataset.

2   Related Work

Existing water scenes semantic segmentation networks are mainly applied to specific applications with shallow understanding, such as salient foreground segmentation [23], water body segmentation [6], oil spill monitoring [4], and obstacles segmentation [1,3]. Therefore, semantic segmentation of water scenes containing multi-categories of floating objects for water environmental assessment is rarely explored in depth at present. As water scene segmentation faces the challenges of imbalanced learning and the weak discriminability of features among categories that are close to the blurred edge, we will introduce related works that focus on these challenges.

For the imbalanced problem, a large number of researches proposed weighted balanced loss to balance the influence of different samples. The difference between these methods is the way to calculate the weighted matrix. Some studies [7,10,20] assign weights of categories according to the sample statistics of categories. But they ignore the sample imbalance problem in different instances that are in the same category. Focal Loss [16] and GHM-C Loss [13] respectively use the confidence and gradient predicted by the model to calculate an instance-wise weighted matrix that assigns different weights to instances. However, as they focus on learning hard instances, it is easy to learn invalid noise samples, especially in small-scale datasets. In this paper, we use Distance Field Loss to calculate pixel-wise weighted matrix by distance field, which can correctly represent the small object and avoid learning to invalid noise samples.

In semantic segmentation, most methods [18,22] fuse multi-level representations to restore detailed features, including edge feature, to generate better segmentation results. In addition, several methods [14,19] use normal binary edge representations as prior knowledge to constrain the learning process and improve the segmentation performance. However, their knowledge exploration in edge representation is still shallow. Lee et al. [12] propose an edge key point selection algorithm to encode shape knowledge which is used to learn structure information indicated by experts and preserve edge structure in the intermediate representation. However, shape knowledge encoded by an edge key point is hard to fit the complex shape. Different from the above methods, this paper learns and explores multi-level edge features of each category to enhance edge features.

3   Distance Field Loss for Imbalance Problem

Distance field loss is used to solve the imbalance problem of the segmentation task. It is a cross entropy loss with a pixel-wise weighted matrix $\hat{d} \in \mathbb{R}^{W \times H}$, where the weight is calculated from the distance field $df \in \mathbb{R}^{W \times H}$, as shown below:

$$L_{df}\left(\hat{d}, p, y\right) = \frac{1}{N} \sum\nolimits_{j=1}^{N} \left(\hat{d}_j L_{ce}\left(p_j, y_j\right)\right) \tag{1}$$

where N is the total number of pixel samples in the image, and j is the index of the pixel of the image. $p \in \mathbb{R}^{C \times W \times H}$ represents the output ranking score map and $y \in \mathbb{R}^{C \times W \times H}$ represents the corresponding non-overlap semantic label, where C refers to the number of categories.

Distance field $df \in \mathbb{R}^{W \times H}$ represents the nearest distance for each pixel to the object's boundary. Specifically, the pixels of small objects are close to boundary. Thus, the smaller the value of df_j is, the greater the probability that the pixel j is located in the small object. To improve the ability of the network to learn small objects, it is necessary to increase their weight in the loss function. We use an inverse function to calculate the weighted matrix $\hat{d}$. The weight $\hat{d}_j$ of the pixel j is calculated as follows:

$$\hat{d}_j = \log \frac{max(df)}{df_j + \alpha} \tag{2}$$

where α is set to 0.0001 to avoid the 0 denominator. To make the model easier to converge, we use a logarithmic function to compress the difference between weights to avoid excessive attention to difficult samples. In addition, df_j is calculated by the ground truth $y \in \mathbb{R}^{C \times W \times H}$ of the image as follows:

$$df_j = \min_{1 \leq c \leq C}(df_j^c)$$
$$= \min_{1 \leq c \leq C}\left(\begin{cases} |j, q|, & y_j^c \in 1 \\ \infty, & y_j^c \in 0 \end{cases}\right) \tag{3}$$

where $c \in \{1, 2, \cdots, C\}$ is the c-th category. df^c is the distance field matrix of c-th category. As shown in Eq. 3, df^c is generated from the c-th channel of the ground truth $y \in \mathbb{R}^{C \times W \times H}$. For each foreground pixel in y^c, df_j^c is set to the distance $|j, q|$ between j and q, where q is the nearest pixel of object's boundary from pixel j. For each background pixel in y^c, we set df_j^c as infinity.

4   Detail Perception Network

In this section, we introduce the proposed Detail Perception Network (DPNet) in detail. First, we describe the overall structure of DPNet in Subsect. 4.1. Then we introduce the proposed Category Edge Perception Attention (CEPA) module in Subsect. 4.2. Finally, we present the loss function used to supervise the proposed network in Subsect. 4.3.

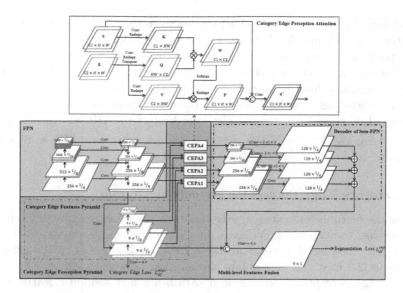

Fig. 2. The model structure of the Detail Perception Network. The numeric expression on the quadrilateral represents the size of the features, whose unified format is $C \times R$. C refers to the channel number while R refers to the scaling ratio of the features relative to the original image size. The format like $(Conv \rightarrow T\times) \times N$ refers to the sequence, consisting of a convolution operation and a T times up-sampling operation, is iterated N times.

4.1 Model Structure

The overall structure of DPNet is shown in Fig. 2. The network can be divided into three parts: FPN [15], Category Edge Perception Pyramid (CEPP), and Multi-level Features Fusion (MFF).

In Category Edge Perception Pyramid (CEPP), a Category Edge Feature Pyramid (CEFP) is used to explore multi-level edge representations of each category. Moreover, Multiple Category Edge Perception Attention (CEPA) modules are used to reduce the semantic gap between CEFP and the Semantic Feature Pyramid (SFP) generated by FPN. As shown in Fig. 2, the highest-resolution feature in the SFP is first fed to a convolution layer to generate a high-resolution category edge feature that is supervised by category edge loss L_{df}^{edge}. The CEFP is then generated by iterative downsampling from the high-resolution category edge feature. Finally, multiple CEPA modules fuse CEFP into SFP to generate a new SFP with category edge representations.

Multi-level Feature Fusion (MMF) is used to restore the resolution of semantic features accordingly. As shown in Fig. 2, the decoder of Sem-FPN is utilized to fuse the new SFP and generate high-resolution semantic features firstly. Then the high-resolution and complete category edge feature is cascaded to the high-resolution semantic feature to enhance the constraint of category edge features to semantic features. The final semantic feature is supervised by semantic segmentation loss L_{df}^{seg}. In summary, the final loss of the DPNet consists of the category edge loss L_{df}^{edge} and the semantic segmentation loss L_{df}^{seg} as described in Subsect. 4.3.

4.2   Category Edge Perception Attention Module

As semantic features and category edge features are supervised by different target functions, they have a large semantic gap, which makes the model difficult to converge. To solve this problem, the CEPA module refers to the attention mechanism [21] that can (1) explore the relation between features, (2) extract category edge perception features, and (3) obtain new semantic features with category edge features.

As shown in the green box in the upper part of Fig. 2, the semantic feature $S \in \mathbb{R}^{C1 \times W \times H}$ and the category edge feature $E \in \mathbb{R}^{C2 \times W \times H}$ of each layer are fed into CEPA and three transformed features $K \in \mathbb{R}^{C1 \times WH}, Q \in \mathbb{R}^{WH \times C2}, V \in \mathbb{R}^{C2 \times WH}$ are generated with reference to the attention mechanism. The feature $W \in \mathbb{R}^{C1 \times C2}$ explores the relationship between the transformed semantic feature K and the transformed category edge feature Q by matrix multiplication. Then, another transformed category edge feature V is mapped into the feature space of the semantic feature to generate the category edge perception feature $P \in \mathbb{R}^{C1 \times W \times H}$, according to the relationship feature W. Finally, the output feature $S' \in \mathbb{R}^{C1 \times W \times H}$ of CEPA, is transformed from P by the sequence of $Concatenation \rightarrow Conv1 \times 1 \rightarrow BN \rightarrow ReLU$.

4.3   Total Loss of DPNet

During training, DPNet jointly supervises a multi-categories segmentation task and a multi-categories edge detection task, which have the same category. In order to improve the ability of the model to learn detailed features, we use DF loss both in the segmentation task and the edge detection task. The multi-categories edge detection task is decoupled to multiple binary classification tasks to explore the edge representation of each category. The total loss L is calculated as follows:

$$L = L_{df}^{edge}(\hat{d}, E, \hat{e}) + L_{df}^{seg}(\hat{d}, O, \hat{s})$$
$$= L_{df}(\hat{d}, E, \hat{e}) + L_{df}(\hat{d}, O, \hat{s}) \tag{4}$$

where $L_{df}(\bullet)$ is calculated by Eq. 1. According to the GT of image's semantic mask $\hat{s} \in \mathbb{R}^{C \times W \times H}$, the pixel-wise weighted matrix $\hat{d}$, calculated by Eqs. 2–3, is not only used in the semantic segmentation loss L_{df}^{seg} but also used in category edge loss L_{df}^{edge}. In the category edge loss, $E \in \mathbb{R}^{C \times W \times H}$ represents the multi-category edge detection task's ranking score map that is transformed from the highest-resolution category edge feature in CEFP. The set of the GT of binary edge detection masks $\hat{e} \in \mathbb{R}^{C \times W \times H}$ is generated by $\hat{s}$. In the semantic segmentation loss, $O \in \mathbb{R}^{C \times W \times H}$ is the ranking score map of multi-categories segmentation task in the MFF module.

5 Experiments

5.1 Datasets

We evaluate our method on two datasets, including our proposed ColorWater dataset and the public Aeroscapes [17] dataset, which both have a large number of small objects and objects with blurred edges.

ColorWater is a segmentation dataset that contains 1279 water scene images with pixel-wise semantic annotation for 9 categories. Object categories include plant debris, duckweeds, domestic waste, sewage, and hybrid, which are primary pollution sources influencing the water environment. Besides, normal non-pollution object categories include background, water, navigation equipment, and landscapes to better distinguish pollution sources and non-pollution sources. The training set, validation set, and test set contain 1029, 137, and 113 images respectively.

Aeroscapes [17] dataset contains 3269 urban scenes images with pixel-wise semantic annotation for 12 categories, including background, person, bikes, cars, drones, boats, animals, obstacles, construction, vegetation, roads, and sky. The training set and validation set contain 2621 and 648 images respectively.

5.2 Ablation Study

Ablation of DPNet. To verify the impact of each key part of DPNet, we conducted ablation experiments for CEPP, CEF-C, and DF Loss respectively, as shown in Table 1.

Table 1. Ablation study of three key parts in DPNet

Method	Backbone	CEPP	CEF-C	DF loss	mIoU	mAcc
Sem-FPN [11]	Resnet50	-	-	-	71.38	83.21
DPNet	Resnet50	✓	-	-	71.53	84.64
DPNet	Resnet50	✓	✓	-	72.86	82.90
DPNet	Resnet50	-	-	✓	73.00	82.94
DPNet	Resnet50	✓	✓	✓	**73.97**	**85.08**

CEPP extracts CEFP and fuses CEFP to SFP through CEPA. Compared with the baseline, the mAcc of the model contained with CEPP is improved by 1.43, indicating that CEPP can effectively improve the classification ability of the model. However, because category edge features will be weakened in the fusion process, the prior knowledge of category edge has limited constraints on the model, and mIoU is not significantly improved.

CEF-C is one of the operations in MFF, which is used to cascade the complete category edge features to semantic features. When incorporating CEF-C, the mIoU of the model is increased by 1.48. This operation can retain the complete category edge features and effectively improve the segmentation performance of the model under the constraints of these features.

DF Loss can replace the cross entropy loss and allow the model to focus on the learning of small objects, which can better supervise the learning of semantic segmentation tasks. In Table 1, the mIoU of the model was significantly improved from 71.38 to 73.00 after using DF Loss alone. It can be seen that DF Loss can effectively improve the segmentation ability of the model. Under the combined effect of CEPP, CEF-C, and DF Loss, the model's mIoU and mAcc reached 73.97 and 85.08, which increased 2.59 and 1.87 respectively compared with the baseline Sem-FPN [11]. Obviously, DPNet can effectively improve the segmentation and classification performance and is suitable for datasets containing many small objects and blurred edge objects, like ColorWater.

DPNet with the above three key parts can effectively enhance the detailed features representations and improve the segmentation performance of small objects and blurred edge regions. Specifically, DPNet can reduce the misclassification of the blurred edge region, thus the segmentation of the blurred edge region is more precise than Sem-FPN [11], as shown in the red rectangle in the first row of Fig. 3. In addition, DPNet can recognize more small objects than Sem-FPN, as shown in the red rectangle in the last row of Fig. 3.

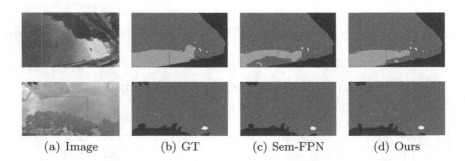

(a) Image   (b) GT   (c) Sem-FPN   (d) Ours

Fig. 3. Examples of semantic segmentation results on ColorWater

Ablation of DF Loss. DF Loss is portable and can be easily migrated to other networks to effectively improve performance. As shown in Table 2, several methods combined with DF Loss can improve the model mIoU when tested on the ColorWater validation set. In addition, we compare DF Loss with several loss functions [13, 20] that deal with data imbalance learning problems. As shown in Table 3, we train Sem-FPN with 4 different loss function on ColorWater and Aeroscapes [17] dataset. Sem-FPN with DF Loss gets the highest mIoU both in ColorWater and Aeroscapes as it can accurately improve the ability of the model to learn small objects. Sem-FPN with GHM-C loss [13] gets a low mIoU. The main reason is that GHM-C Loss is easy to learn noisy data that makes it difficult for the model trained with GHM-C loss to converge. Sem-FPN trained with recall loss [20] also has a low mIoU in ColorWater and Aeroscapes as it only calculates the weight of the category dimension that is too rough to attend to the region with few pixel samples, such as small objects.

Table 2. Comparison of several methods with and without using DF loss

Methods	DeepLabv3+ [2]	HRNet [18]	Sem-FPN [11]
w/o DF loss	70.20	71.43	71.38
w/ DF loss	**71.98**	**72.79**	**73.00**

Table 3. Comparison of several losses for imbalance problem

Datasets	CE	GHM-C loss [13]	Recall loss [20]	DF loss
ColorWater	71.38	67.07	66.30	**73.00**
Aeroscapes [17]	62.31	54.48	61.19	**63.36**

5.3 Comparison with State-of-the-Arts

We further report the comparison of performances of six state-of-the-art methods [2,8,9,11,18,21] on ColorWater in Table 4. In the validation set and test set of ColorWater, our DPNet achieves state-of-the-art performance. The mIoU and mAcc reached 73.97 and 85.08 in the validation set and 70.18 and 81.03 in the test set, respectively. For the mIoU of the validation set, our DPNet is 1.89 higher than APCNet, which has the highest mIoU among the other six methods. In Table 5, we compare the IoU of each category between APCNet [9] and DPNet, where DPNet outperforms APCNet [9] in most categories. DPNet significantly improves the model's ability to segment various floating objects in the water scene, especially small objects like plant debris, duckweed, domestic waste, and navigation equipment.

We also show the research gaps between small and big objects on 6 methods in Table 6. It should be noted that an object whose scale is bigger than the average scale is defined as a big object, otherwise as a small object. As the influence of small objects on the loss function is weakened by big objects, the learning of small objects is more difficult than that of big objects. Thus, the mIoU of small objects $mIoU_s$ is always lower than the mIoU of big objects $mIoU_b$. Comparing the performance of 6 methods for small objects, our method gets the highest

Table 4. Comparison of the performance of several methods on ColorWater

Methods	val		test	
	mIoU	mAcc	mIoU	mAcc
DMNet [8]	68.98	83.30	68.32	78.58
DeepLabv3+ [2]	70.20	79.80	66.25	78.61
Sem-FPN [11]	71.38	83.21	70.16	78.83
HRNet [18]	71.43	80.03	65.38	77.23
OCRNet [21]	71.98	83.80	68.27	77.41
APCNet [9]	72.08	81.44	66.85	76.81
DPNet (Ours)	**73.97**	**85.08**	**70.18**	**81.03**

Table 5. Comparison of the IoU of APCNet and DPNet on the ColorWater

Methods	bg	water	plant debris	duckweeds	domestic waste
APCNet [9]	96.74	94.54	48.75	46.73	50.83
DPNet (Ours)	**96.84**	**94.75**	**53.91**	**53.22**	**52.75**

Methods	sewage	navigation equipment	hybrid	landscapes	mIoU
APCNet [9]	**63.25**	84.54	70.58	92.72	72.08
DPNet (Ours)	62.06	**85.80**	**71.69**	**94.68**	**73.97**

Table 6. Quantification results of segmentation performance for different scale objects

Indexs*	Deeplabv3+ [2]	APCNet [9]	HRNet [18]	OCRNet [21]	Sem-FPN [11]	Ours
$mIoU_s$	55.79	57.49	57.68	61.32	63.54	**66.54**
$mIoU_b$	71.04	71.77	**72.40**	69.61	69.18	70.12

* $mIoU_s = \frac{TP_s}{TP_s+FP_s+FN_s}$, $mIoU_b = \frac{TP_b}{TP_b+FP_b+FN_b}$

Table 7. Comparison of the IoU of several methods on the Aeroscapes [17]

Methods	bg	person	bikes	cars	drones	boats	animals
Sem-FPN [11]	78.82	**51.89**	20.98	83.34	48.68	57.18	44.26
DeepLabv3+ [2]	79.73	51.13	20.48	82.97	51.72	**73.03**	37.54
APCNet [9]	81.15	50.85	24.20	83.58	**56.54**	71.59	44.17
DPNet (Ours)	**81.99**	49.91	**29.37**	**85.71**	52.57	71.33	**44.27**

Methods	obstacles	construction	vegetation	roads	sky	mIoU	mAcc
Sem-FPN [11]	15.70	73.04	93.06	90.67	90.08	62.31	71.01
DeepLabv3+ [2]	18.83	73.40	93.09	90.80	89.05	63.56	71.46
APCNet [9]	17.09	74.98	94.06	90.83	91.30	65.03	**73.87**
DPNet (Ours)	**19.95**	**75.46**	**96.61**	**92.82**	**92.88**	**65.82**	73.55

mIoU, which further verifies that DPNet significantly improves the ability of the model to learn small objects. Furthermore, the performances of DPNet for small and big objects are higher than our baseline Sem-FPN, showing that our proposed method can better balance the segmentation of multi-scale objects.

To verify the generalization ability of our method, we test the result of several methods on the validation set of the Aeroscapes [17] dataset. As shown in Table 7, DPNet achieves the best performance. The mIoU of our proposed method is 65.82, which is 3.51 higher than baseline Sem-FPN. In addition, our method is superior to other methods when segmenting the stuff of irregularly shaped such as roads, sky, and vegetation, as well as most small things such as bikes, animals, and obstacles.

6   Conclusion

In order to provide diverse and high-quality annotated data for water environment monitoring and analysis, this paper establishes the ColorWater dataset with multi-categories semantic labels and proposes DPNet for semantic segmentation in the water scene. The key innovations lie in the DF Loss and the learning of category edge features. We propose DF Loss to solve the sample imbalance among objects of different scales, which could precisely improve the ability of the model to learn small objects. We propose a Category Edge Perception Pyramid (CEPP) module to learn category edge features, which can be regarded as prior knowledge to improve the discriminability of features among categories that are close to the blurred edge. Extensive ablation experiments demonstrate the effectiveness of the proposed methods. Comparisons with state-of-the-art models show that DPNet achieves outstanding performance on our ColorWater dataset and a public Aeroscapes dataset. We hope these efforts could facilitate new advances in the field of semantic segmentation in the water scene. In the future work, we will publish and introduce our dataset in detail at https://github.com/L-cuixiao/ColorWater.

References

1. Bovcon, B., Kristan, M.: WaSR-a water segmentation and refinement maritime obstacle detection network. IEEE Trans. Cyberneti. 1–14 (2021)
2. Chen, L.C., Zhu, Y., Papandreou, G., Schroff, F., Adam, H.: Encoder-decoder with Atrous separable convolution for semantic image segmentation. In: Proceedings of the European Conference on Computer Vision (ECCV), pp. 801–818 (2018)
3. Chen, X., Liu, Y., Achuthan, K.: WODIS: water obstacle detection network based on image segmentation for autonomous surface vehicles in maritime environments. IEEE Trans. Instrum. Measur 70, 1–13 (2021)
4. Chen, Y., Li, Y., Wang, J.: An end-to-end oil-spill monitoring method for multi-sensory satellite images based on deep semantic segmentation. Sensors 20(3), 725 (2020)
5. Cheng, F., et al.: Learning directional feature maps for cardiac MRI segmentation. In: Martel, A.L., et al. (eds.) MICCAI 2020. LNCS, vol. 12264, pp. 108–117. Springer, Cham (2020). https://doi.org/10.1007/978-3-030-59719-1_11

6. Cheng, Y., Jiang, M., Zhu, J., Liu, Y.: Are we ready for unmanned surface vehicles in inland waterways? The USVInland multisensor dataset and benchmark. IEEE Robot. Autom. Lett. **6**(2), 3964–3970 (2021)
7. Cui, Y., Jia, M., Lin, T.Y., Song, Y., Belongie, S.: Class-balanced loss based on effective number of samples. In: Proceedings of the IEEE/CVF Conference on Computer Vision and Pattern Recognition, pp. 9268–9277 (2019)
8. He, J., Deng, Z., Qiao, Y.: Dynamic multi-scale filters for semantic segmentation. In: Proceedings of the IEEE/CVF International Conference on Computer Vision, pp. 3562–3572 (2019)
9. He, J., Deng, Z., Zhou, L., Wang, Y., Qiao, Y.: Adaptive pyramid context network for semantic segmentation. In: Proceedings of the IEEE/CVF Conference on Computer Vision and Pattern Recognition. pp. 7519–7528 (2019)
10. Huang, C., Li, Y., Loy, C.C., Tang, X.: Deep imbalanced learning for face recognition and attribute prediction. IEEE Trans. Pattern Anal. Mach. Intell. **42**(11), 2781–2794 (2019)
11. Kirillov, A., Girshick, R., He, K., Dollár, P.: Panoptic feature pyramid networks. In: Proceedings of the IEEE/CVF Conference on Computer Vision and Pattern Recognition, pp. 6399–6408 (2019)
12. Lee, H.J., Kim, J.U., Lee, S., Kim, H.G., Ro, Y.M.: Structure boundary preserving segmentation for medical image with ambiguous boundary. In: Proceedings of the IEEE/CVF Conference on Computer Vision and Pattern Recognition, pp. 4817–4826 (2020)
13. Li, B., Liu, Y., Wang, X.: Gradient harmonized single-stage detector. In: Proceedings of the AAAI Conference on Artificial Intelligence, vol. 33, pp. 8577–8584 (2019)
14. Li, X., et al.: Improving semantic segmentation via decoupled body and edge supervision. In: Vedaldi, A., Bischof, H., Brox, T., Frahm, J.-M. (eds.) ECCV 2020. LNCS, vol. 12362, pp. 435–452. Springer, Cham (2020). https://doi.org/10.1007/978-3-030-58520-4_26
15. Lin, T.Y., Dollár, P., Girshick, R., He, K., Hariharan, B., Belongie, S.: Feature pyramid networks for object detection. In: Proceedings of the IEEE Conference on Computer Vision and Pattern Recognition, pp. 2117–2125 (2017)
16. Lin, T.Y., Goyal, P., Girshick, R., He, K., Dollár, P.: Focal loss for dense object detection. In: Proceedings of the IEEE International Conference on Computer Vision, pp. 2980–2988 (2017)
17. Nigam, I., Huang, C., Ramanan, D.: Ensemble knowledge transfer for semantic segmentation. In: 2018 IEEE Winter Conference on Applications of Computer Vision (WACV), pp. 1499–1508. IEEE (2018)
18. Sun, K., Xiao, B., Liu, D., Wang, J.: Deep high-resolution representation learning for human pose estimation. In: Proceedings of the IEEE/CVF Conference on Computer Vision and Pattern Recognition, pp. 5693–5703 (2019)
19. Takikawa, T., Acuna, D., Jampani, V., Fidler, S.: Gated-SCNN: gated shape CNNs for semantic segmentation. In: Proceedings of the IEEE/CVF International Conference on Computer Vision, pp. 5229–5238 (2019)
20. Tian, J., Mithun, N.C., Seymour, Z., Chiu, H.P., Kira, Z.: Striking the right balance: recall loss for semantic segmentation. CoRR abs/2106.14917 (2021). https://arxiv.org/abs/2106.14917
21. Yuan, Y., Chen, X., Wang, J.: Object-contextual representations for semantic segmentation. In: Vedaldi, A., Bischof, H., Brox, T., Frahm, J.-M. (eds.) ECCV 2020. LNCS, vol. 12351, pp. 173–190. Springer, Cham (2020). https://doi.org/10.1007/978-3-030-58539-6_11

22. Zhang, Z., Zhang, X., Peng, C., Xue, X., Sun, J.: ExFuse: enhancing feature fusion for semantic segmentation. In: Proceedings of the European Conference on Computer Vision (ECCV), pp. 269–284 (2018)
23. Štricelj, A., Kačič, Z.: Detection of objects on waters' surfaces using CEIEMV method. Comput. Electr. Eng. **46**, 511–527 (2015)

Learning Discriminative Representation Base on Attention for Uplift

Guoqiang Xu[1], Cunxiang Yin[1], Yuchen Zhang[1(✉)], Yuncong Li[1], Yancheng He[1], Jing Cai[1], and Zhongyu Wei[2]

[1] Tencent Inc., Shenzhen, China
{chybotxu,jasonyin,ericyczhang,yuncongli,collinhe,samscai}@tencent.com
[2] School of Data Science, Fudan University, Shanghai, China
zywei@fudan.edu.cn

Abstract. Uplift modeling aims to estimate Conditional Average Treatment Effects (CATE) for a given factor, such as a marketing intervention or a medical treatment. Given covariates of a single subject under different treatment indicators, most of existing approaches, especially those based on deep learning, either learn the same representation from a single model or learn different representations from two separate models. Thus, these methods could not learn discriminative representations or could not utilize both information of control and treatment group. In this paper, we develop an attentive neural uplift model to alleviate the above shortcomings by utilizing attention mechanisms to map the original covariate space $\mathcal{X}$ into a latent space $\mathcal{Z}$ in a single model. Given covariates of a subject, the learned representations in space $\mathcal{Z}$ which we called *after-treatment representation* are discriminative under different treatment indicators, thus can model potential outcomes more effectively. Moreover, the model is trained on a single neural network so that the information shared by treatment and control group is utilised. Experiments on synthetic and real-world datasets show our proposed method is competitive with the state-of-the-art.

Keywords: Uplift model · Causal inference · Attention mechanism

1 Introduction

Uplift models refer to techniques that model the incremental impact of treatment on an individual's behaviour. For example, in the domain of marketing, uplift models produce for each customer an uplift score that indicates how susceptible (or persuadable) they are by the marketing interventions. This score is then used to rank customers in the order to maximize gains and reduce costs [6]. The main difficulty to estimate uplift score is that it is not directly measurable: for a single subject, we can only observe the outcome of the subject treated or not treated, but cannot simultaneously observe both outcomes of the subject treated and not

---

G. Xu and C. Yin—Equal contribution.

© The Author(s), under exclusive license to Springer Nature Switzerland AG 2022
J. Gama et al. (Eds.): PAKDD 2022, LNAI 13282, pp. 200–211, 2022.
https://doi.org/10.1007/978-3-031-05981-0_16

treated, commonly known as *The Fundamental Problem of Causal Inference* [13]. To make the individual treatment effect (ITE) estimation identifiable, uplift models are achieved in randomized controlled trial (RCT) settings, with both the treatment and outcome as binary random variables, where prediction power is the most important issue.

In recent years, various uplift modeling techniques have been developed and applied in a wide range of domains [6,29], including health care [14], digital marketing [12] and public policies [11]. Most of the existing uplift modeling methods are based on Potential Outcome Paradigm [22] and can be categorized into two major categories [29]. The first category consists of methods that extend existing supervised learning methods for Conditional Average Treatment Effect (CATE) estimation, including Single Model Approach [16], Two Model Approach [3] and Transformed Outcome Approach [1], etc. The second category consists of tailored methods for uplift modelling, including Tree-based Method [27], Learn to rank based Method [7] and so on. In practice, Single Model Approach and Two Model Approach are two kinds of widely used approaches because they are easy to implement and have the flexibility of being able to use any off-the-shelf supervised learning algorithm [29]. The Two Model Approach is constructed by fitting a separate model for the control and treatment observations and then taking the difference between the predicted treatment outcome and the predicted control outcome to estimate the CATE. Different from the Two Model Approach, the Single-model Approach uses the concatenation of treatment and covariates as the features and outcome as the target to train a single supervised model.

However, given covariates of a single subject under different treatment indicators, most of existing approaches, especially those based on deep learning, either learn the same representation from a single model or learn different representations from two separate models. Thus, these methods may not learn discriminative representations or utilize both information of control and treatment group. Intuitively, a subject receiving different treatments should have different representations. For example, after patients take different drugs (i.e. treatments in our model), their bodies (i.e. subject representations in our model) will have different inner changes.

In this paper, we propose an attentive neural uplift model to alleviate above shortcomings through attention mechanism. First, we utilize attention mechanisms to explicitly model the process that treatments influence subjects' inside based on the intuition that the effect of different drugs on the same part of patients may be different. The model maps the original covariate space $\mathcal{X}$ into a latent space $\mathcal{Z}$. Given covariates of a subject, the learned representations in space $\mathcal{Z}$ which we called the *After-treatment Representation* are discriminative under different treatment indicators, thus can model potential outcomes more effectively. Second, the model is trained on a single neural network so that the information shared by treatment and control group is utilised.

The main contributions of this work are as follows.

- To the best of our knowledge, this is the first attempt to learn discriminative representations from covariates under different treatments in a single model.

The representation is generated by attention mechanism so that it can extract relevant feature of the given treatment indicator, helps in reducing generalization error and estimating potential outcome more effectively.

- The proposed model can take advantages of both Single Model Approach and Two Model Approach. It takes a special role of treatment indicator by applying attention mechanism and utilise information shared by treatment and control group.
- We evaluate proposed attentive neural uplift model through extensive experiments on synthetic and real-world datasets and achieve state-of-the-art results.

2   Related Work

The methods for uplift modelling can be categorized into two major categories [29]. The first category consists of methods that extend existing supervised learning methods for CATE estimation and the second category consists of tailored methods for uplift modelling.

Most of exist uplift models belong to the first category, including Meta-Learners [15] and Transformed Outcome Approach [1]. There are many uplift models belong to Meta-Learners. For example: The Two Model Approach, known as T-Learner, is constructed by fitting a separate model for the control and treatment observations and then taking the difference between the predicted treatment outcome and the predicted control outcome to estimate the CATE. The Single-model Approach [16], also known as S-leaner, is another Meta-Learner. Different from T-learner, S-Learner uses the concatenation of treatment and covariates as the features and outcome as the target to train a single supervised model. The advantages of T-learner and S-learner are simplicity and easy to implement. However, both S-Learner and T-leaner have their own drawbacks and may not perform well in practice. More complex meta-learners include X-Learner proposed by [15] and R-Learner proposed by [18]. Transformed Outcome Approach models the CATE by creating a new binary target variable of which the expectation conditional mean is equal to uplift. A main advantage of the transformed outcome approach is that the CATE can be modelled directly after transformation. But the drawback of Transformed Outcome Approach is that relies heavily on the accurate estimation of the propensity score.

Methods in the second category are Tree-based Methods, Learn-to-rank based Method and so on. The most recent Tree-based Methods are based on regression trees. These proposed methods view the forests as an adaptive neighborhood metric and estimate the treatment effect at the leaf node [27,30]. The criteria used for choosing each split during the growth of the uplift trees is based on maximization of the difference in uplifts between the two child nodes. However, in practice, these approaches are prone to over-fitting. Learn-to-rank based Method is proposed in [7], which employs the learning-to-rank technique LambdaMART to optimize the ranking according to *promoted cumulative gain* (PCG). This method focuses on modeling the rank of uplift instead of true uplift score.

In the recent years, several deep learning-based uplift modelling algorithms have been proposed. Belbahri introduced a twin neural model for uplift based on a loss function defined by leveraging a connection with the Bayesian interpretation of the relative risk. Betlei proposed a uplift model which directly optimizes an upper bound on AUUC [2].

3 Preliminaries

In this section, we discuss and unify the definitions, assumptions and objectives for uplift modelling under the Rubin-Neyman potential outcomes framework [23,25]. The letters in upper case are used to denote random variables and the letters in lower case their realizations.

- Let $X \in \mathbb{R}^d$ denote the treatment covariates and x its realization. The number of covariates of a subject is d. Superscript is used to indicate specific covariate.
- Let $T = \{0, 1\}$ denote a binary treatment variable, T = 0 for receiving no treatment and T = 1 for receiving treatment.
- Let Y denote the observed outcome and y its realization. Let $Y(0)$ and $Y(1)$ be the binary potential outcomes under control and treatment respectively.

Following the potential outcomes framework, every subject i has two potential outcomes: $Y_i(0)$, the potential outcome if the subject had received no treatment; and $Y_i(1)$, the potential outcome if the subject had received the treatment. The individual treatment effect (ITE) of a treatment T, is defined as the difference between the two potential outcomes:

$$\tau_i := Y_i(1) - Y_i(0). \tag{1}$$

Since one can never simultaneously observe both potential outcomes for a single subject, ITE is not identifiable in real world. However, under some assumptions including Stable Unit Treatment Value Assumption (SUTVA) [19], Consistency, Ignorability and Population Overlapping [9], we can estimate the Conditional Average Treatment Effect (CATE) defined as $CATE = \mathbb{E}[Y(1) - Y(0)|X = x]$. It has been proved that the CATE is the best estimator for the ITE in terms of the mean squared error. For uplift modeling, the target quantity is to infer CATE from data that are obtained from experiments with randomized treatment assignment, which have the following objective:

$$\mathbb{U}(x) = \mathbb{E}(Y|T = 1, X = x) - \mathbb{E}(Y|T = 0, X = x). \tag{2}$$

The objective involves two conditional expectations of the observed outcomes and can be estimated from randomized experiment data for unbiased estimation.

To measure and compare the performance among uplift models, a popular metric named Area Under the Uplift Curve (AUUC) [2,7] is used in this paper. The AUUC is learned on qini curves [7]. Qini curves are built by sorting users from the best to the worst in segments based on predictions of the trained uplift model. The AUUC is the area under the qini curve. Greater AUUC represents better performance of the uplift model. In practice, the AUUC can be calculated as follows:

$$g(p) = Y_p^T/N^T - Y_p^C/N^C \tag{3}$$

$$AUUC = \sum_p g(p) \tag{4}$$

where p represents the proportion of the users, Y_p^T is the number of positive samples in treatment group at p-proportion, N^T is the total number of samples in treatment group, Y_p^C is the number of positive samples in control group at p-proportion and N^C is the total number of samples in control group.

4 Proposed Method

In this section, we introduce the proposed attentive neural uplift model (ANU) based on deep representation learning. The key idea of ANU is to map the original covariate space $\mathcal{X}$ into a latent space $\mathcal{Z}$. Particularly, given the covariates x_i of a special subject, the learned representations in latent space $\mathcal{Z}$ are discriminative under different treatment indicators. Let $z_i(x_i, t = 0)$ and $z_i(x_i, t = 1)$ denote the representations of subject x_i when it is not treated or treated, respectively. We call this kind of representation *the after-treatment representation*.

The framework of ANU is shown in Fig. 1, which contains three major components: the input representation, the attention network and the outcome prediction network. The following subsections explain each component in detail.

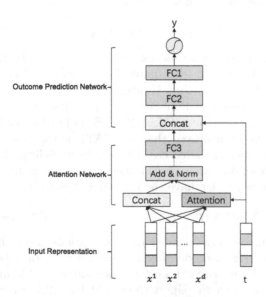

Fig. 1. Attentive neural uplift model.

4.1 Input Representation

The input of ANU contains two categories of features: the covariates x of subject and the observed treatment indicator t. Let x^i denote ith field of covariates x. For each field of covariates x, an embedding layer is built to learn the representations. Another embedding layer is built to learn the representations of treatment indicator t. We use learned embeddings to convert the input features to vectors of dimension d_{model}.

4.2 Attention Network

Intuitively, a subject receiving different treatments should have different representations. For example, after patients take different drugs (i.e. treatments in our model), their bodies (i.e. subject representations in our model) will have different inner changes. However, to predict the outcomes of different treatments, previous single-model approaches generate representations of subjects without considering the influence of treatments, which is unreasonable. In this work, we utilize attention mechanisms to explicitly model the process that treatments influence subjects' inside based on the intuition that the effect of different drugs on the same part of patients may be different. Next, we detail the attention mechanisms used in our model to generate subject representations.

There are two sub-layers in the attention network: attention layer and feed-forward network. For stability and superior performance, we add a residual connection for the sub-layer of feed-forward networks.

Attention mechanisms have been used in large numbers of applications in Natural Language Processing (NLP) [10], Speech [5], Computer Vision (CV) [28]. An attention function maps a query and a set of key-value pairs to an output, where the query and output are vectors (denoted by Q and O respectively) and the keys and values are matrices (denoted by K and V respectively) [26]. The output is computed as a weighted sum of the values, where the weight assigned to each value is computed by a compatibility function of the query with the corresponding key. In our proposed ANU model, the query refers to the embedding of treatment T, while both keys and values refer to vectors of field embeddings of covariates X. Although there are various attention models in history, in practice, we use three most classic attention models in our framework. They are Bahdanau Attention, Scaled Dot-Product Attention and Multi-Head Attention, described as follow:

Bahdanau Attention. As shown in Fig. 2 (left), a bahdanau attention computes the compatibility function using a feed-forward network with a single hidden layer, *tanh* is used as activate function. The output is computed as:

$$Attention(Q, K, V) = softmax(tanh(W[Q, K])V \tag{5}$$

Scaled Dot-Product Attention. As shown in Fig. 2 (middle), a scaled dot-product attention computes the dot products of the query with all keys, divide each by

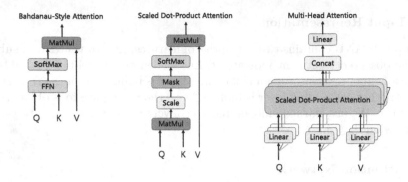

Fig. 2. (left) Bahdanau attention. (middle) Scaled dot-product attention. (right) Multi-head attention.

$\sqrt{d_k}$ and apply a *softmax* function to obtain the weights on the values. The output is computed as:

$$Attention(Q, K, V) = softmax(\frac{QK^T}{\sqrt{d_k}})V \tag{6}$$

Multi-head Attention. As shown in Fig. 2 (right), a multi-head attention linearly projects the query, keys and values h times with different, learned linear projections to d_k, d_k and d_v dimensions, respectively. On each of these projected versions of queries, keys and values, scaled dot-product attentions are performed in parallel, yielding d_v-dimensional output values. These are concatenated and once again projected, resulting in the final output, which is computed as:

$$MultiHead(Q, K, V) = Concat(head_1, ..., head_h)W^O \tag{7}$$

$$head_i = Attention(QW_i^Q, KW_i^K, VW_i^V) \tag{8}$$

Let $f(\cdot)$ denote the function learned by attention network, the after-treatment representation of covariates x_i for subject i under treatment t_i can be computed as $z_i(t_i) = f(x_i, t_i)$. With attention mechanism, the representation of $z_i(t_i)$ pays more attention to covariates relate to outcome under treatment t_i, thus can estimate potential outcome more effectively.

4.3   Outcome Prediction Network

Finally, the outcome prediction network with a feed-forward network is employed to estimate the outcome $y_i(t_i)$ for subject i under treatment t_i. We concatenate the representation of z_i and the representation of treatment t_i and take it as input to the outcome prediction network. Let $g(\cdot)$ denote the function learned by the outcome prediction network. We have $\hat{y}_i(t_i) = g(z_i, t_i) = g(f(x_i, t_i), t_i)$. The loss function is as follows:

$$\mathcal{L}(y_i(t_i), \hat{y}_i(t_i)) = -\sum_{i=1}^{n}(y_i(t_i)log\hat{y}_i(t_i) + (1 - y_i(t_i))log(1 - \hat{y}_i(t_i))). \tag{9}$$

5   Experiments

Table 1. Summary statistics of data sets.

Data set	Treatment	Positive samples	Negative samples	#Features
Hillstrom	No E-Mail	2.2k	19k	7
	Womens E-Mail	3.2k	18.1k	
Criteo	No targeting advertising	80k	2M	11
	Targeting advertising	577k	11.3M	
Zenodo	$Treatment_0$	124k	376k	36
	$Treatment_1$	179k	321k	

5.1   Experimental Data

In order to ensure reproducibility, the experiments are performed on three public datasets. The first dataset is *Hillstrom* [20] which contains results of an e-mail campaign for an Internet retailer. Like the previous works [2,7], the 'visit' target variable was selected as outcome and the 'Womens E-Mail' variable is selected as treatment in our experiments. The second dataset is *Criteo-UPLIFT v2* which is a large scale dataset donated by Criteo AI Lab. It is constructed from several incrementality tests in advertising [8]. The last dataset is a synthetic data set provided by Zenodo. Table 1 summarizes the statistics of the control group and the treatment groups of each dataset. For each dataset, we sample randomly 70% of data as training set and the remaining 30% as test set. All models use the same training set and test set in the spirit of fair comparison.

5.2   Implementation Details

We implement our model through the Keras framework, with tensorflow as the backend. The input embedding size d_{model} is set to 8 and the dimensions in the dense layer FC_1, FC_2, FC_3 are 1, 64, 128 respectively. The optimizer is set to Adam ($lr = 1e - 4, \beta_1 = 0.9, \beta_2 = 0.999$). During training, the batch size is set to 256 for all datasets.

5.3   Comparison Methods

We compare our ANU model with a wide range of baseline uplift models in literature. These models are **(1) Meta-learner-based methods** including S-learner, T-learner, X-learner [15]; **(2) Tree-based methods** including random forests on KL divergence, Euclidean Distance, Chi-Square and Conditional Treatment Selection; **(3) Learn to Rank based methods** including PCG [7]; **(4) Neural-network-based models** including CEVAE [17], DragonNet [24], AUUC-max [2]. Most of baseline models are implement by CausalMl package [4] except PCG and AUUC-max because their codes are not available.

Our ANU model uses the Scaled Dot-Product Attention as the default attention mechanism. The comparisons of four variants of our ANU model are also provided. The difference between of these variants is the attention mechanism that they used. Specifically, the variants are ANU-bahdanau, ANU-multi-head-2, ANU-multi-head-4 and ANU-multi-head-8, which use Bahdanau Attention, Multi-Head Attention with 2 heads, Attention with 4 heads and Multi-Head Attention with 8 heads, respectively. In addition, ANU-FC (replace the attention network in ANU with FC layer) used as the baseline for the whole ANU models. For fair comparison all models except PCG and AUUC-max are run 10 times and the average results on the test data are reported. We report results of PCG and AUUC-max taken from the original papers.

Table 2. Comparison of baselines and ANU. The bold scores are the best scores among ANU and the baselines. The underlined scores are the best scores among ANU and its variants.

Model	Hillstrom	Criteo	Zenodo
	AUUC ± S.E.	AUUC ± S.E.	AUUC ± S.E.
S-learner (xgboost) [15]	$0.02990 \pm 9.72e{-}4$	$0.01500 \pm 1.99e{-}5$	$0.10397 \pm 8.99e{-}5$
T-learner (xgboost) [15]	$0.02969 \pm 1.08e{-}3$	$0.01502 \pm 1.06e{-}5$	$0.10408 \pm 5.04e{-}5$
X-learner (xgboost) [15]	$0.02988 \pm 1.01e{-}3$	$0.01507 \pm 1.37e{-}5$	$0.10411 \pm 7.31e{-}5$
CEVAE [17]	$0.02408 \pm 1.88e{-}3$	$0.01501 \pm 1.17e{-}5$	$0.10388 \pm 1.75e{-}4$
DragonNet [24]	$0.02633 \pm 2.76e{-}3$	$0.01499 \pm 3.82e{-}5$	$0.10401 \pm 9.55e{-}5$
AUUC-max (deep, s_{log}) [2]	$0.02999 \pm 3.25e{-}3$	$0.00924 \pm 1.0e{-}5$	N/A
PCG [7]	0.03055	0.01601^{*}	N/A
UpliftTree-KL [21]	$0.02998 \pm 4.12e{-}4$	$0.01479 \pm 1.10e{-}5$	$0.10389 \pm 2.06e{-}5$
UpliftTree-ED [21]	$0.02994 \pm 2.95e{-}4$	$0.01492 \pm 2.51e{-}5$	$0.10392 \pm 1.99e{-}5$
UpliftTree-Chi [21]	$0.02989 \pm 8.75e{-}4$	$0.01500 \pm 1.48e{-}5$	$0.10354 \pm 2.49e{-}5$
UpliftTree-CTS [21]	$0.02478 \pm 1.03e{-}3$	$0.01455 \pm 2.09e{-}5$	$0.10358 \pm 7.65e{-}5$
ANU	$\underline{\mathbf{0.03072 \pm 2.44e{-}4}}$	$\mathbf{0.01514 \pm 1.21e{-}5}$	$\mathbf{0.10591 \pm 7.25e{-}6}$
ANU-FC	$0.02998 \pm 5.54e{-}4$	$0.01510 \pm 2.79e{-}5$	$0.10415 \pm 3.02e{-}5$
ANU-bahdanau	$0.03046 \pm 2.86e{-}4$	$0.01514 \pm 8.85e{-}6$	$0.10564 \pm 3.12e{-}5$
ANU-multi-head-2	$0.03025 \pm 3.17e{-}4$	$0.01516 \pm 1.40e{-}5$	$0.10585 \pm 1.43e{-}5$
ANU-multi-head-4	$0.03038 \pm 2.72e{-}4$	$0.01516 \pm 1.04e{-}5$	$\underline{0.10592 \pm 8.55e{-}6}$
ANU-multi-head-8	$0.03060 \pm 2.13e{-}4$	$\underline{0.01518 \pm 7.54e{-}6}$	$0.10591 \pm 7.96e{-}6$

5.4   Experimental Results

Table 2 shows the AUUC achieved by all the tested models. Note that the results of AUUC-max and PCG are taken from [2,7] respectively and the result of PCG on Criteo is not comparable to the results of other methods because it is obtained on a random subset of Criteo dataset while the results of other methods are got on the whole dataset. First of all, it is obviously that ANU has achieved the best performance on multiple data sets. Compared to the previous state-of-the-art uplift model (i.e. PCG for Hillstrom, X-learner for Criteo and Zenodo), the

AUUC values of our proposed method on the three data sets has increased by 0.56%, 0.73% and 1.74% respectively. Furthermore, the variances in multiple tests indicate that our ANU model has achieved more robust results. Second, to investigate how attention mechanism impact representation $z(x, t)$ and the performance of uplift prediction, we compared three different attention style to illustrate the improvement of attention mechanism on the estimated CATE of the uplift model. Comparing the results of multiple attention models, our ANU model has achieved **state-of-the-art** on each data set. Yet diverse attention style shows varied effects in three data sets. No such attention style can achieve the best results on the whole multiple data sets. For sufficient data set (i.e. criteo), the three attention modes achieved similar results, but on smaller data sets (i.e. Hillstrom), ANU model uses a simple and effective attention structure to get the best results. Third, in order to further show the performance of multi-head attention, we also compared the effects of head numbers. Multi-head attention can focus on different aspects of information on different heads. Comparing the number of heads, ANU-multi-head-8 shows the best performance on the whole data sets. As the number of heads grows, the AUUC is gradually getting better.

5.5 Effect of Attention Mechanism

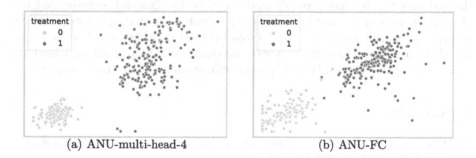

(a) ANU-multi-head-4   (b) ANU-FC

Fig. 3. Comparison between ANU-multi-head-4 and ANU-FC. Dark color dots and light color dots represent the representations of samples from the Zenodo dataset when they are treated (i.e. $t = 1$) or not treated (i.e. $t = 0$), respectively.

We design an experiment to analyze the effect of attention mechanism on after-treatment representation. The experiment is conducted on Zenodo, because among the three data sets, the true uplift values are known only on Zenodo. We select top-300 samples ordered by true uplift values in the test set denoted as D_{top}. For those samples with relatively high uplift values, the predicted outcomes of them that are treated (i.e. $t = 1$) or not treated (i.e. $t = 0$) should be very different. That means that, for a subject, its representations produced by FC_2 in ANU should be discriminative when it is treated or not treated. Therefore, we project the output dimension of $FC_2(x, t)$ in our ANU model to 2 for conveniently visualizing the discriminant in space $\mathbb{R}^2$. Specifically, We first train the ANU-multi-head-4 model and the ANU-FC model, the output dimension of $FC_2(x,t)$ of which are set to 2. Then, samples in D_{top} associated with $t = 0$ and $t = 1$ are fed into the two well-trained models and their representations generated by $FC_2(x,t)$ are obtained. These representations produced by ANU-multi-head-4 and ANU-FC are illustrated in Fig. 3(a) and 3(b), respectively.

The corresponding points of a subject given $t = 1$ or $t = 0$ in the figure should be farther apart for a good model. Compare Fig. 3(a) with Fig. 3(b), it is obvious that the distance between point sets of $t = 1$ and $t = 0$ of model ANU-multi-head-4 is larger than ANU-FC, which means that our proposed after-treatment representation is more effective in estimate potential outcome and uplift value.

6   Conclusion

We have presented an attentive neural uplift model (ANU), the first uplift model to predict CATE with attention mechanism, which generates after-treatment representation with relevant feature of the given treatment. ANU utilizes attention mechanisms to map the original covariate space $\mathcal{X}$ into a latent space $\mathcal{Z}$. Given different treatment indicators and the covariates of a subject, the learned representations in space $\mathcal{Z}$ are discriminative.

Experiments on a number of uplift data sets and existing methods, especially those data sets with more complicated features (i.e. Zenodo), suggest that our attentive neural uplift model by leveraging attention mechanisms for covariate x and treatment indicator, resulting a state-of-the-art performance when compared to other baselines. Through the visual analysis, we have further explained the effect of attention mechanisms on covariate x when given diverse treatment.

References

1. Athey, S., Imbens, G.W.: Machine learning methods for estimating heterogeneous causal effects. Stat **1050**, 1–26 (2015)
2. Betlei, A., Diemert, E., Amini, M.R.: Uplift modeling with generalization guarantees. In: Proceedings of KDD (2021)
3. Cai, T., Tian, L., Wong, P.H., Wei, L.: Analysis of randomized comparative clinical trial data for personalized treatment selections. Biostatistics **12**, 270–282 (2011)
4. Chen, H., Harinen, T., Lee, J.Y., Yung, M., Zhao, Z.: Causalml: Python package for causal machine learning. arXiv preprint arXiv:2002.11631 (2020)
5. Cho, K., Courville, A., Bengio, Y.: Describing multimedia content using attention-based encoder-decoder networks. IEEE Trans. Multimedia **17**, 1875–1886 (2015)
6. Devriendt, F., Moldovan, D., Verbeke, W.: A literature survey and experimental evaluation of the state-of-the-art in uplift modeling: a stepping stone toward the development of prescriptive analytics. Big Data **6**, 13–41 (2018)
7. Devriendt, F., Van Belle, J., Guns, T., Verbeke, W.: Learning to rank for uplift modeling. IEEE Trans. Knowl. Data Eng. (2020)
8. Diemert, E., Betlei, A., Renaudin, C., Amini, M.R.: A large scale benchmark for uplift modeling. In: Proceedings of the KDD (2018)

9. D'Amour, A., Ding, P., Feller, A., Lei, L., Sekhon, J.: Overlap in observational studies with high-dimensional covariates. J. Econom. **221**, 644–654 (2021)
10. Galassi, A., Lippi, M., Torroni, P.: Attention in natural language processing. IEEE Trans. Neural Netw. Learn. Syst. **32**, 4291–4308 (2020)
11. Grimmer, J., Messing, S., Westwood, S.J.: Estimating heterogeneous treatment effects and the effects of heterogeneous treatments with ensemble methods. Polit. Anal. **25**, 413–434 (2017)
12. Gubela, R., Bequé, A., Lessmann, S., Gebert, F.: Conversion uplift in e-commerce: a systematic benchmark of modeling strategies. Int. J. Inf. Technol. Decis. Mak. **18**, 747–791 (2019)
13. Holland, P.W.: Statistics and causal inference. J. Am. Stat. Assoc. **81**, 945–960 (1986)
14. Jaskowski, M., Jaroszewicz, S.: Uplift modeling for clinical trial data. In: ICML Workshop on Clinical Data Analysis (2012)
15. Künzel, S.R., Sekhon, J.S., Bickel, P.J., Yu, B.: Metalearners for estimating heterogeneous treatment effects using machine learning. In: Proceedings of the National Academy of Sciences (2019)
16. Lo, V.S.: The true lift model: a novel data mining approach to response modeling in database marketing. ACM SIGKDD Explor. Newsl **4**, 78–86 (2002)
17. Louizos, C., Shalit, U., Mooij, J., Sontag, D., Zemel, R., Welling, M.: Causal effect inference with deep latent-variable models. arXiv preprint arXiv:1705.08821 (2017)
18. Nie, X., Wager, S.: Quasi-oracle estimation of heterogeneous treatment effects. Biometrika **108**, 299–319 (2021)
19. Pearl, J.: Causal inference in statistics: an overview. Stat. Surv. **3**, 96–146 (2009)
20. Radcliffe, N.J.: Hillstrom's MineThatData Email Analytics Challenge: An Approach Using Uplift Modelling. Stochastic Solutions Limited, Edinburgh (2008)
21. Radcliffe, N.J., Surry, P.D.: Real-world uplift modelling with significance-based uplift trees. White Paper TR-2011-1, Stochastic Solutions (2011)
22. Rubin, D.B.: Estimating causal effects of treatments in randomized and nonrandomized studies. J. Educ. Psychol. **66**, 688 (1974)
23. Rubin, D.B.: Causal inference using potential outcomes: design, modeling, decisions. J. Am. Stat. Assoc. **100**, 322–331 (2005)
24. Shi, C., Blei, D.M., Veitch, V.: Adapting neural networks for the estimation of treatment effects. arXiv preprint arXiv:1906.02120 (2019)
25. Splawa-Neyman, J., Dabrowska, D.M., Speed, T.: On the application of probability theory to agricultural experiments. Essay on principles. section 9. Statistical Science (1990)
26. Vaswani, A., et al.: Attention is all you need. In: Proceedings of the NeurIPS (2017)
27. Wager, S., Athey, S.: Estimation and inference of heterogeneous treatment effects using random forests. J. Am. Stat. Assoc. **113**, 1228–1242 (2018)
28. Wang, F., Tax, D.M.: Survey on the attention based RNN model and its applications in computer vision. arXiv preprint arXiv:1601.06823 (2016)
29. Zhang, W., Li, J., Liu, L.: A unified survey of treatment effect heterogeneity modelling and uplift modelling. ACM Comput. Surv. (CSUR) **54**, 1–36 (2021)
30. Zhao, Y., Fang, X., Simchi-Levi, D.: A practically competitive and provably consistent algorithm for uplift modeling. In: Proceedings of the ICDM (2017)

Mental Health Treatments Using an Explainable Adaptive Clustering Model

Usman Ahmed[1], Jerry Chun-Wei Lin[1(✉)], and Gautam Srivastava[2]

[1] Western Norway University of Applied Sciences, Bergen, Norway
usman.ahmed@hvl.no, jerrylin@ieee.org
[2] Brandon University, Brandon, Canada
srivastavag@brandonu.ca

Abstract. In this paper, we propose a model to help psychologists identify the most important aspects of emotions in mentally ill people. As a first step, we created emotional lexicon embeddings using natural language processing, followed by deep clustering based on the attention mechanism. The generated representation is used to evaluate patient-written text from an emotional perspective. To increase the patient authored emotional lexicon, we apply synonymous semantic expansion. The EANDC method is used to classify latent semantic representations according to context. This is an explainable attention network based adaptive deep clustering approach. To increase the explainability of the learning, we used similarity metrics to select the instances that label the text to optimize the following selection using the curriculum-based optimization technique. Experimental results show that increasing the number of synonyms based on emotion lexicons improves accuracy without negatively affecting performance.

Keywords: NLP · Explainable · Text clustering · Healthy treatments

1 Introduction

Mental impairment is a widespread problem around the world, according to WHO (World Health Organization)[1]. Both the increase and decrease in incarceration as a result of the pandemic harm mental health more than physical health. The mental health of the population is further exacerbated by social isolation and limited contact at work and school. Frontline health workers suffer from anxiety and depression caused by isolation from family, lack of protective equipment, stressful working conditions, and social isolation in addition to patients. Stress, worry, and despair are all emotional states that affect a person's psychological well-being. A substantial and growing number of publications contribute

---

[1] https://www.who.int/campaigns/connecting-the-world-to-combat-coronavirus/healthyathome/healthyathome---mental-health.

J. Gama et al. (Eds.): PAKDD 2022, LNAI 13282, pp. 212–222, 2022.
https://doi.org/10.1007/978-3-031-05981-0_17

to the diverse body of knowledge. Because of conflicting reports, it remains not easy to obtain meaningful and up-to-date information [7].

According [18], a patient's mental state can be considered as a primary consequence of a series of events that a single urgent situation/event can trigger. According [15], the process of determining the cause of each emotion is complicated. The person may be able to recognize critical signs and seek medical help. According to [1], individuals can connect and express their grief anonymously through the growing data hub of social media as well as online forums; as stated in [18, 19], sharing personal ideas often leads to an exposed circumstance, as is the case here. Preventing and recognizing such circumstances through the use of texts can contribute to the quick resolution of conflicts and the overall well-being of communities [20].

Globally, depression leads to the most common diagnoses by healthcare professionals [9] and affects approximately 264 million people worldwide. The consequences of untreated depression are even more severe, according to [16]. With more than 800, 000 cases of suicide attempts per year, depression is one of the leading causes of death among 15–29 year olds. Middle- and low-income people suffering from mental illnesses that cause personal (individual) or social problems do not receive treatment. Other problems include misdiagnosis, a lack of appropriately qualified health care providers, and the presence of emotional bias among people who have a mental illness, according [9]. People try to avoid medical care because they are afraid or ashamed of the physiological treatment they will receive [19]. As a result, many people try to ignore or deny their mental illness.

Adoption-based interventions can lead to positive effects. User behavior is helpful to the intervention system, and analysis of individual user behavior should lead to tailored recommendations for the user. The user's interests, hobbies, social environment, and psychological symptoms are considered in the behavior analysis [18]. In this paper, we are interested in considering the psychological symptoms from a text written by a patient by using important indicators of depression. We created an emotional embedding using the latent representation approach. We then used this embedding to cluster the empty sentences using the clustering approach used in the deep attention-based model. Communication is the most important tool for controlling one's mental health and emotional well-being. The text contained several important emotional cues that served as an early warning to the patient when the crucial triggering signs were identified. The main purpose of this study lies in the data labeling tasks relevant to the extraction and consideration of depression-related emotional vital indicators for use in psychotherapy. Most of our contributions are listed here:

1. We focus on important emotional signs to extract essential contextual information presented to psychologists to make preventive suggestions.
2. The term lexicon formation was developed and used to describe the process of expanding a synonym based on its occurrence in context, followed by the use of that lexicon as a latent representation.

3. Using the attention-based strategy, the learned representation is used to train and create a cluster in the input text. The final embedding visually represents the vital emotional indicators. The approach is batch-based, with the curriculum-based learning system controlling each batch.

2   Related Work

Dinakar et al. [6] examined the online community and discussed a stacked generalization modeling strategy for its analysis. The ensemble classifier was developed using an SVM and GBDT in the designed framework, where SVM stands for Support Vector Machine and GBDT stands for stochastic gradient boosted decision trees. The text is then divided into 23 sections. The features were extracted using unigrams, bigrams, part-of-speech bigrams as well as TF-IDF filters. The result is then predicted using the stacking base classifier. The model evaluated 7,147 personal narratives on a well-known youth support website. However, data-driven deep learning (D-DL) is believed to be the approach that significantly outperforms previous models, i.e., feature engineering, as it can more accurately account for both the more general data and the sequence-to-sequence connectivity that both unigram and bigram models lack. Feature engineering is domain dependent and complicates modeling. D-DL provides unique solutions to these problems.

Also, another technique based on Twitter feeds is presented and explored [3]. Depression, bipolar disorder, and seasonal affective disorder are topics covered in the text feed. The Linguistic Inquiry Word Count (LIWC) identifies and classifies each disorder. To regularize the response, the approach evaluates character occurrences in order. The classifier can then distinguish between groupings. The group's Twitter feed is analyzed to discover the relationship between each group's class distribution and the Twitter feed [17]. Lin et al. [12] compared the analysis of microblogs using the four-layer network with more classical statistical classifiers such as SVM, Random Forest as well as Naïve Bayes. They investigated different pooling techniques, including maximum pooling, mean-over-instance, as well as mean-over-time. The optimal approach for the DNN was mean-over-time pooling.

The attention mechanism improves RNN encoder and decoder models by using the alignment approach. Compared to multi-hop sequences, the values of the neighborhood features have an impact on the sequence [5]. The attention approach assigns weights to certain inputs. The inputs and their positions help the decoder to determine the correct context vector weights for the higher feature representation. Then, the developed approach is improved using the RNN weight optimization method. The trained attention vector helps in the representation of context and features [13]. The attention mechanism is proposed to use soft attention, hard attention, and global attention. Soft attention minimizes the amount of contextual information available [1]. Then learn the representations of the hidden features using the averaging strategy. For hard attention, Xu et al. calculated the vector transferred from the context using the sampling method [22]. It was decided to use both the intermediate version of soft and hard attention for local and global attention [14]. Using the prediction function, each attention point is considered from the stack of inputs, and then the position of each attention point is predicted using the selection function. The approach contributes to the enrichment of domain-specific data.

Henry et al. [8] examine NLP-based techniques that have been used to treat mental health. They conclude that NLP may be a valuable tool for addressing mental health in the coming years. Le Glaz et al. [11] explore the combination of machine learning and NLP techniques in a review article collecting their treatments in mental health. The authors show that the topic itself has not produced many strong solutions in real-world situations or settings. Kulkarni et al. [10] give a thorough examination of the methods of cluster analysis used for mental health discourse. The authors strongly believe that the cluster method has a long way to go to be a viable tool for health practitioners to address mental health.

3 Deep Adaptive Clustering Based on an Explainable Attention Network (EAN)

For the depression symptom recognition model, we used an emotion lexicon-based embedding. The architecture is shown in Fig. 1, where we evaluated the PHQ-9 symptom score using the cosine similarity distance method. The knowledge of the lexicon is enhanced by using the word vector model. In the following sections, the specifics of the proposed model are explained.

3.1 Psychometric Questionnaires (PQ)

The psychiatrist's model used in this paper is called ICD10[2] classification. Depending on the patient's psychological difficulties, the standard questions help the psychiatrist recommend the most reliable diagnosis based on the answers. In the questionnaire, you are asked to rate the frequency of various vital signs, as well as the rating is based on the occurrence frequency multiplied by the established threshold. The Clinical Symptom Elicitation Process (CSEP) (see footnote 2) categorizes each disease into mild, moderate, and severe situations based on the severity of the symptom. First, we enrich the cadaver's knowledge with new words before training the deep attention-based clustering approach, which is then automated using the embedded words from the enrichment phase. The severity of the symptom and the frequency with which it occurs are estimated to obtain a score for patient evaluation.

The PHQ-9 approach is used to extract the DSM-V behavioral types (https://www.psychiatry.org/psychiatrists/practice/dsm). Each of the nine symptoms is classified as a condition, such as sleep, interest, eating, or social problems, as described in Table 1 as well as in a sample document[3]. The questions are scored to

---

[2] https://www.who.int/classifications/icd/en/GRNBOOK.pdf.

[3] https://www.uspreventiveservicestaskforce.org/Home/GetFileByID/218.

help classify the patient. Each sentence is lowercase, non-meaningful full symbols $(\#, +, -, *, =, \text{HTTP}, \text{HTTPS})$ are eliminated, and colloquial or text-based terms are converted to whole words during preprocessing.

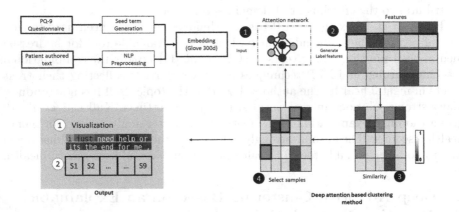

Fig. 1. A flowchart depicting the training as well as adaptation processes associated with attention-based domain adoption.

Table 1. PHQ-9 questionnaire as well as seed terms for each symptoms.

Symptoms	PHQ-9	Seed terms
S1	Little interest or pleasure in doing things	Interest
S2	Felling down depressed or hopeless	Feeling, depressed, hopeless
S3	Trouble failing or staying asleep or sleeping too much	Sleep, asleep
S4	Feeling tired or having little energy	Tired, energy
S5	Poor appetite or over eating	Appetite, overeating
S6	Feeling bad about yourself or that you are a failure or have let yourself or your family down	Failure, family
S7	Trouble concentrating on things such as reading the newspaper or watching television	Concentration, reading, watching
S8	Moving or speaking so slowly that other people could have noticed or the opposite being or restless that you have been moving around a lot more than usual	Moving, speaking, restless
S9	Thoughts that you would be better off dead or of hurt yourself	Dead, hurt suicide

3.2   Emotional Lexicon Used to Embed Words in Sentences

Current research on knowledge-based dynamic systems is scanty. In this experiment, we used the word sense lexicon to manipulate the contextual embedding taking place. The emotional data used to create the word sense lexicon came from an Internet discussion group. Each tokenized word in the patient-authored text is represented by a 300-dimensional pre-trained model, which can also be referred to as a global vector for word representation (Glove) [21]. It is possible to teach the structural embedding of a phrase using the idea of semantic composition as well as the hypothesised [2].

Linguistic patterns can estimate the frequency of co-occurrence of the vectorized terms. The output of the model is a single word with a fixed vector representation, called a unique word. Overall, the pre-trained model outperforms its assigned task. On the other hand, they lack emotional analysis. Therefore, we use transfer learning to extend the meaning of words in terms of emotions. The use of *happy* as well as *sad* to identify the emotions associated with *feelings* was appropriate because most of the models are based on publicly available data, namely *(Twitter data)* as well as *(Twitter data)* sentiment information (Twitter data). However, the two terms denote different mental states. Therefore, their word lexicon is extended to include texts written by patients. Below are some initial definitions for classifying the posts.

Definition 1 (Corpus). *A corpus D is a collection of texts, $D = \{t_1, t_2, \ldots, t_n\}$.*

Definition 2 (Emotion set). *It is used to find synonyms, antonyms, hypernyms, as well as the physical meaning of each extracted speech segment. For each document, a sentiment word is formed from the set $W = \{w_1, w_2, \ldots, w_K\}$.*

Definition 3 (Vocabulary set). *The vocabulary is created using the W dataset used to train the model. The resulting embedding is the learned vector V, i.e., $V = \{v_1, v_2, \ldots, v_m\} \in \mathbb{R}^{m \times \delta}$, where δ is the dimension of the word vector.*

Based on the given definition, the designed deep learning-based clustering method is then shown in Algorithm 1.

3.3   Dataset

An online forum, website, and social media site were used to obtain the data used in this study. The Amazon Mechanical Turk[4] service was used for the texts to be labeled. The EANDC was used for the rest of the labeling. According to [18], the degree of depression is indicated by the number of symptoms (nine in total), where 0 means the person is not depressed, one means the person is mildly depressed, two means the person is moderately depressed, and three means the person is severely depressed. We move to a multiple-assignment procedure in which the presence of each symptom is scored as 1 and the absence of each symptom is scored as 0, rather than a single assignment. The information collected is shown in Table 2.

---

[4] https://www.mturk.com/.

Table 2. Summary of statistical data from training and testing sets

Type	Statistics
# of collected posts (corpus size)	15,044
# of sentences	133,524
Avg. sentences per post	8.87
Avg. words per post	232
# of posts (Training set size)	14,944
# of posts (Testing set size)	100

Algorithm 1. Deep learning based clustering method.

Require: Corpus, m instances per batch; n, total instances; d, documents; C, set of instances in the batch; as well as T, set of training samples.

Ensure: Cluster label c_i.

1: **for** $d \in corpus$ **do**
2:     $d \leftarrow preprocessing(d)$;
3:     **for** term $t \in d$ **do**
4:         synonyms $\leftarrow Extract_{synonyms}(t)$;
5:         $z \leftarrow Extract_{hyperonym, hyponym, antonyms}(synonyms)$;
6:         **for** $w \in z$ **do**
7:             terms $\leftarrow$ WordNet (synonym);
8:         **end for**
9:         vocabulary $\leftarrow$ terms;
10:     **end for**
11: **end for**
12: Embed $\leftarrow attention_{BiLSTM}(vocabulary, corpus)$;
13: **while** $K \leq \{1, 2, ; \frac{n}{m}\}$ **do**
14:     Sample batch C from $corpus$;
15:     $Embed(C)$;
16:     Select training samples from T;
17:     Calculate similarity using the Cosine similarity;
18:     Update the gradient descent algorithm;
19: **end while**
20: **while** $C_i \in corpus$ **do**
21:     $\{c_i\} = argmax_h(l_{ih})$;
22: **end while**
23: **Return** Cluster label c_i.

3.4 Developed EANDC Model

The studies were performed using a feed-forward neural network as a basis. EANDC embedding is used to extract features in all models while averaging is used to extract phrase vectors. The buried layer is a $(30, 20, 10)$ structure with a ReLU activation function [1]. We used a loss function based on cross-entropy. To measure performance, we used the rate of true positives TPR $= \frac{TP}{TP+FN}$ and the false positive rate FPR $= \frac{FP}{FP+TN}$. We developed a model with an RNN gated architecture that is well suited for sequential workloads. Thus, the memory app-roach of bidirectional LSTM (Bi-LSTM) structure can support the learning of class representation and labeling. In the hidden layer, we used the element-wise averag-ing approach. The design uses the two given inputs for each token, which are then merged and concatenated to produce the output. We introduced a dropout ratio of **0:5** to avoid an overfitting problem across generations. In addition, we applied the attention approach to consider the emotional content of words in the predic-tion [1]. Thus, the additional layer may be beneficial for learning classes and vector representations. A larger network with a diversified set of data points can achieve generalization. The cosine and casual learning methods start by identifying the easy examples in each category and then remove the difficult examples. Using vec-tor similarity and input feature labeling, the model continues to train and converge on the position of the significant word. As the number of labeled instances grows, performance improves by fitting to the data distribution. In addition, the learning process becomes more extensive over time.

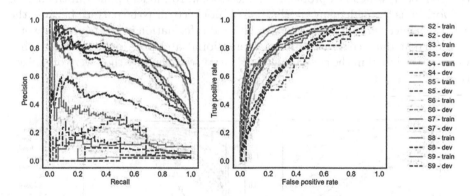

Fig. 2. The baseline feed-forward neural network model.

4 Experimental Analytics

We extracted the emotional lexicon from the text written by the patient. We then trained and labeled the input text using the PHQ-9 questionnaire. We used the 300-dimensional Glove vector to vectorize the transfer learning. Each of the nine symptoms represents a separate vector corresponding to a separate embedding. Each instance is labeled using cosine similarity. We compared both feed-forward networks and bidirectional LSTM. The receiver operating charac-teristic curve (ROC) and the precision-recall curve were used to evaluate the developed framework in the experiments. The hyper-tuning model and Adam optimizer are then applied in the developed model with a learning rate of 0.05%. The results can be seen in Tables 3 and 4. We ran the model for longer epochs and gradually recovered the model by stopping early. We also used the gradient pruning approach to circumvent gradient problems [4].

Table 3. ROC-AUC values for both training and testing sets.

Compared model	Train	Test
Baseline	0.76	0.73
Bi-LSTM	0.91	0.80

Table 4. Performance results of precision and recall.

Compared model	Precision	Recall
Baseline	0.82	0.80
Bi-LSTM	0.89	0.88

Figure 2 shows the performance of the baseline model in detail. The values in the upper left corner of the precision retrieval curve tend to be over-performed by the model, which is a problem. The feed-forward model applied in the neural network runs through many epochs, but cannot recognize the pattern because the word sequence was not preserved as relational information in the neural network. The current data contains significant emotional and positional meanings that were not included in the simpler network because of the way they were organized (Fig. 3).

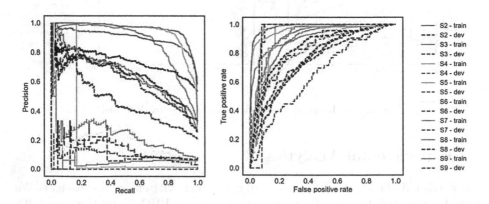

Fig. 3. Results for Bi-LSTM model.

For clinical text analysis, we used the clustering approach. A deep neural network can be built by analyzing large amounts of training data and automatically searching for features that distinguish between different classes. A neural network is a collection of interconnected nodes. On the other hand, deep learning algorithms require a large amount of training data and do not scale well in clinical settings. In addition, models based on a single data source are prone to bias or "overfitting". They may not adapt perfectly to the new instances that are added, limiting their use in different clinical scenarios and reducing their overall accuracy. Based on knowledge development, we proposed a strategy to classify data more efficiently than the baseline model.

5   Conclusion

The technique investigated in this study focuses on emotional text expansion, which is used to expand expressive words in patient-written texts (or self-written texts). We began by expanding and training word embeddings and subsequent clustering based on cosine similarity. The developed model starts with basic patterns and gradually introduces more sophisticated patterns as the cluster size increases. Bidirectional LSTM (Long Short Term Memory) with weighted word layers is used to determine the probability distribution of symptoms. Clustering allows both label generation and representation learning. The results are used to develop adaptive interventions based on individual patient input.

References

1. Ahmed, U., Mukhiya, S.K., Srivastava, G., Lamo, Y., Lin, J.C.W.: Attention-based deep entropy active learning using lexical algorithm for mental health treatment. Front. Psychol. **12**, 471 (2021)
2. Charles, W.G.: Contextual correlates of meaning. Appl. Psycholinguist. **21**(4), 505–524 (2000)
3. Chen, E., Lerman, K., Ferrara, E.: Tracking social media discourse about the COVID-19 pandemic: development of a public coronavirus twitter data set. JMIR Public Health Surveill. **6**(2), e19273 (2020)
4. Chen, X., Wu, S.Z., Hong, M.: Understanding gradient clipping in private SGD: a geometric perspective. Adv. Neural. Inf. Process. Syst. **33**, 13773–13782 (2020)
5. Cho, K., et al.: Learning phrase representations using RNN encoder-decoder for statistical machine translation. In: Moschitti, A., Pang, B., Daelemans, W. (eds.) The Conference on Empirical Methods in Natural Language Processing, pp. 1724–1734 (2014)
6. Dinakar, K., Weinstein, E., Lieberman, H., Selman, R.: Stacked generalization learning to analyze teenage distress. In: Proceedings of the International AAAI Conference on Web and Social Media, pp. 1–8 (2014)
7. Ebadi, A., Xi, P., Tremblay, S., Spencer, B., Pall, R., Wong, A.: Understanding the temporal evolution of COVID-19 research through machine learning and natural language processing. CoRR abs/2007.11604 (2020)
8. Henry, S., Yetisgen, M., Uzuner, O.: Natural language processing in mental health research and practice. In: Tenenbaum, J.D., Ranallo, P.A. (eds.) Mental Health Informatics, pp. 317–353. Springer, Cham (2021). https://doi.org/10.1007/978-3-030-70558-9_13
9. James, S.L., et al.: Global, regional, and national incidence, prevalence, and years lived with disability for 354 diseases and injuries for 195 countries and territories, 1990–2017: a systematic analysis for the global burden of disease study 2017. Lancet **392**(10159), 1789–1858 (2018)

10. Kulkarni, A., Hengle, A., Kulkarni, P., Marathe, M.: Cluster analysis of online mental health discourse using topic-infused deep contextualized representations. In: Proceedings of the 12th International Workshop on Health Text Mining and Information Analysis, pp. 83–93 (2021)
11. Le Glaz, A., Berrouiguet, S., et al.: Machine learning and natural language processing in mental health: systematic review. J. Med. Internet Res. **23**(5), e15708 (2021)
12. Lin, H., et al.: User-level psychological stress detection from social media using deep neural network. In: Proceedings of the 22nd ACM International Conference on Multimedia, pp. 507–516 (2014)
13. Lu, J., Yang, J., Batra, D., Parikh, D.: Hierarchical question-image co-attention for visual question answering. In: Lee, D.D., Sugiyama, M., von Luxburg, U., Guyon, I., Garnett, R. (eds.) Advances in Neural Information Processing Systems 29: Annual Conference on Neural Information Processing Systems, pp. 289–297 (2016)
14. Luong, T., Pham, H., Manning, C.D.: Effective approaches to attention-based neural machine translation. In: Màrquez, L., Callison-Burch, C., Su, J., Pighin, D., Marton, Y. (eds.) The Conference on Empirical Methods in Natural Language Processing, pp. 1412–1421 (2015)
15. Losada, D.E., Gamallo, P.: Evaluating and improving lexical resources for detecting signs of depression in text. Lang. Resour. Eval. **54**(1), 1–24 (2018)
16. Mazza, M.G., et al.: Anxiety and depression in COVID-19 survivors: role of inflammatory and clinical predictors. Brain Behav. Immun. **89**, 594–600 (2020)
17. McDonnell, M., Owen, J.E., Bantum, E.O.: Identification of emotional expression with cancer survivors: validation of linguistic inquiry and word count. JMIR Form. Res. **4**(10), e18246 (2020)
18. Mukhiya, S.K., Ahmed, U., Rabbi, F., Pun, K.I., Lamo, Y.: Adaptation of IDPT system based on patient-authored text data using NLP. In: International Symposium on Computer-Based Medical Systems (CBMS), pp. 226–232. IEEE (2020)
19. Mukhiya, S.K., Wake, J.D., Inal, Y., Pun, K.I., Lamo, Y.: Adaptive elements in internet-delivered psychological treatment systems: systematic review. J. Med. Internet Res. **22**(11), e21066 (2020)
20. Neuraz, A., et al.: Natural language processing for rapid response to emergent diseases: case study of calcium channel blockers and hypertension in the covid-19 pandemic. J. Med. Internet Res. **22**(8), e20773 (2020)
21. Pennington, J., Socher, R., Manning, C.D.: Glove: global vectors for word representation. In: The Conference on Empirical Methods in Natural Language Processing, pp. 1532–1543 (2014)
22. Xu, K., et al.: Show, attend and tell: neural image caption generation with visual attention. In: Bach, F.R., Blei, D.M. (eds.) The International Conference on Machine Learning. JMLR Workshop and Conference Proceedings, vol. 37, pp. 2048–2057 (2015)

S²QL: Retrieval Augmented *Zero-Shot* Question Answering over Knowledge Graph

Daoguang Zan[1,2(✉)], Sirui Wang[3], Hongzhi Zhang[3], Yuanmeng Yan[4],
Wei Wu[3], Bei Guan[2,5], and Yongji Wang[2,5,6]

[1] Cooperative Innovation Center, Institute of Software,
Chinese Academy of Sciences, Beijing, China
`daoguang@iscas.ac.cn`
[2] University of Chinese Academy of Sciences, Beijing, China
`beiguan@iscas.ac.cn`, `ywang@itechs.iscas.ac.cn`
[3] Meituan Inc., Beijing, China
`{wangsirui,zhanghongzhi03,wuwei30}@meituan.com`
[4] Beijing University of Posts and Telecommunications, Beijing, China
[5] Integrative Innovation Center, Institute of Software, Chinese Academy of Sciences,
Beijing, China
[6] State Key Laboratory of Computer Science, Institute of Software,
Chinese Academy of Sciences, Beijing, China

Abstract. Knowledge Graph Question Answering (KGQA) is a challenging task that aims to obtain the entities from the given Knowledge Graph (KG) to answer the user's natural language questions. Most existing studies are focused on the traditional KGQA task, where the test distribution is the same as the training distribution over questions. In contrast, few efforts have been made to explore the *zero-shot* KGQA task. Logically, the existing models for the traditional KGQA task naturally show poor performance on the *zero-shot* setting. It is a non-trivial task to migrate the off-the-shelf *zero-shot* solutions in other common tasks to KGQA since an intrinsical gap exists between other common tasks and the KGQA task under the *zero-shot* settings. Furthermore, we observed that **S**imilar **Q**uestions tend to have **S**imilar **L**ogic forms. Motivated by this, we propose a simple yet effective framework S²QL. In detail, we first elaborately devise three similarity measurement units to category the user's questions. Then based on the Similarity Relation Graph (SRG) constructed by the above similarity measurement units, we devise a retrieval augmented strategy to further answer arduous *zero-shot* questions with its retrieved similar questions. Extensive experiments on the GrailQA and WebQSP benchmarks demonstrate that our approach is more effective than a number of competitive KGQA baselines on the *zero-shot* setting.

Keywords: Question answering · Knowledge graph · Information retrieval

D. Zan and Y. Yan—Work done during internship at Meituan Inc. The first two authors contribute equally.

J. Gama et al. (Eds.): PAKDD 2022, LNAI 13282, pp. 223–236, 2022.
https://doi.org/10.1007/978-3-031-05981-0_18

1   Introduction

Knowledge Graph Question Answering (KGQA) is a challenging but meaningful task that aims to reason the answer entity to the user's query over a given Knowledge Graph (KG). Over recent years, the KGQA task has attracted a great deal of research attention from academia and industry. Though remarkable performance has been achieved, most of these works assume that the training and test questions are from the same data distribution. With the increasing diversity and complexity of user's requirements, the assumption does not hold enough in practical business scenarios. At the same time, *zero-shot KGQA*, where the entities or relations involved in the test questions are unseen in the training phase, is becoming more and more common. However, it is almost unexplored [16]. Inevitably, most existing solutions for the traditional KGQA task have poor performance on the *zero-shot* setting since the traditional KGQA models themselves lack semantic modeling of unseen entities or relations, further affecting the correct semantic reasoning over the KG.

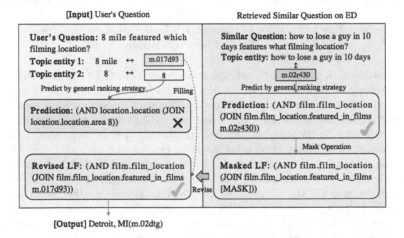

Fig. 1. A practical GrailQA [16] example to illustrate that similar queries tend to have similar logic forms. Therefore, we leverage the retrieved similar question (right) that answers correctly to guide the user's query (left) that answers wrongly until the correct answer is obtained.

To combat with the above problems, an intuitive idea is to leverage the off-the-shelf *zero-shot* solutions based on semantic similarity technology in other common tasks (*e.g.*, the classification task) to directly migrate to the KGQA task for dealing with the *zero-shot* setting. Although the idea is appealing, it is challenging to migrate the general *zero-shot* method to the KGQA task since an intrinsical gap exists between other common tasks and the KGQA task under the *zero-shot* setting. We concretize other common tasks into the classification task to illustrate the gap. In the classification task, the *zero-shot* solutions be summarized into two types. One type [27] is to align the vector spaces of the input sample and its output label. Another [1] is to leverage the generative models (*e.g.,* Autoencoder, GAN) to generate samples of the test set to enhance the robustness of the unseen domain. However, the above two solutions can not be directly applied to the KGQA task. For the former, the semantic space corresponding to KGQA's entities or relations is ambiguous because the mentions of the entities or relations in the user's question are not always available, further exacerbating the difficulty of aligning the vector spaces [2]. For the latter, compared with generating input samples (*e.g.,* image) based on the label description in the classification task, it is extremely challenging for the SQL2Text task [32] to generate the question based on the intricate logic form containing unseen entities or relations. Thus employing the generative models to the KGQA task is a troublesome job.

We further observed that it is potential for KGQA to benefit from referring to similar questions from practical KGQA scenarios in Fig. 1. For example, the question `how to lose a guy in 10 days features what filming location?` and the *zero-shot* question `8 mile featured which filming location?` are highly consistent in the correct logic form except for their topic entity[1]. The example inspires our main idea: referring to its corresponding similar questions when answering the arduous *zero-shot* question. To fill the gap between other common tasks and the KGQA task under the *zero-shot* setting, we propose a simple but effective framework S²QL to solve out-of-distribution questions based on the above idea. The method poses two problems: 1) how to measure the similarity between multiple queries? 2) how to utilize the measured similarity scores with other queries to improve the performance of *zero-shot* KGQA? The following will separately introduce how to solve the two problems.

For the first problem, we first devise three different units to measure the similarity of the questions in the KGQA task: 1) Entity Difference (ED) that divides the questions corresponding to logic forms that only differ in the topic entity into one category, 2) Relation Difference (RD) that classifies the questions corresponding to logic forms with different relations into the same category, and 3) None Difference (ND) that categorizes these questions that are entirely consistent in logic form into one category. Then based on the three different similarity measurement methods, we devise a training strategy to calculate the similarity and categorize all the questions involved in the inference phase.

For the second problem, we first employ the static score of similarity by ED, RD and ND to construct a Similarity Relationship Graph (SRG). Then based on the SRG, we design a retrieval augmented KGQA training strategy for catering to our main idea. More detailed, in addition to model the correlation between the user's question and the corresponding candidate answers, the strategy also considers the retrieved similar questions that guide the user's question to answer correctly by the masked logic form (as Masked LF in Fig. 1).

In general, S²QL can effectively weaken the model's dependence on unseen elements (*e.g.* entities or relations), allowing the model to only focus on the seen

---

[1] Follow the [16], the logical form used is s-expression in our experiment.

part of the logic form. Therefore, the unseen elements only require extraction from the question and filling in the masked logic form. The contributions can be summarized as follows:

- We propose a simple but effective framework S^2QL to solve *zero-shot* KGQA. To the best of our knowledge, it is the first approach to leverage similar questions to guide the *zero-shot* questions in the KGQA task.
- Under the framework, we elaborately devise three similarity measurement units (ED, RD, ND) to classify all questions. Based on the Similarity Relationship Graph (SRG) constructed by calculating ED, RD and ND, we devise a retrieval augmented strategy to answer arduous *zero-shot* questions with its retrieved similar questions.
- Through a series of experiments on GrailQA [16] and WebQSP [36], we demonstrate the effectiveness of our approach, especially in the *zero-shot* setting.

2   Preliminaries

Knowledge Graph (KG): A KG typically organizes factual information as a set of triple fact, denoted by $\mathcal{G} = \{\langle h, r, t \rangle | h, t \in \mathcal{E}, r \in \mathcal{R}\}$, where $\mathcal{E}$ and $\mathcal{R}$ represent all entities and relations, respectively. A triple fact $\langle h, r, t \rangle$ denotes that relation r exists between head entity h and tail entity t. A KG is given as the available resource.

Knowledge Graph Question Answering (KGQA): Knowledge Graph Question Answering (KGQA) is a task that aims to map user's questions to logic forms, *e.g.*, s-expression, programs, or SPARQL, which can be directly executed on the KG to yield answers. In this paper, we focus on solving the *zero-shot* KGQA task, where the entities or relations involved in the test questions is unseen in the training phase.

Masked Logic Form (MLF): MLF is a concept we proposed. We classify the questions according to our proposed three similarities of ED, RD, and ND. Concretely, we first obtain the logic form with the highest probability of being correct in each category, and then mask its entity, relation, or none with the token [MASK]. Thus the MLF is generated. As an example in Fig. 1, we assume that the generalization level of the user's question (8 mile featured which filming location?) is *zero-shot*, the similarity measurement unit we utilized is Entity Difference (ED), and the general BERT-based ranking strategy cannot reach the correct answer. Therefore, we first extract its similar question (how to lose a guy in 10 days features what filming location?) that has the highest probability of correctness in the same similarity category. Then replace the entity in the logic form corresponding to the extracted question with the token [MASK] to generate MLF (AND film.film_location (JOIN film.film_location.featured_in_films [MASK])).

3   Method

3.1   The General Pipeline of the KGQA Model

As shown in the left part of Fig. 2, we summarize the general pipeline of the KGQA model into three steps. First, we extract the topic entity over the given $\mathcal{G}$ for the user's question. Then, the candidate logic forms are generated. Next, the BERT-based ranking strategy ranks all the candidate logic forms to select top-1 as the predicted logic form. Finally, we execute the predicted logic form over $\mathcal{G}$ to obtain the final answers for the user's question.

Entity Linking. We train a BERT-based model[2] to conduct Name Entity Recognition (NER). We choose the most popular one(s) based on FACC1 for entity disambiguation from the detected mentions.

Candidate Logic Form Generation. We first extract all reasoning paths within k-hop to the topic entity over $\mathcal{G}$ and then encapsulate the paths into candidate logic forms according to the pre-defined grammar-based framework. We take GrailQA as an example to illustrate it. Since the limitation of the enormous search space, previous works [16] only set up 12 common grammar-based schema types for the inference phase (*e.g.,* (ARGMAX), (AND (le), (AND(JOIN)))). To enhance the diversity of candidate logic forms, we increase the number of schema

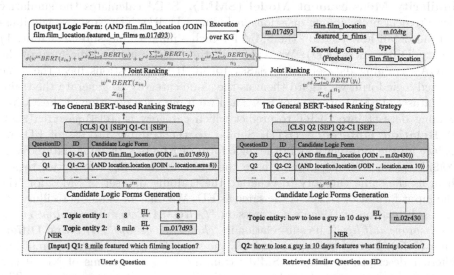

Fig. 2. An overview of S²QL. We only leverage ED as examples to illustrate the pipeline of our model. For brevity, we abbreviate *Named Entity Recognition*, *Entity Linking* and the path of candidate logic forms as *NER*, *EL* and ellipsis.

---

[2] https://github.com/kamalkraj/BERT-NER.

types to 148 according to GrailQA's training set. To deal with the explosion of candidate logic forms, we use the neural model to prune these irrelevant candidate logic forms.

The General BERT-Based Ranking Strategy. As shown in Fig. 2, for the user's question q, we obtain its topic entity e_{topic} by the entity linking. Then, the module of candidate logic form generation is used to generate the candidate logic forms, and we name each logic form as $\mathcal{F}_{in}^v$. Next, we concatenate the question q and its each candidate logic form $\mathcal{F}_{in}^v$ to make an BERT's input sample $x_{in} =$ [CLS] q [SEP] $\mathcal{F}_{in}^v$ [SEP], where we determine the semantic match score of the input sample according to the output of the [CLS] token. Finally, we select the $\mathcal{F}_{in}^v$ with the highest score as the prediction of logic form, which can be executed over $\mathcal{G}$ to obtain the final predicted answer for q.

3.2 S²QL

The general pipeline of the KGQA model in Sect. 3.1 only models the user's question, ignoring the context-based knowledge of the retrieved similar questions. Therefore, we first propose the Similarity Measurement Model (SMM) to build the Similarity Relationship Graph (SRG). Then we devise a retrieval augmented ranking strategy to facilitate modeling of similar questions based on the SRG, as displayed in the right part of Fig. 2.

Similarity Measurement Model (SMM). SMM calculates the similarity measurement units among questions including ED, RD, ND. Next, we will illustrate in detail how to train the SMM for each similarity measurement unit. **1) Entity Difference (ED):** We replace all the entities in the logic form of the KGQA's training dataset with [MASK] token. If two questions have the same masked logic form, we regard them as the same category to generate the training dataset of SMM in terms of ED. Then we input each question of the SMM's training dataset into BERT to predict which category is: [CLS] question [SEP]. **2) Relation Difference (RD):** Similar to the training process of ED, we replace all the sub-relations in the logic form of KGQA's training dataset with [MASK] token. If two questions have the same masked logic form, we regard them as the same category to generate the training dataset of SMM for RD. The input format of BERT is the same as ED. We take an example to illustrate what sub-relation is. Given a s-expression *"(ARGMAX chemistry.isotope chemistry.isotope.half_life)"*, its sub-relation is *"chemistry.isotope"*. **3) None Difference (ND):** In real business scenarios, the same logic form often has multiple different descriptions. For the SMM's training dataset in terms of ND, if two questions have the same logic form, we regard them as the same category. The input format of BERT is also the same as ED.

Similarity Relationship Graph (SRG). It is proposed by this paper, a graph composed of questions and the relationships among questions. The nodes of the graph are user's question, and the edges of the graph consist of ED, RD, and ND. Our experiment uses SRG to store the similarity relationship among the questions to facilitate the loading of the questions similar to the user's question in the KGQA model later. **1) The process of constructing SRG.** We first obtain all categories on the training set. Then we utilize the SMM model to classify the questions on the inference phase. Finally, SRG is constructed according to the above category results. **2) How to efficiently retrieve similar questions for user's questions?** To efficiently retrieve similar questions based on SRG, we load SRG as a graph into a graph database engine, gStore System^3 [37] where the graph's node is the user's question, and the graph's edge is one of three similarity measurement types (ED, RD and ND).

A Retrieval Augmented Ranking Strategy with SRG. Based on the general ranking strategy in Sect. 3.1, to model the similar questions of q as shown in the right part of Fig. 2, we retrieve all similar questions (Q_{ed}, Q_{rd} and Q_{nd}) by SRG. Then for each question q_{ed} in Q_{ed}, we extract all candidate logic forms that are consistent with $\mathcal{F}_{in}^v$ except for the topic entity e_{topic}, its i-th candidate logic form is named $\mathcal{F}_{ed}^i$. Similar to Q_{ed}, we also extract the candidate logic forms that are consistent with $\mathcal{F}_{in}^v$ besides the sub-relation for each question q_{rd} in Q_{rd}, its j-th candidate logic form is named $\mathcal{F}_{rd}^j$. In the same way, we can find all candidate logic forms that are absolutely the same as $\mathcal{F}_{in}^v$ for each question q_{nd} in Q_{nd}, its k-th candidate logic form is named $\mathcal{F}_{nd}^k$. Finally, we make input samples for the similar questions q_{ed}, q_{rd} and q_{nd}: $y_i = $ [CLS]q_{ed}[SEP]$\mathcal{F}_{ed}^i$[SEP], $z_j = $ [CLS]q_{rd}[SEP]$\mathcal{F}_{rd}^j$[SEP] and $p_k = $ [CLS]q_{nd}[SEP]$\mathcal{F}_{nd}^k$[SEP]. Our ultimate goal is to model the semantic evaluation score of the user's question q and each candidate logic form, where similar questions are also considered to facilitate the above process. Thus, we feed the above samples to BERT and take the representation corresponding to the [CLS] token for binary classification:

$$s_i = \sigma(w^{in}BERT(x_{in}) + w^{ed}\frac{\sum_{i=0}^{n_1}BERT(y_i)}{n_1} +$$
$$w^{rd}\frac{\sum_{j=0}^{n_2}BERT(z_j)}{n_2} + w^{nd}\frac{\sum_{k=0}^{n_3}BERT(p_k)}{n_3}), \tag{1}$$

where $\sigma(.)$ is the sigmoid function and w^{in}, w^{ed}, w^{rd} and w^{nd} represent the weight coefficient corresponding to q, q_{ed}, q_{rd} and q_{nd}, and thus our loss function $\mathcal{L}_i$ is defined as the negative log-likelihood of the logistic model:

$$\mathcal{L}_i = -(g \times \log s_i + (1 - g) \times \log(1 - s_i)), \tag{2}$$

where g is the correct label indicating if $\mathcal{F}_{in}^v$ is the ground truth logic form of q or not.

---

^{3} https://github.com/pkumod/gStore.

4   Experiments and Results

4.1   Setup

Datasets and Metrics. We adopt two benchmark datasets for the KGQA task: **GrailQA** [16] provides evaluation settings for all three levels of generalization: *i.i.d.*, *compositional*, and *zero-shot*, which account for 25%, 25%, and 50% of test dataset, respectively. **WebQuestionSP (WQSP)** [36] consists of 4,737 natural language questions that are answerable through Freebase KG. To make a fair comparison, we strictly keep the same metrics as the baselines. We report the Exact Match (EM) and F1 score evaluated on the GrailQA test set. At the time of writing, S^2QL achieves the second position on the GrailQA leaderboard in both the *overall* setting and *zero-shot* setting[4]. For the WebQSP benchmark, we use hits@1 and F1 score as evaluation metrics.

Baseline Models. We compare our model against the previous competitive methods on GrailQA and WebQSP benchmarks, including QGG [26], Ranking [16], Transduction (alias Trans) [16] and RetraCk [7] for GrailQA, and GRAFT-Net [34], EmbedKGQA [31], PullNet [33], Ranking [16], QGG [26], NSM+h [20] and RetraCt [7] for WebQSP. In general, All the above baselines are limited to model the user's question, while S^2QL models similar questions with the user's question.

Implementation Details. We implement our proposed framework S^2QL with PyTorch, AllenNLP and Transformers library. In our experiment, we use a single Tesla V100 for the training and inferencing phase, and the logic form is s-expression. In particular, if more than 10,000 candidate paths correspond to the question, we randomly select 10,000 paths. In training, we employ the Adam optimizer. The KGQA task's batch size is set to 8. The value of K in k-hop is set to 2 for both GrailQA and WebQSP.

4.2   Results

The main experiment in left part of Table 1 presents the exact matching (EM) accuracy and F1 score of S^2QL and various baselines on the test and development set of GrailQA benchmark. In detail, S^2QL clearly outperforms all the baselines by a substantial margin on the *zero-shot* setting. It obtains 6.5% and 7.9% absolute improvement over the previous state-of-art model on the EM accuracy and F1 score. On the *overall* setting of GrailQA, our proposed approach also surpass all baselines on the F1 score.

To better study whether S^2QL has a continuous and significant generalization performance, we also conduct plenty of experiments on the WebQSP benchmark where we integrate ED, RD, and ND into NSM_{+h} proposed by [20]

---

[4] https://dki-lab.github.io/GrailQA.

Table 1. The performance on GrailQA (left) and WebQSP (right). For GrailQA, all baseline results with ‡ are taken from [7], with ‡ are taken from the leaderboard of GrailQA. Specifically, S²QL = General Ranking + ED, RD, ND. For WebQSP, all baseline results are from [7] except for NSM$_{+h}$ taken from [20]. † denotes our re-implemented since the unpublished source code in [7].

dataset / model	Overall EM	Overall F1	Zero-shot EM	Zero-shot F1
Baseline (test)				
QGG‡	-	36.7	-	36.6
Glove+Trans‡	17.6	18.4	3.0	3.1
BERT+Trans‡	33.3	36.8	25.7	29.3
Glove+Ranking‡	39.5	45.1	28.9	33.8
BERT+Ranking‡	50.6	58.0	48.6	55.7
RetraCk‡	**58.1**	65.3	44.6	52.7
Our implementation				
S²QL (ours)	57.5	**66.2**	**55.1**	**63.6**
Ablation Experiments (dev)				
General Ranking	51.0	58.0	51.7	58.5
w/ ED	53.6	60.8	53.8	60.7
w/ RD	52.8	60.4	52.5	60.1
w/ ND	51.6	58.6	52.3	59.2
w/ ED,RD,ND	**57.7**	**65.0**	**58.3**	**65.0**

dataset / model	F1	Hits@1
Baseline		
GRAFT-Net	62.8	67.8
QGG	74.0	-
EmbedKGQA	-	66.6
GrailQA Ranking†	67.0	-
RetraCt	74.7	**74.6**
NSM$_{+h}$	74.3	67.4
Ablation Experiments		
NSM$_{+h}$	74.3	67.4
w/ ED	74.6	67.6
w/ RD	75.0	68.1
w/ ND	74.1	66.9
w/ ED,RD,ND	**75.2**	68.5
GrailQA Ranking†	67.0	-
w/ ED,RD,ND	68.2	-

and GrailQA Ranking proposed by [16] respectively. The results are shown in right part of Table 1. The similarity measurement strategy (w/ ED, RD, ND) we devised can indeed bring +0.9% and +1.2% performance improvements on F1 score compared with both NSM$_{+h}$ and GrailQA Ranking. More importantly, our model reaches 75.2% on the F1 score based on NSM$_{+h}$, outperforming all baselines.

4.3 Ablation Study

To verify the effectiveness of our proposed similarity measurement methods (ED, RD, and ND) and our devised pre-training models, we conduct ablation studies on the GrailQA and WebQSP benchmarks to analyze the contribution of each design choice. The left part of Table 1 also presents the ablation study results on the development set of GrailQA since the leaderboard of GrailQA does not allow ablation study or parameter study on the test dataset. Specifically, we use the general BERT-based ranking model (Sect. 3.1) as the basic model. By attaching one or more similarity measurement modules (ED, RD, and ND), the performance improves a large margin of both EM and F1 score, which also demonstrate that the knowledge of similarity between user's queries can indeed promote the performance of the Q&A over $\mathcal{G}$.

To prove that our proposed approach also performs well on the traditional KGQA task, we conduct plenty of analytical experiments on the WebQSP benchmark. To make our method more convincing, we regard the NSM$_{+h}$ proposed

Table 2. Ablation study on the dev set of GrailQA for both BERT+TRANSDUCTION and GloVe+TRANSDUCTION.

Model	Dataset							
	Overall		I.I.D.		Comp		Zero-shot	
	EM	F1	EM	F1	EM	F1	EM	F1
Baseline (TRANSDUCTION)								
BERT-based (dev)	33.1	35.2	50.5	52.7	31.4	34.0	26.3	28.1
Ablation experiments								
w/ ED	36.6	40.1	53.4	55.1	36.5	40.6	29.3	33.4
w/ RD	36.7	38.8	52.2	53.9	35.1	37.6	30.6	32.8
w/ ND	34.0	35.4	50.9	51.6	33.2	34.5	26.9	28.7
w/ ED, RD, ND	39.8	43.6	56.1	59.2	40.5	44.9	32.5	36.3
Baseline (TRANSDUCTION)								
GloVe-based (dev)	16.5	17.6	48.2	49.4	16.1	18.9	2.9	3.2
Ablation experiments								
w/ ED	18.5	19.3	50.3	51.6	19.4	20.9	4.2	4.5
w/ RD	18.1	19.1	49.4	51.2	19.7	21.0	3.8	4.4
w/ ND	16.6	17.6	46.4	48.1	17.3	19.0	3.3	3.7
w/ ED,RD,ND	20.0	20.7	52.4	53.1	21.7	22.3	5.1	5.9

by [20] as the basic model to implement the similarity measurement methods (ED, RD, ND). As shown in the right part of Fig. 1, using the ED (w/ ED), RD (w/ RD), ND (w/ ND) or all of the above operation (w/ ED, RD, ND) bring about 0.3%, 0.7%, −0.2% and 0.9% absolute improvement in F1 score, 0.2%, 0.7%, −0.5% and 1.1% increment in EM accuracy. Note that the performance of applying ND on the WebQSP benchmark is descending because there are relatively few instances of the same logic form in WebQSP. All the above experiments assume that the basic model is the type of query-graph generation KGQA solution. To illustrate that our model also has high generalization on another type of solution (*e.g.,* TRANSDUCTION-based models), we conduct the experiments based on both BERT+TRANSDUCTION and GloVe+TRANSDUCTION proposed by [16] as shown in Table 2. Such above detailed results demonstrate that similar questions indeed boost the performance of the *zero-shot* user's question with various KGQA models.

4.4   Case Study

The major novelty of our approach lies in the retrieved similar question. Next, we present a case study for illustrating how it helps and directs the user's question to obtain the correct answer. As shown in Fig. 3, given the user's question q and the ground truth logic form $\mathcal{F}$. However, the original model without ED, RD and ND is not good at reasoning on unseen elements (*e.g.,* Card game→m.01mtt),

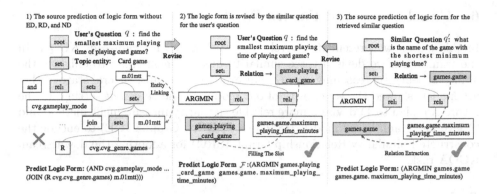

Fig. 3. A case study from the GrailQA dataset. To facilitate the detailed illustration of the revising process, we also drew the logic form's grammar tree. In the figure, all red items denote that the predicted logic form is wrong, and the green ones represent the correct logic form. In general, the predicted logic form of the user's question (left) is revised by a similar question (right), and thus the correct logic form can be correctly formed to answer the given question (middle).

and thus it makes the wrong prediction (left part). We observe that the general BERT-based ranking strategy is challenging to solve the question q, it yet makes a perfect prediction for its retrieved similar question q' (right part). According to our idea: similar questions tend to have similar logic forms. Thus, the similar retrieved question q' is utilized to revise the question q to obtain the correct logic form $\mathcal{F}$.

5   Related Work

The earliest KGQA system focused on factoid question answering over $\mathcal{G}$. Widely-used datasets of simple KGQA include SimpleQuestion [5], WebQSP [36] and so on, while the existing baselines [30] have extraordinary performance on the mentioned above factoid Q&A benchmarks. With the complexity of real business requirements, researchers are gradually focusing on the complex and multi-hop KGQA task, including ComplexWebQuestion [35], ComplexQuestion [3] and so on. Since the construction process of the KGQA dataset is time-consuming and laborious, the use of few-shot and zero-shot samples [16] for training has become an urgent problem. However, the existing models [26,31] of KGQA have poor generalization in actual application with *zero-shot* setting, and thus how to devise a KGQA system's architecture to enhance the generalization is of great significance. In general, KGQA's solutions [16] can be roughly categorized into information-retrieval methods and semantic-parsing-based methods. The former method ranks several candidate answers or paths to select the top-1 as the prediction result [9,16,20,26,31,33,34]. While the latter generates an executable logic form from the given question to get the final answer [7,16,36].

6   Conclusions and Future Works

In this paper, we presented a novel framework S^2QL for *zero-shot* KGQA. In the framework, we devised three similarity measurement units to enhance the ability of semantic reasoning over $\mathcal{G}$. Our evaluation against several state-of-the-art models showed the superiority of our model on the *zero-shot* setting. In the future, we plan to 1) devise a more reliable similarity measurement method and 2) leverage graph-based models [18,24] (*e.g.*, GCN, GraphSAGE) to encode the constructed Similarity Relationship Graph (SRG) for adopting the downstream KGQA task.

References

1. Annadani, Y., Biswas, S.: Preserving semantic relations for zero-shot learning. In: Proceedings of the IEEE Conference on Computer Vision and Pattern Recognition, pp. 7603–7612 (2018)
2. Banerjee, D., Chaudhuri, D., Dubey, M., Lehmann, J.: PNEL: pointer network based end-to-end entity linking over knowledge graphs. In: Pan, J.Z., et al. (eds.) ISWC 2020. LNCS, vol. 12506, pp. 21–38. Springer, Cham (2020). https://doi.org/10.1007/978-3-030-62419-4_2
3. Bao, J., et al.: Constraint-based question answering with knowledge graph. In: Proceedings of COLING 2016, the 26th International Conference on Computational Linguistics, Technical Papers, pp. 2503–2514 (2016)
4. Bollacker, K., et al.: Freebase: a collaboratively created graph database for structuring human knowledge. In: Proceedings of the ACM SIGMOD International Conference on Management of Data, pp. 1247–1250 (2008)
5. Bordes, A., et al.: Large-scale simple question answering with memory networks. arXiv preprint arXiv:1506.02075 (2015)
6. Chen, J., et al.: Knowledge-aware Zero-shot learning: survey and perspective. arXiv preprint arXiv:2103.00070 (2021)
7. Chen, S., et al.: ReTraCk: a flexible and efficient framework for knowledge base question answering. In: ACL: System Demonstrations, pp. 325–336 (2021)
8. Chen, X., Jia, S., Xiang, Y.: A review: knowledge reasoning over knowledge graph. Expert Syst. App. **141**, 112948 (2020)
9. Chen, Z.Y., et al.: UHop: an unrestricted-hop relation extraction framework for knowledge-based question answering. In: ACL: Human Language Technologies, vol. 1 (Long and Short Papers), pp. 345–356 (2019)
10. Dinu, G., Lazaridou, A., Baroni, M.: Improving zero-shot learning by mitigating the hubness problem. arXiv preprint arXiv:1412.6568 (2014)
11. Do, T., et al.: Compact trilinear interaction for visual question answering. In: Proceedings of the IEEE/CVF International Conference on Computer Vision, pp. 392–401 (2019)
12. Frome, A., et al.: DeViSE: a deep visual-semantic embedding model. In: NIPS (2013)
13. Fuglede, B., Topsoe, F.: Jensen-Shannon divergence and Hilbert space embedding. In: Proceedings of International Symposium on Information Theory: ISIT 2004, p. 31. IEEE (2004)
14. Furlanello, T., et al.: Born again neural networks. In: International Conference on Machine Learning, pp. 1607–1616. PMLR (2018)

15. Gabrilovich, E., Ringgaard, M., Subramanya, A.: FACC1: Freebase annotation of ClueWeb corpora (2013)
16. Gu, Y., et al.: Beyond IID: three levels of generalization for question answering on knowledge bases. In: Proceedings of the Web Conference 2021, pp. 3477–3488 (2021)
17. Guo, Y., et al.: Synthesizing samples from zero-shot learning. In: IJCAI (2017)
18. Hamilton, W.L., Ying, R., Leskovec, J.: Inductive representation learning on large graphs. In: Proceedings of the 31st International Conference on Neural Information Processing Systems, pp. 1025–1035 (2017)
19. Hayashi, T., Fujita, H.: Cluster-based zero-shot learning for multivariate data. J. Ambient Intell. Human. Comput. **12**(2), 1897–1911 (2021)
20. He, G., et al.: Improving multi-hop knowledge base question answering by learning intermediate supervision signals. In: Proceedings of the 14th ACM International Conference on Web Search and Data Mining, pp. 553–561 (2021)
21. Hinton G, Vinyals O, Dean J. Distilling the knowledge in a neural network. arXiv preprint arXiv:1503.02531 (2015)
22. Hu, M., et al.: Attention-guided answer distillation for machine reading comprehension. In: EMNLP, pp. 2077–2086 (2018)
23. Huang, S., et al.: Learning hypergraph-regularized attribute predictors. In: Proceedings of the IEEE Conference on Computer Vision and Pattern Recognition, pp. 409–417 (2015)
24. Kipf ,T.N., Welling, M.: Semi-supervised classification with graph convolutional networks. arXiv preprint arXiv:1609.02907 (2016)
25. Kullback, S., Leibler, R.A.: On information and sufficiency. Ann. Math. Statist. **22**(1), 79–86 (1951)
26. Lan, Y., Jiang, J.: Query graph generation for answering multi-hop complex questions from knowledge bases. In: Association for Computational Linguistics (2020)
27. Li, Y., et al.: Zero-shot recognition using dual visual-semantic mapping paths. In: Proceedings of the IEEE Conference on Computer Vision and Pattern Recognition, pp. 3279–3287 (2017)
28. Nguyen, H.V., Gelli, F., Poria, S.: DOZEN: cross-domain zero shot named entity recognition with knowledge graph. In: Proceedings of the 44th International ACM SIGIR Conference on Research and Development in Information Retrieval, pp. 1642–1646 (2021)
29. Noh, J., Kavuluru, R.: Joint learning for biomedical NER and entity normalization: encoding schemes, counterfactual examples, and zero-shot evaluation. In: Proceedings of the 12th ACM Conference on Bioinformatics, Computational Biology, and Health Informatics, pp. 1–10 (2021)
30. Petrochuk, M., Zettlemoyer, L.: Simple questions nearly solved: a new upperbound and baseline approach. In: EMNLP, pp. 554–558 (2018)
31. Saxena, A., Tripathi, A., Talukdar, P.: Improving multi-hop question answering over knowledge graphs using knowledge base embeddings. In: Proceedings of the 58th Annual Meeting of the Association for Computational Linguistics, pp. 4498–4507 (2020)
32. Shu, C., et al.: Logic-consistency text generation from semantic parses. In: Findings of the Association for Computational Linguistics: ACL-IJCNLP 2021, pp. 4414–4426 (2021)
33. Sun, H., Bedrax-Weiss, T., Cohen, W.: PullNet: open domain question answering with iterative retrieval on knowledge bases and text. In: EMNLP-IJCNLP, pp. 2380–2390 (2019)

34. Sun, H., et al.: Open domain question answering using early fusion of knowledge bases and text. In: EMNLP, pp. 4231–4242 (2018)
35. Talmor, A., Berant, J.: The web as a knowledge-base for answering complex questions. In: ACL: Human Language Technologies, vol. 1 (Long Papers), pp. 641–651 (2018)
36. Yih, W., et al.: semantic parsing via staged query graph generation: question answering with knowledge base. In: ACL, pp. 1321–1331 (2015)
37. Zou, L., et al.: gStore: a graph-based SPARQL query engine. VLDB J. **23**(4), 565–590 (2014)

Improve Chinese Spelling Check
by Reevaluation

Shuai Wang[1,2] and Lin Shang[1,2(✉)]

[1] State Key Laboratory for Novel Software Technology, Nanjing University,
Nanjing 210023, China
wangshuai@smail.nju.edu.cn
[2] Department of Computer Science and Technology, Nanjing University,
Nanjing 210023, China
shanglin@nju.edu.cn

Abstract. Chinese Spelling Check (CSC) aims to detect and correct
the spelling errors in Chinese. Most Chinese spelling errors are misused semantically, phonetically or graphically similar characters. Previous state-of-the-art works on the CSC task pursue transitions from
misspelled sentences to correct sentences directly. However, the spelling
errors, especially the continuous incorrect characters, usually confuse the
meaning of the semantic context. It is difficult to make correct modifications for CSC models based on the error contextual information. To
address this issue, we propose a simple but effective pipeline for CSC
by searching the most appropriate candidate sentences as the original
correct sentence. Specifically, candidate sentences are generated based
on possible error characters with the confusion set. Then we reevaluate
the candidate sentences to find the best in terms of character probabilities and similarity compared to the original error characters. Besides, we
extend the widely used confusion set (The code and data are available
at https://github.com/zuoyecihua/CSC.). Simply applying the confusion
set as a filter will bring large performance improvement. The experimental results show that our approach outperforms previous methods and
performs well on bi-gram errors.

Keywords: Chinese spelling check · Text process · Natural language
process

1 Introduction

Spelling check plays an important role in many natural language applications,
such as machine translation [2], search query correction [7,15], part-of-speech
tagging [17], optical character recognition [1]. The goal of Chinese spelling check
(CSC) is to identify and correct typos in Chinese, so that the grammar of the

This work is supported by the National Natural Science Foundation of China (No.
61672276, No. 51975294).

modified text is correct and the modified text is consistent with the meaning of the original sentence. Chinese is hieroglyphic. Each Chinese character has its own font and pronunciation and can be seen as a word in alphabetic languages such as English. Most Chinese spelling errors are misused semantically, phonetically or graphically similar characters.

Recently, methods based on pre-trained language models have achieved state-of-art performance in the CSC task [8,18,24,25]. These methods are mainly divided into two categories: 1) based on greedy strategy in decoding stage. During the training phase, the input is Chinese sentences with error characters and the labels are sentences without any error. During the inference stage on test datasets, the model gives probability distribution at each position of the input. If the most likely character is different from the input character at the same position, the most likely character would be considered as a corresponding correction. 2) based on well designed decoding strategy. Multiple candidates are generated for each location and dependencies between adjacent Chinese characters are modeled by the attention network.

To obtain accurate prediction results, CSC models have to understand the input text as accurate as possible. However, the input sentences in the CSC task are always with error characters and the true meaning of the sentence may change greatly because of the errors. Besides, it is almost impossible for models to capture the accurate meaning when the error characters are adjacent, e.g., bi-gram errors. When the model makes predictions, the accuracy may suffer from noisy information from error characters.

In this paper, we firstly proposed an algorithm based on maximum posterior probability to alleviate the influence of error characters, which works well with bi-gram errors. A Pinyin enhanced language model with confusion set is adopted to generate candidates. Possible error characters will be replaced by the candidates to generate modified sentences. All modified sentences will be evaluated again to obtain more accurate probability distributions, which are posterior probabilities. The sentence with maximum posterior probability and maximum similarity to the original error characters will be selected as the final best answer. Besides, we extend the widely used confusion set [22] according to widely used training materials in the CSC task. Simply applying filter strategy guided by the extended confusion set brings considerable benefits on correction performance.

To summarize, our contributions are as follows:

- We propose a reevaluation strategy to avoid noisy information from error characters. This strategy works well on bi-gram errors which is usually difficult for previous CSC methods.
- We extend the widely used confusion set according to the common training materials in the CSC task. Simply applying filter strategy guided by the extended confusion will bring benefits on CSC performance.
- Experimental results on the benchmark SIGHAN datasets show that our method outperforms the others on SIGHAN 15 and achieved competitive F1 scores on SIGHAN 13/14. This demonstrates the effectiveness of our method.

2   Related Work

The CSC task has attracted lots of research attention. Early works on CSC mainly focus on rule based unsupervised methods. Various rules are designed to deal with different kinds of spelling errors [3]. A character would be considered as a spelling error if its probability in n-gram language model is lower than the pre-defined threshold [13]. When it comes to error correction, confusion set is widely used for candidates generation. With the development of neural networks, some work regards the CSC task as a sequence-labeling task with RNN models [21, 23]. Copy mechanism in sequence-to-sequence framework is also introduced to the CSC task to copy the possible correction words from the confusion set [20].

With the great success of large pre-trained language models [5], many BERT-based CSC models are proposed and have made tremendous progress in the CSC task. In [9], a language model is employed as candidates generator and a confidence-similarity curve is employed to select the best candidate considering word confidence and character similarity. Soft-Masked-Bert [26] consists of error detection module based on GRU [6] and error correction module based on BERT. With the development of graph neural networks[11,27], SpellGCN [4] utilizes a graph convolution neural network to model similarity relations between error characters and the candidates. [18] proposed Dynamic Connected Networks (DCN) to model the dependencies between two adjacent Chinese characters. Multimodal information, etc. glyph and pronunciation feature, is incorporated in BERT and achieved good performance. ReaLiSe [24] makes correction by directly leveraging multimodal information of Chinese characters. MLM-phonetics [25] propose a novel end-to-end CSC model that integrates phonetic features into a language model by leveraging the powerful pre-training and fine-tuning method. Recently, some studies begin to pay attention to the problem of noisy information caused by error characters. [8] proposed a global attention decoder (GAD) to alleviate the impact of noisy information. GAD learns the global relationships of the potential correct input characters and the candidates of potential error characters.

3   The Proposed Approach

In this section, we introduce the P-BERT-RF model, which is the Pinyin enhanced BERT with reevaluation and filter strategy. Firstly, the problem formulation is elaborated. Then we describe the Pinyin Enhanced BERT (P-BERT) model. Finally, we introduce our proposed strategy of reevaluation and filter.

3.1   Problem

Given an input Chinese text sequence $X = (x_1, x_2, ..., x_N)$ which contains error characters, the CSC task aims to generate a correct sequence $Y = (y_1, y_2, ..., y_N)$. N denotes the number of characters. The input qequence and the generated sequence are the same length in the CSC task, which is different from Grammatical Error Correction (GEC).

3.2 The Pinyin Enhanced BERT

Pinyin Embeddings for Phonetic Features of Chinese Characters. According to [12], more than 80% of Chinese spelling errors are from phonologically similar characters. Since phonological errors account for large portion of Chinese spelling errors, we adopt Pinyin embeddings to model phonetic features of Chinese characters so that the model can achieve better performance on phonological errors. Pinyin is an official romanization system for Chinese characters and is widely used in China. Pinyin provides phonetic information for all Chinese characters, which would be of great help to find out the best one from many candidate characters. The conversion from a Chinese character to its Pinyin is certain while the conversion from Pinyin to Chinese characters is uncertain. In other words, a single Pinyin code is shared by many different Chinese characters. For example, the Pinyin for Chinese characters "户禿" and "糊涂" are both "hutu". Since "户禿" and "糊涂" are both pronounced as "hutu", we have more confidence to convert error characters "户禿" into the correct version "糊涂". We ignored the tone and the final (vowel) of Pinyin to obtain more phonological candidates for a single Pinyin. The Pinyin embeddings are randomly initialized.

The Pinyin Enhanced BERT. Figure 1 shows our model architecture, which is a variant of BERT with additional Pinyin embeddings. The initialized Pinyin embeddings are summed up with the original token embeddings as the input of BERT. In the training process, the model would predict the correct characters based on the input sequence characters and their Pinyin codes. In the inference process, the model is employed as both the detector and evaluator for Chinese spelling errors. The details of inference will be described in the following sections.

3.3 The Reevaluation Strategy

The difficulty of correcting Chinese spelling errors largely depends on the number of error characters in same sentence. It is easier for CSC models to understand the true meaning of the input text and make a successful correction when there is only one error character. However, when the number of error characters increases, the accuracy of CSC models usually decreases sharply. Because the CSC models may suffer from noisy information from error characters, especially continuous error characters. To address the problem, we propose a simple but effective method to handle bi-gram errors based on maximum posterior probability. The optimization consists of three steps:

1. Locate the possible continuous errors. Considering the continuous characters are mostly bi-grams in CSC datasets and to simplify the difficulty of implementation, we only make optimization for bi-gram errors in the experiments. To locate the possible error bi-grams, we obtain probabilities for each token of the input sequence by the Pinyin Enhanced BERT. If two adjacent characters both have probabilities lower than 0.8 or the product of two adjacent probabilities lower than 0.64, we assume them a possible bi-gram error. Besides, if

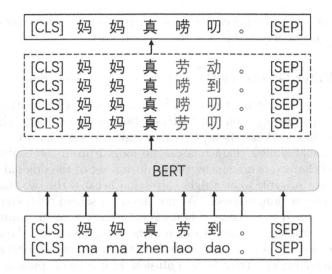

Fig. 1. Architecture overview of the Pinyin Enhanced BERT. The Pinyin codes of Chinese characters are also the input of BERT which is summed up with token embeddings of BERT. In the example input, for the bi-gram error "劳到", several candidate sentences are generated and the correct one is selected according to the reevaluation strategy.

two adjacent characters are modified by the Pinyin Enhanced BERT, we also assume them a possible bi-gram error.

2. Generate candidate sentences. To generate candidate bi-grams, we retrieve top characters from the extended confusion set by probability. Specifically, we sample top k characters with similar pronunciation and top k other characters in the confusion set at each position. Then there are k^2 candidate bi-grams composed of the sampled characters. To narrow the search space, we drop bi-grams with extremely low word frequency in training materials.

3. Select the best candidate. By replacing the possible bi-gram errors with candidate bi-grams, many new candidate sequences are generated. We obtain probabilities for the replaced bi-gram again from the Pinyin Enhanced BERT. To select the best candidate bi-gram, we retain those with product of two adjacent probabilities higher than 0.9 as high confident candidates. Then we select a best one by computing a selective score which is mainly reflecting similarity between candidate bi-gram and the original bi-gram. The principle is to find the most similar candidate from the high confident candidates.

The selective score is used to find the best candidate from all high confident candidates. Given the original possible error bi-gram $span_x = (x_i, x_j)$ and the corresponding candidate bi-gram $span_y = (y_i, y_j)$, we compute the number of similar characters between $span_x$ and $span_y$ as the selective score. If the word frequency of a candidate bi-gram is extremely high in training materials, e.g., more than 1000, its selective score will get additional one point. All high confident candidates are sorted by selective scores and reevaluated probabilities. Finally, the first high confident candidate is the selected answer.

3.4   The Filter Guided by the Extended Confusion Set

Confusion set is widely used in the CSC task. Many CSC methods such as [4,8,18] leveraged a confusion set, which contains a set of similar characters in terms of phonological and visual. Given a single Chinese character, its phonological and visual similar characters can be looked up in the confusion set. If the corrected characters are among the confusion set of the original error character, it is more possible to be a right correction because the two characters are similar in glyph or pronunciation. We use the open source confusion set [22] as a candidate filter. If the generated best candidate is not in the confusion set of the original character, we discard this modification. To extend the coverage of confusion set, we retrieve character pairs from the training materials as the additional similar characters. Traditional confusion set consists of phonologically and visually similar characters, while the retrieved pairs may be semantically similar characters.

4   Experiments

4.1   Datasets

The training sets of SIGHAN 13/14/15 are combined up as a whole training set. Similar to previous works, we use the large automatically generated pseudo corpus [19] as additional training data. A few incorrectly labeled samples in the automatically generated corpus are removed by us. We evaluate our methods on test sets from SIGHAN 13/14/15. Following the previous works, we use OpenCC tool[1] to convert SIGHAN datasets to simplified Chinese.

4.2   Implementation Details

The Pinyin Enhanced BERT is implemented using Pytorch [16]. We resize the size of BERT embeddings from 21128 to 21954 to store the randomly initialized Pinyin embeddings and a few unknown tokens in the training materials. For candidates generation in reevaluation strategy, we retrieve top-5 phonetically similar characters and top-5 other characters in the confusion set based on probability, which means we may generate $(5+5)^2 = 100$ candidate sentences one time. To narrow the search space, we remove the generated bi-grams if their word frequency is lower than 3 in the training materials. We train the Pinyin enhanced model with the AdamW [14] optimizer with stochastic gradient descent strategy [10] with for 5 epochs. The learning rate is set to $5e-5$ and the batch size is set to 32.

---

[1] https://github.com/BYVoid/OpenCC.

4.3   Baselines

We compare our methods with several strong baseline methods.

- **FASPELL** [9] generates candidates for each character in the input and filter them with visual and phonetic similarity features to select the best candidate.
- **SpellGCN** [25] builds two similarity graphs from the confusion set and incorporates the graphs into a pre-trained language model for the CSC task.
- **BERT** [24] is the vanilla BERT finetuned for the CSC task. The input in sentences with error characters and the target is to predict the correct sentence directly.
- **MLM-phonetics** [25] integrates phonetic features into a language model by leveraging the powerful pre-training and fine-tuning method. MLM-phonetics uses an adaptive weighted objective to jointly train error detection and correction in a unified framework.
- **DCN** [18], the Dynamic Connected Networks, generates the candidate Chinese characters via a Pinyin Enhanced Candidate Generator and then utilizes an attention-based network to model the dependencies between two adjacent Chinese characters.
- **Global Attention Decoder (GAD)** [8] learns the global relationship of the potential correct input characters and the candidates of potential error characters. Rich global contextual information is obtained to alleviate the impact of the local error contextual information.
- **REALISE** [24] directly leverages the multimodal information of the Chinese characters. The REALISE model tackles the CSC task by (1) capturing the semantic, phonetic and graphic information of the input characters, and (2) selectively mixing the information in these modalities to predict the correct output.

4.4   Main Results

Table 1 shows sentence level detection and correction scores on the SIGHAN 13/14/15 test sets. It shows that our method, P-BERT-RF, achieves the state-of-the-art performance on SIGHAN15 and achieves competitive results on SIGHAN 13/14 test sets. The proposed P-BERT-RF method has the best detection performance on SIGHAN 14/15. The correction F1-scores on SIGHAN 15 test set has 1.3 point improvement ($77.8 \rightarrow 79.1$).

MLM-phonetics achieves best correction score on SIGHAN14. However, MLM-phonetics utilizes an additional large scale corpus of 0.3 billion sentences for pre-training. REALISE achieves best scores on SIGHAN 13 test set. It is a multimodal model which consists of the semantic encoder, phonetic encoder, graphical encoder and modality fusion module. The architecture of the semantic encoder is same as the $BERT_{base}$. Graphical encoder and modality fusion module are based on 4 and 3 layers transformers respectively. Our P-BERT-RF uses much less parameters, which is only a variant of $BERT_{base}$.

Table 1. The performance of our method and all baseline models on SIGHAN test sets. Similar to previous work [24], "的地得" are removed from model outputs of SIGHAN 13 because "的地得" are not annotated in SIGHAN 13. We also report the rank sorted by correction level F1 scores of different models.

Dataset	Method	Detection level			Correction level			Rank
		Pre	Rec	F1	Pre	Rec	F1	
SIGHAN13	Faspell (2019)	76.2	63.2	69.1	73.1	60.5	66.2	8
	SpellGCN (2020)	80.1	74.4	77.2	78.3	72.7	75.4	7
	BERT	85.0	77.0	80.8	75.2	83.0	78.9	5
	REALISE (2021)	**88.6**	**82.5**	**85.4**	**87.2**	**81.2**	**84.1**	1
	MLM-phonetics (2021)	82.0	78.3	80.1	79.5	77.0	78.2	6
	DCN (2021)	86.8	79.6	83.0	84.7	77.7	81.0	4
	GAD (2021)	85.7	79.5	82.5	84.9	78.7	81.6	3
	P-BERT-RF (ours)	86.7	79.1	82.7	86.3	78.7	82.3	2
SIGHAN14	Faspell (2019)	61.0	53.5	57.0	59.4	52.0	55.4	8
	SpellGCN (2020)	65.1	69.5	67.2	63.1	67.2	65.3	6
	BERT	64.5	68.6	66.5	62.4	66.3	64.3	7
	REALISE (2021)	67.8	71.5	69.6	66.3	70.0	68.1	3
	MLM-phonetics (2021)	66.2	**73.8**	69.8	64.2	**73.8**	**68.7**	1
	DCN (2021)	67.4	70.4	68.9	65.8	68.7	67.2	5
	GAD (2021)	66.6	71.8	69.1	65.0	70.1	67.5	4
	P-BERT-RF (ours)	**70.4**	69.9	**70.1**	**68.7**	68.2	68.4	2
SIGHAN15	Faspell (2019)	67.6	60.0	63.5	66.6	59.1	62.6	8
	SpellGCN (2020)	74.8	80.7	77.7	72.1	77.7	75.9	5
	BERT	74.2	78.0	76.1	71.6	75.3	73.4	7
	REALISE (2021)	77.3	**81.3**	79.3	75.9	**79.9**	77.8	2
	MLM-phonetics (2021)	77.5	83.1	80.2	74.9	80.2	77.5	3
	DCN (2021)	77.1	80.9	79.0	74.5	78.2	76.3	4
	GAD (2021)	75.6	80.4	77.9	73.2	77.8	75.4	6
	P-BERT-RF (ours)	**80.5**	80.0	**80.3**	**79.5**	78.9	**79.1**	1

Nevertheless, our proposed method achieves either the best or the second performance on both detection and correction F1 scores. The mean rank of our method on SIGHAN 13/14/15 based on F1 scores is the smallest. This proves the effectiveness of our proposed "reevaluation and filter" strategy.

5   Ablation Study

We explore the contribution of each component in P-BERT-RF by conducting ablation studies with the following settings: 1) removing the filter, 2) removing the reevaluation strategy, 3) removing the reevaluation and filter strategy.

Table 2. Ablation results of the Pinyin enhanced model on SIGHAN 15 test set. We apply the following changes: removing the reevaluation and filter strategy (P-BERT), removing the reevaluation strategy (P-BERT-F), removing the filter strategy (P-BERT-R).

Dataset	Method	Detection level			Correction level		
		Pre	Rec	F1	Pre	Rec	F1
SIGHAN15	P-BERT	77.4	80.4	78.9	75.3	78.1	76.7
	P-BERT-F	80.1	79.1	79.6	79.2	78.2	78.7
	P-BERT-R	77.8	**80.9**	79.3	75.9	**78.9**	77.4
	P-BERT-RF	**80.5**	80.0	**80.3**	**79.5**	**78.9**	**79.1**

Table 3. Performance of using different k value when retrieving top k candidates for error characters.

Dataset	Top K	Detection level			Correction level		
		Pre	Rec	F1	Pre	Rec	F1
SIGHAN15	k = 1	80.6	79.8	80.2	79.3	78.5	78.9
	k = 3	80.7	80.0	80.4	79.6	78.9	79.3
	k = 5	80.5	80.0	80.3	79.5	78.9	79.1
	k = 10	80.4	79.8	80.1	79.3	78.7	79.0

As shown in Table 2, when we removed the reevaluation and filter strategy, the Pinyin Enhanced BERT achieved a sentence level correction score of 76.7, which improves 3.3 points compared to the vanilla BERT. The Pinyin Enhanced BERT with the filter improves 2.0 point correction F1 score compared to P-BERT. The correction precision jumps by a massive 3.9 point which leads to an increase on the F1 score. The filter strategy avoids many wrongly corrected characters and brings improvements both on sentence level precision and recall scores. The Pinyin Enhanced BERT with the reevaluation strategy brings improvement on recall scores, which also leads to an increase on the F1 score.

Table 3 shows sentence level performance using different k value. When k is set to 3, the detection and correction scores get a slight increase. We think it is because top 3 candidates can cover most ground truth characters without bringing too many choices.

6   Case Study

As shown in Table 4, the P-BERT without the reevaluation and filter strategy can detect the position of error characters. However, P-BERT fails to find the correct characters because of the noisy information from other error characters. For example, in the original sentence "你不会说很表尊(biao zun)都没有用", "表尊" should be corrected as "标准(biao zhun)". P-BERT predicts "表情(biao

Table 4. Examples of the input and output of our model. We highlight the wrong/correct characters in red/blue color.

In:	中文对我说是非常很难，不论你会学一千多个字，你不会说很表尊都没有用。
Pinyin:	zhong wen dui wo shuo shi fei chang nan, bu lun ni hui xue yi qian duo ge zi, ni bu hui shuo hen biao zun dou mei you yong.
P-BERT Out:	中文对我说是非常很难，不论你会学一千多个字，你不会说很表情都没有用。？
Pinyin:	zhong wen dui wo shuo shi fei chang nan, bu lun ni hui xue yi qian duo ge zi, ni bu hui shuo hen biao qing dou mei you yong.
P-BERT-FR Out:	中文对我说是非常很难，不论你会学一千多个字，你不会说很标准都没有用。？
Pinyin:	zhong wen dui wo shuo shi fei chang nan, bu lun ni hui xue yi qian duo ge zi, ni bu hui shuo hen biao zhun dou mei you yong.
In:	你可以告诉我那家书店有中文数马？
Pinyin:	ni ke yi gao su wo na jia shu dian you zhong wen shu ma ?
P-BERT Out:	你可以告诉我那家书店有中文课吗？
Pinyin:	ni ke yi gao su wo na jia shu dian you zhong wen ke ma ?
P-BERT-FR Out:	你可以告诉我那家书店有中文书吗？
Pinyin:	ni ke yi gao su wo na jia shu dian you zhong wen shu ma ?

qing)" to replace "表尊(biao zun)", which is not a good correction. According to the reevaluation strategy guided by the confusion set, P-BERT-FR retrieves top-k candidates at the corresponding positions and generate $k * k$ candidate sentences. The ground truth candidate sentences are usually among the $k * k$ candidate sentences. By reevaluating the generated sentence and compute the probabilities at the detected error positions, we find the sentence with maximum posterior probability and similarity compared to the original error characters.

7 Conclusion

In this paper, we propose the simple and effective reevaluation and filter strategy for the CSC task. For bi-gram errors, instead of predicting the correct characters based on noisy information from error characters, we evaluate the probabilities at detected error positions based on the modified sentences and select the most possible and similar one. Besides, we extend the confusion set by retrieving pair characters from the training materials. Simply applying the extended confusion set will bring considerable improvements in the CSC task. Experiments on benchmark datasets show that the reevaluation and filter strategy achieved state-of-the-art on SIGHAN 15 and competitive performance on SIGHAN 13/14 compared to previous methods. In the future, we are going to improve the reevaluation strategy for better performance on continuous error characters, e.g. performance on tri-gram and four-gram errors.

References

1. Afli, H., Qiu, Z., Way, A., Sheridan, P.: Using SMT for OCR error correction of historical texts. In: Proceedings of the Tenth International Conference on Language Resources and Evaluation (LREC 2016), pp. 962–966 (2016)
2. Belinkov, Y., Bisk, Y.: Synthetic and natural noise both break neural machine translation. arXiv preprint arXiv:1711.02173 (2017)
3. Chang, T.H., Chen, H.C., Yang, C.H.: Introduction to a proofreading tool for Chinese spelling check task of SIGHAN-8. In: Proceedings of the Eighth SIGHAN Workshop on Chinese Language Processing, pp. 50–55 (2015)
4. Cheng, X., et al.: SpellGCN: incorporating phonological and visual similarities into language models for Chinese spelling check. In: Proceedings of the 58th Annual Meeting of the Association for Computational Linguistics, pp. 871–881. Association for Computational Linguistics, Online, July 2020. https://doi.org/10.18653/v1/2020.acl-main.81, https://aclanthology.org/2020.acl-main.81
5. Devlin, J., Chang, M.W., Lee, K., Toutanova, K.: BERT: pre-training of deep bidirectional transformers for language understanding. In: Proceedings of the 2019 Conference of the North American Chapter of the Association for Computational Linguistics: Human Language Technologies, Volume 1 (Long and Short Papers), pp. 4171–4186. Association for Computational Linguistics, Minneapolis, Minnesota, June 2019. https://doi.org/10.18653/v1/N19-1423, https://aclanthology.org/N19-1423
6. Dey, R., Salem, F.M.: Gate-variants of gated recurrent unit (GRU) neural networks. In: 2017 IEEE 60th International Midwest Symposium on Circuits and Systems (MWSCAS), pp. 1597–1600. IEEE (2017)
7. Gao, J., et al.: A large scale ranker-based system for search query spelling correction (2010)
8. Guo, Z., Ni, Y., Wang, K., Zhu, W., Xie, G.: Global attention decoder for Chinese spelling error correction. In: Findings of the Association for Computational Linguistics: ACL-IJCNLP 2021, pp. 1419–1428 (2021)
9. Hong, Y., Yu, X., He, N., Liu, N., Liu, J.: FASPell: a fast, adaptable, simple, powerful Chinese spell checker based on DAE-decoder paradigm. In: Proceedings of the 5th Workshop on Noisy User-generated Text (W-NUT 2019), pp. 160–169 (2019)
10. Izmailov, P., Podoprikhin, D., Garipov, T., Vetrov, D., Wilson, A.G.: Averaging weights leads to wider optima and better generalization. arXiv preprint arXiv:1803.05407 (2018)
11. Kipf, T.N., Welling, M.: Semi-supervised classification with graph convolutional networks. In: 5th International Conference on Learning Representations, ICLR 2017, Toulon, France, April 24–26, 2017, Conference Track Proceedings. OpenReview.net (2017). https://openreview.net/forum?id=SJU4ayYgl
12. Liu, C.L., Lai, M.H., Chuang, Y.H., Lee, C.Y.: Visually and phonologically similar characters in incorrect simplified Chinese words. In: Coling 2010: Posters, pp. 739–747 (2010)
13. Liu, X., Cheng, K., Luo, Y., Duh, K., Matsumoto, Y.: A hybrid Chinese spelling correction using language model and statistical machine translation with reranking. In: Proceedings of the Seventh SIGHAN Workshop on Chinese Language Processing, pp. 54–58 (2013)
14. Loshchilov, I., Hutter, F.: Fixing weight decay regularization in Adam (2018)

15. Martins, B., Silva, M.J.: Spelling correction for search engine queries. In: Vicedo, J.L., Martínez-Barco, P., Muñoz, R., Saiz Noeda, M. (eds.) EsTAL 2004. LNCS (LNAI), vol. 3230, pp. 372–383. Springer, Heidelberg (2004). https://doi.org/10.1007/978-3-540-30228-5_33

16. Paszke, A., et al.: Pytorch: an imperative style, high-performance deep learning library. Adv. Neural Inf. Process. Syst. **32**, 8026–8037 (2019)

17. Sakaguchi, K., Mizumoto, T., Komachi, M., Matsumoto, Y.: Joint English spelling error correction and POS tagging for language learners writing. In: Proceedings of COLING 2012, pp. 2357–2374 (2012)

18. Wang, B., Che, W., Wu, D., Wang, S., Hu, G., Liu, T.: Dynamic connected networks for chinese spelling check. In: Findings of the Association for Computational Linguistics: ACL-IJCNLP 2021, pp. 2437–2446 (2021)

19. Wang, D., Song, Y., Li, J., Han, J., Zhang, H.: A hybrid approach to automatic corpus generation for Chinese spelling check. In: Proceedings of the 2018 Conference on Empirical Methods in Natural Language Processing, pp. 2517–2527 (2018)

20. Wang, D., Tay, Y., Zhong, L.: Confusionset-guided pointer networks for Chinese spelling check. In: Proceedings of the 57th Annual Meeting of the Association for Computational Linguistics, pp. 5780–5785 (2019)

21. Wang, Y.R., Liao, Y.F.: Word vector/conditional random field-based Chinese spelling error detection for SIGHAN-2015 evaluation. In: Proceedings of the Eighth SIGHAN Workshop on Chinese Language Processing, pp. 46–49 (2015)

22. Wu, S.H., Liu, C.L., Lee, L.H.: Chinese spelling check evaluation at SIGHAN bake-off 2013. In: Proceedings of the Seventh SIGHAN Workshop on Chinese Language Processing, pp. 35–42 (2013)

23. Xiong, J., Zhang, Q., Zhang, S., Hou, J., Cheng, X.: Hanspeller: a unified framework for Chinese spelling correction. In: International Journal of Computational Linguistics and Chinese Language Processing, vol. 20, No. 1, June 2015-Special Issue on Chinese as a Foreign Language (2015)

24. Xu, H.D., et al.: Read, listen, and see: leveraging multimodal information helps Chinese spell checking. In: Findings of the Association for Computational Linguistics: ACL-IJCNLP 2021, pp. 716–728. Association for Computational Linguistics, Online, August 2021. https://doi.org/10.18653/v1/2021.findings-acl.64, https://aclanthology.org/2021.findings-acl.64

25. Zhang, R., et al.: Correcting Chinese spelling errors with phonetic pre-training. In: Findings of the Association for Computational Linguistics: ACL-IJCNLP 2021, pp. 2250–2261 (2021)

26. Zhang, S., Huang, H., Liu, J., Li, H.: Spelling error correction with soft-masked BERT. In: Proceedings of the 58th Annual Meeting of the Association for Computational Linguistics, pp. 882–890. Association for Computational Linguistics, Online, July 2020. https://doi.org/10.18653/v1/2020.acl-main.82, https://aclanthology.org/2020.acl-main.82

27. Zou, H., et al.: On embedding sequence correlations in attributed network for semi-supervised node classification. Inf. Sci. **562**, 385–397 (2021)

Extreme Multi-label Classification with Hierarchical Multi-task for Product Attribute Identification

Jun Zhang[1(✉)], Menqian Cai[1], Chenyu Zhao[1], Xiaowei Zhang[1],
Zhiqian Zhang[1], Haiheng Chen[1], and Sulong Xu[2]

[1] JD.com, Shenzhen, China
{zhangjun35,caimengqian,zhaochenyu8,zhangxiaowei9,zhangzhiqian1,
chenhaiheng}@jd.com
[2] JD.com, Beijing, China
xusulong@jd.com

Abstract. Identification of product attributes (product type, brand, color, gender, etc.) from a query is critically important for e-commerce search systems, especially the identification of brand intent. Recently, Named Entity Recognition (NER) method has been used to address this issue. However, the limitation of NER method is that it can only identify brand intent specified by terms of a query and cannot work appropriately if brand terms are not provided explicitly. To overcome this limitation, we propose a novel **E**xtreme **M**ulti-label based hierarchical **M**ulti-t**A**sk (EMMA) framework, where we treat the brand identification as an issue of extreme multi-label classification; thereafter, a deep learning model is also developed to jointly learn query's product intent and brand intent in a coarse-to-fine approach. The results from both online A/B test and offline experiment on real industrial dataset demonstrate the effectiveness of our proposed framework. Additionally, this framework may be extended potentially from e-commerce system to other search scenarios.

Keywords: Product attributes · Multi-task · Multi-label classification

1 Introduction

Query understanding is mainly to identify a query's product intent with the corresponding product attributes (product type, gender, brand, color, etc.). These product attributes can be used by e-commerce search engine to return items that better match the query's product intent. Most recent studies have applied Named Entity Recognition (NER) task [1,2] to annotate individual terms in a query. In the NER process, terms in a query are classified individually into the predefined named entities, then a sequence labeling task is trained on the probabilistic output of the base classifier, and lastly the sequential label prediction is obtained using Viterbi decoding [1]. For example, there is no explicit brand term in query "mate30", and it can not be identified with "HUAWEI" brand by NER

© The Author(s), under exclusive license to Springer Nature Switzerland AG 2022
J. Gama et al. (Eds.): PAKDD 2022, LNAI 13282, pp. 249–260, 2022.
https://doi.org/10.1007/978-3-031-05981-0_20

method. Therefore, the limitation of NER method is that it can not identify the product's brand characteristic with no specific brand terms in the query, but it is necessary for search engine to leverage the implicit brand "HUAWEI" to return items that better match user's intent. To overcome the above limitation, we treat the brand intent identification as an issue of extreme multi-label classification. In order to solve the problem of extreme brand categories, we propose a novel Extreme Multi-label based hierarchical Multi-tAsk (EMMA) framework, in which three hierarchical tasks are built to identify a query's brand intent in a coarse-to-fine approach. Moreover, we employ category attention to capture the semantic relation between product and brand in a query. Besides, we proposed a framework to jointly learn brand intent and product intent.

To summarize, the contributions of this work are as follows:

- To the best of our knowledge, it is the first work to treat the brand intent detection as an issue of extreme multi-label classification compared to the traditional NER task.
- We propose a deep learning model to jointly learn query's product intent and brand intent in a unified framework.
- We conduct both online A/B test and offline experiment on real industrial dataset, the results of which demonstrate the effectiveness of our proposed model.

2   Related Works

Query understanding is one of critical challenges in search engine module, especially in e-commerce [3]. Queries are usually short, vague, data insufficiency and suffer from the lack of textual evidence [4]. Moreover, queries have multiple intent [5] and product category often has large scale label set [6]. In this section, we review the related techniques proposed in our method and put our work in proper context.

2.1   Multi-task Learning

Multi-task Learning (MTL) [7,8] is an approach to inductive transfer learning that improves the performance of a target task by using information of related tasks. Recent MTL has been proved to be effective in Natural Language Processing [9,10]. Ahmadvand et al. [11] propose a JointMap model to predict a query's "commercial" vs. "non-commercial" intent and product category. Gao et al. [12] jointly learn users' different types of behaviors (e.g., views, clicks and purchasing) in recommender systems. In [13], Wang et al. use a Pointer Network to automatically select the key words as compressed title and the auxiliary task is a seq2seq model to generate user's search query. Zhang et al. [14] proposed a multi-task framework that jointly learn the intent detection and entity linking tasks. In our work, we propose a hierarchical multi-task framework that jointly learn the query's product intent and product intent in a coarse-to-fine approach.

2.2 Extreme Multi-label Classification

Extreme multi-label classification (XMC) is the problem of finding the relevant labels for each input from an extremely large-scale label set. In general, XMC approaches fall into three categories: one-vs-all methods, tree-based methods and embedding-based methods. First, one-vs-all methods [15,16] simply treat each label as binary classification problem and learns a classifier for each label. Moreover, one-vs-all methods have been proved to achieve high accuracy, but still suffer from high computational complexity. Second, tree-based methods aim to reduce computational complexity in one-vs-all methods. These methods use a hierarchical tree structure by partitioning labels [17] or samples [18]. Third, embedding-based methods [19,20] reduce the effective number of labels by projecting high dimensional label space into low dimensional space. In our work, we do negative sampling dynamically based on hierarchical label structure for high-performance extreme multi-label classification.

3 Problem Formulation

In this section, we first describe connection between three classification tasks, and provide a brief describe of extreme multi-label classification (XMC), which help readers to understand our work.

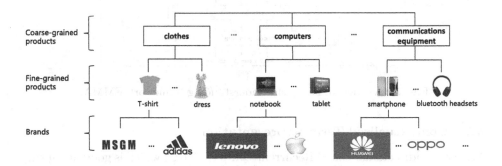

Fig. 1. Hierarchical tree labels for three tasks.

Given a query, we aim to predict its product intent and brand intent. Due to the design of products in JD's dataset, we predict the product intent based on coarse-to-fine predicting mechanism. Specifically, product intent detection is divided into two tasks: coarse-grained product intent detection and fine-grained product intent detection. Figure 1 shows the relation between coarse-grained products, fine-grained products and brands. As the picture shows, there are three layers, which are all linked together within a predefined hierarchy. We cast query product and brand intent detection as XMC problem. For each query, we aim to build a classifier that recognizes the most related product labels and brand labels from a large number of label sets.

4   The Proposed Model

In this section, we describe our network architecture of EMMA as shown in Fig. 2, which consists of multiple hierarchical multi-label classification layers, the first layer for predicting the coarse-grained product intent, the second layer for predicting the fine-grained product intent, and the third layer for predicting the brand intent. The later layer has larger number of labels. Specifically, there are about 560 coarse-grained product labels, 6,184 fine-grained product labels and 160,000 brand labels.

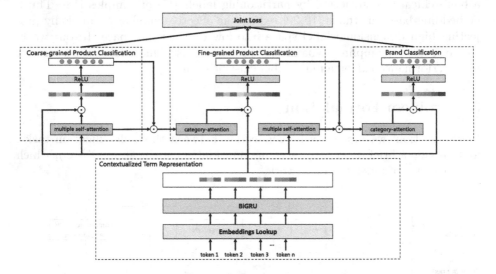

Fig. 2. Overview of our query understanding framework EMMA.

4.1   Contextualized Term Representation

We use a Bidirectional Gated Recurrent Units (BiGRU) which is good at capturing sequential information to extract each term's contextualized representation in a sequential input. The sequence of term vectors in query q is denoted by $\mathbf{E} = \{\mathbf{e}_1, \mathbf{e}_2, \ldots, \mathbf{e}_N\}$. The BiGRU encodes the query q as:

$$\mathbf{h}_t = [\mathrm{GRU}^f(\mathbf{h}_{t-1}^f, \mathbf{e}_t); \mathrm{GRU}^b(\mathbf{h}_{t-1}^b, \mathbf{e}_t)], \tag{1}$$

where GRU^f encodes the query q in a forward direction and GRU^b encodes in backward direction. $[;]$ denotes the concatenation of hidden states from two directions. The output of each term at the position t in the sequential input is $\mathbf{h}_t$.

4.2   Coarse-Grained Product Classification

In this layer, we introduce how to predict queries' coarse-grained product intent. After obtaining the contextualized term representation from BiGRU model, we can obtain semantic vectors $\mathbf{H} = [\mathbf{h}_1, \mathbf{h}_2, \ldots, \mathbf{h}_N] \in \mathbb{R}^{2D \times N}$ for N terms. To learn the importance of each term towards multiple view, we apply multiple self-attention mechanism [9]. The multiple self-attention matrix $\mathbf{A}^{s_1}$ is computed by:

$$\mathbf{A}^{s_1} = \mathrm{softmax}(\mathbf{W}_a^c \tanh(\mathbf{W}_t^c \mathbf{H})), \tag{2}$$

where $\{\mathbf{W}_t^c \in \mathbb{R}^{2D \times 2D}, \mathbf{W}_a^c \in \mathbb{R}^{C \times 2D}\}$ are learnable parameters and C is the number of coarse-grained product labels. And $\mathbf{A}^{s_1} \in \mathbb{R}^{C \times N}$ is a multiple self-attention matrix. Then, we compute the query embedding vector $\mathbf{Q}^c$ as follows:

$$\mathbf{Q}^c = \mathrm{ReLU}(\mathrm{Avg}(\mathbf{A}^{s_1} \mathbf{H})), \tag{3}$$

where $\mathrm{Avg}(\cdot)$ is the average pooling in the first dimension and ReLU is a fully connected layer with activation function Rectified Linear Units (ReLU). Then, the output $\mathbf{Q}^c$ is fed into the classifier layer with C output units. The prediction probability is computed by:

$$\hat{\mathbf{y}}^c = \sigma(\mathbf{W}^c \mathbf{Q}^c + \mathbf{b}^c), \tag{4}$$

where $\mathbf{W}^c \in \mathbb{R}^{C \times 2D}$ is the classifier weight matrix with C labels, $\mathbf{b}^c \in \mathbb{R}^C$ is the bias vector, and σ is the sigmoid activation. We use the sigmoid cross-entropy loss for multi-label classification.

$$L_c = -\sum_{i=1}^{C} \mathbf{y}_i^c \log(\hat{\mathbf{y}}_i^c) + (1 - \mathbf{y}_i^c) \log(1 - \hat{\mathbf{y}}_i^c), \tag{5}$$

where $\mathbf{y}^c$ is the multiple one-hot label of coarse-grained product.

4.3   Fine-Grained Product Classification

In this layer, we introduce how to predict queries' fine-grained product intent. To leverage the hierarchy between coarse-grained product and fine-grained product, motivated by [21], we use prediction label scores of former layer to adjust the multiple self-attention $\mathbf{A}^{s_1}$ to get category attention. Let $\hat{\mathbf{y}}^c$ repeat N times to form the matrix $\hat{\mathbf{Y}}^c = [\hat{\mathbf{y}}^c, \ldots, \hat{\mathbf{y}}^c] \in \mathbb{R}^{C \times N}$, we can transform the multiple self-attention matrix $\mathbf{A}^{s_1}$ to the category attention $\mathbf{A}^{c_2} \in \mathbb{R}^{C \times N}$ as follows:

$$\mathbf{A}^{c_2} = \mathbf{A}^{s_1} \odot \hat{\mathbf{Y}}^c, \tag{6}$$

where $\odot$ denotes the element-wise product. Each row of $\mathbf{A}^{c_2}$ represents the weight of terms corresponding to the certain coarse-grained product label. To learn global weight of each term on all coarse-grained product categories, we sum up the term weights across all coarse-grained product categories to get the attention weight of this layer:

$$\mathbf{a}^f = \mathrm{softmax}\left(\sum_{i=1}^{C} \mathbf{A}_i^{c_2}\right), \tag{7}$$

where $\mathbf{a}^f \in \mathbb{R}^N$ is a weight vector for the sequential input terms and $\mathbf{A}_i^{c_2}$ is the i-th row of the category attention matrix $\mathbf{A}^{c_2}$. The query embedding $\mathbf{Q}^f$ in this layer is obtained by:

$$\mathbf{Q}^f = \text{ReLU}(\sum_{i=1}^{N} a_i^f \mathbf{h}_i), \tag{8}$$

Same as the former layer, we can get prediction probability $\hat{\mathbf{y}}^f$ and sigmoid cross-entropy loss L_f.

4.4   Brand Classification

In JD's e-commerce search platform, there are a large scale of brand labels and the number of brand labels B is about 160k. In this part, we describe how to sample hierarchical negative labels to reduce computational complexity and speed up convergence. Specifically, we sample negative labels based on the hierarchy between fine-grained product layer and brand layer. In our dataset, we can obtain each product category with some specific brand categories. Let $G = \{G_i\}_{i=1}^{F}$ denotes F groups where each group G_i is composed by some specific brand categories with respect to the i-th product category.

Same as the former layer, we can learn the importance of each term towards the query's fine-grained product intent and the multiple self-attention matrix $\mathbf{A}^{s_2}$ is computed by:

$$\mathbf{A}^{s_2} = \text{softmax}(\mathbf{W}_a^f \tanh(\mathbf{W}_t^f \mathbf{H})), \tag{9}$$

where $\{\mathbf{W}_t^f \in \mathbb{R}^{2D \times 2D}, \mathbf{W}_a^f \in \mathbb{R}^{F \times 2D}\}$ are learnable parameters and F is the number of multiple self-attention and equal to the number of fine-grained product labels. $\mathbf{A}^{s_2} \in \mathbb{R}^{C \times N}$ is a multiple self-attention matrix.

Hierarchical Negative Sampling. Given a query's fine-grained product labels set $Y = \{y_1, y_2, ..., y_m\}$, positive brand labels S_b^+ and the group G. The complete brand label candidates S_b are sampled by:

$$S_b = \{G_{y_1} \cup G_{y_2} \cup ... \cup G_{y_m}\}. \tag{10}$$

Therefore, the negative labels S_b^- are obtained by:

$$S_b^- = S_b - S_b^+. \tag{11}$$

Given the brand labels weight matrix $\mathbf{M}^b \in \mathbb{R}^{B \times 2D}$, we extract embedding of positive labels and negative labels respectively, which can be represented as follows:

$$\mathbf{M}^+ = G_e(\mathbf{M}^b, S_b^+), \tag{12}$$

$$\mathbf{M}^- = G_e(\mathbf{M}^b, S_b^-), \tag{13}$$

where G_e is a function to map labels to its embedding. Same as the fine-grained product layer, we can obtain the category attention $\mathbf{A}^{c_3}$ and attention weight $\mathbf{a}^b$ of each term and query embedding $\mathbf{Q}^b$ as follows:

$$\mathbf{A}^{c_3} = \mathbf{A}^{s_2} \odot \hat{\mathbf{Y}}^f, \tag{14}$$

$$\mathbf{a}^b = \text{softmax}\left(\sum_{i=1}^{F} \mathbf{A}_i^{c_3}\right), \tag{15}$$

$$\mathbf{Q}^b = \text{ReLU}(\sum_{i=1}^{N} a_i^b \mathbf{h}_i), \tag{16}$$

where $\hat{\mathbf{Y}}^f = [\hat{\mathbf{y}}^f, \ldots, \hat{\mathbf{y}}^f] \in \mathbb{R}^{F \times N}$ is the N times copy of $\hat{\mathbf{y}}^f$. In addition, the prediction probability is computed by:

$$\mathbf{P}^+ = \sigma(\mathbf{Q}_b'\mathbf{M}^+ + b_{s_b^+}), \tag{17}$$

$$\mathbf{P}^- = \sigma(\mathbf{Q}_b'\mathbf{M}^- + b_{s_b^-}), \tag{18}$$

where $b_{s_b^+}$ and $b_{s_b^-}$ are the biases.

4.5   Loss Function

In this section, we introduce how to combine the multi-task loss. Multi task learning with homoscedastic uncertainty [22] is adopted. This allows us to simultaneously learn various quantities with different scales in our three classification tasks. The joint loss function is:

$$L = \sum_{j \in \{c,f,b\}} \frac{1}{2\epsilon_j^2} \times L_j + 2 \times \log(\epsilon_j), \tag{19}$$

where L_c is the loss for coarse-grained product classification, L_f is the loss for fine-grained product classification and L_b is the loss for brand classification. There is a noise parameter ϵ to balance each task loss in the overall loss. Our model performs end-to-end training by learning multiple objectives.

4.6   Online Inference

In this section, we introduce a novel strategy to significantly reduce online inference latency. To avoid extreme large label classification computing, we extract the brand embedding from the weights of dense layer of the last trained submodule. The model firstly outputs the coarse-grained product prediction $\hat{\mathbf{y}}^c$, fine-grained product prediction $\hat{\mathbf{y}}^f$ and query embedding $\mathbf{Q}^b$. Then, we recall brand candidates via the fine-grained product to brand mapping dictionary and can easily get the embedding of candidates through indexing. Thus, the manually dot products between brand candidates and query embedding are performed to get scores of brand candidates. The whole online inference process we describe above is shown in Fig. 3.

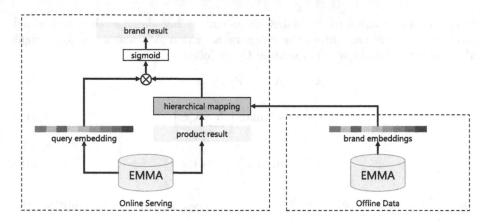

Fig. 3. Online inference process.

5   Experiments

In this section, we evaluate the result of proposed method and analyze the contribution of each component of the method via ablation studies. Furthermore, we apply visualization on attentions and embeddings to illustrate the performance of proposed model.

5.1   Dateset

We collect our training data from clicked-through logs on JD's e-commerce platform. In our dataset, each item has a predefined brand label and product label. So, if a user types a specific query and clicks on the item, we can obtain the query's product and brand labels according to item's label. By this way, we can obtain hundreds of millions of samples to train model. As there is no public dataset for e-commerce query intent classification, we collect evaluation data from JD's e-commerce platform. We randomly collect 1k queries annotated by human as our evaluation set.

5.2   Evaluation Metrics

For extreme multi-label classification, we follow the convention of XMC problems which use both precision at top K (P@K) and the Normalized Discounted Cumulated Gains at top K (N@K) as our evaluation metrics, with $k = 1, 3, 5$. The two metrics are defined as follows:

$$P@k = \frac{1}{k} \sum_{l \in r_k(\hat{\mathbf{y}})} \mathbf{y}_l, \tag{20}$$

$$DCG@k = \frac{1}{k} \sum_{l \in r_k(\hat{\mathbf{y}})} \frac{\mathbf{y}_l}{\log(l+1)}, \tag{21}$$

$$N@k = \frac{DCG@k}{\sum_{l=1}^{\min(k,\|\mathbf{y}\|_0)} \frac{1}{\log(l+1)}}, \tag{22}$$

where $r_k(\hat{\mathbf{y}})$ is the set of rank indices of the ground truth labels among the top-k portion of the predicted ranked list for a test instance, and $\| \mathbf{y} \|_0$ counts the number of relevant labels in the ground truth label vector $\mathbf{y}$. P@K and N@K are calculated for each test query and then averaged over all test set.

5.3   Main Results

The overall comparison of methods on our evaluation dataset is shown in Table 1. We compared EMMA with BiGRU+CRF as baseline model. BiGRU is the most common neural architectures for Named Entity Recognition (NER) and the CRF layer is helpful to predict the most likely entity sequence and to forbid invalid sequence transitions. The first line shows the result of BiGRU+CRF. As we can see in Table 1, the performance of EMMA is statistically better than BiGRU+CRF model. On the evaluation dataset, the evaluation metric precision@1 of the EMMA outperforms the latter method by 8.6%.

Why the proposed EMMA can perform particularly stronger against the NER method? As we mentioned above, the EMMA recognizes the brand intent on classification view. It typically recognizes the brand intent on the case of no explicit brand terms appear in query. For example the query is "iphone 12", which is a famous phone product of brand "APPLE". Notice that the actual brand name does not appear in the query, so the NER method cannot detect the brand intent. However, the EMMA can precisely recognize the brand intent "APPLE" on classification view.

Table 1. Main results in P@K and N@K.

Methods	P@1	P@3	P@5	N@1	N@3	N@5
BiGRU+CRF	0.409	0.206	0.130	0.409	0.427	0.427
EMMA	0.495	0.267	0.180	0.495	0.508	0.507

5.4   Online A/B Tests

To analyze the effectiveness of the proposed method in online experiment, we conduct A/B test on the platform of JD.com, using 10% of the entire site traffic during a period of two weeks. The task to be compared is query brand intent detection. The baseline model based on BiGRU+CRF is a NER method which was deployed online and the test model based on EMMA is a multi-label classification task. Experimental results show that EMMA contributes +0.785% UV value (gross merchandise value/number of users), and +0.098% CTR (click-through rate) improvement compared with baseline.

5.5   Ablation Studies

Given the EMMA model, it would be informative to empirically analyze the impact of some components via ablation test. We conduct some experiments to verify the effect of each component.

Effect of Hierarchical Negative Sampling. We compare the performance on the brand intent detection task. The result is shown in Table 2, where EMMA-w/o-HNS denotes the model without hierarchical negative sampling strategy. We can see the performance of EMMA outperforms EMMA-w/o-HNS by 22.1% on precision@1 and 10.7% on NDCG@3 respectively.

Table 2. Effect of hierarchical negative sampling.

Models	P@1	P@3	P@5	N@1	N@3	N@5
EMMA	0.495	0.267	0.180	0.495	0.508	0.507
EMMA-w/o-HNS	0.267	0.231	0.152	0.267	0.401	0.398

Effect of Multi-task Learning. In order to show the performance of multi-task learning, we also compare the performance on brand intent detection task. Specifically, we ignore the loss of coarse and fine-grained product classification, and only retain brand classification. The result is shown in Table 3. EMMA-w/o-MT denotes the EMMA without multi-task. We can see EMMA outperforms EMMA-w/o-MT on product intent detection task by 17.1% on precision@1 and 6.2% on NDCG@3 respectively.

Table 3. Effect of multi-task learning.

Models	P@1	P@3	P@5	N@1	N@3	N@5
EMMA	0.495	0.267	0.180	0.495	0.508	0.507
EMMA-w/o-MT	0.324	0.255	0.163	0.324	0.446	0.440

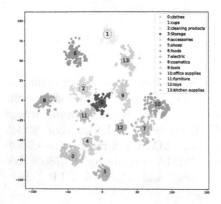

Fig. 4. Visualization of product embedding from popular categories.

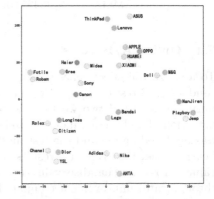

Fig. 5. Visualization of brand embedding.

5.6   Visualization

Product Embedding. For product classification task, the weights of final dense layer can be regarded as the embedding vector of product categories. In Fig. 4, we compute 2-D t-SNE [23] coordinates for some products. We can see the points with same category are gathered. For example, "sweater" and "jacket" are belong to clothes, and categories with similar type are close, such as clothes, accessories and shoes.

Brand Embedding. Similar with product category embedding, we visualize brand embedding in Fig. 5. As picture shows, the distribution of the points is very reasonable, the brands with same product category are close. We can see, ThinkPad, Lenovo, ASUS belong to computers product and clustered together, while Addidas, Nike, ANTA belong to clothes product and clustered together.

6   Conclusion

In this paper, we propose a deep learning model designed for jointly learning of product intent and brand intent in a coarse-to-fine approach, named EMMA. We can identify a query's brand intent without specific brand terms compared to NER method. Our proposed model framework can be potentially extended from e-commerce query understanding to other search scenarios. The results of empirical comparison and the ablation study verify the effectiveness of our method.

References

1. Putthividhya, D., Hu, J.: Bootstrapped named entity recognition for product attribute extraction. In: Proceedings of the 2011 Conference on Empirical Methods in Natural Language Processing, pp. 1557–1567 (2011)
2. Zheng, G., Mukherjee, S., Dong, X.L., Li, F.: OpenTag: open attribute value extraction from product profiles. In: Proceedings of the 24th ACM SIGKDD International Conference on Knowledge Discovery and Data Mining, pp. 1049–1058 (2018)
3. Jansen, B.J., Booth, D.L., Spink, A.: Determining the user intent of web search engine queries. In: Proceedings of the 16th International Conference on World Wide Web, pp. 1149–1150 (2007)
4. Ha, J.W., Pyo, H., Kim, J.: Large-scale item categorization in e-commerce using multiple recurrent neural networks. In: Proceedings of the 22nd ACM SIGKDD International Conference on Knowledge Discovery and Data Mining, pp. 107–115 (2016)
5. Ashkan, A., Clarke, C.L.A., Agichtein, E., Guo, Q.: Classifying and characterizing query intent. In: Boughanem, M., Berrut, C., Mothe, J., Soule-Dupuy, C. (eds.) ECIR 2009. LNCS, vol. 5478, pp. 578–586. Springer, Heidelberg (2009). https://doi.org/10.1007/978-3-642-00958-7_53
6. Yu, W., Sun, Z., Liu, H., Li, Z., Zheng, Z.: Multi-level deep learning based e-commerce product categorization. In: The SIGIR 2018 Workshop On eCommerce co-located with the 41st International ACM SIGIR Conference on Research and Development in Information Retrieval (2018)

7. Caruana, R.: Multitask learning. Mach. Learn. **28**(1), 41–75 (1997)
8. Ruder, S.: An overview of multi-task learning in deep neural networks. arXiv preprint arXiv:1706.05098 (2017)
9. Subramanian, S., Trischler, A., Bengio, Y., Pal, C.J.: Learning general purpose distributed sentence representations via large scale multi-task learning. arXiv preprint arXiv:1804.00079 (2018)
10. Sanh, V., Wolf, T., Ruder, S.: A hierarchical multi-task approach for learning embeddings from semantic tasks. In: Proceedings of the AAAI Conference on Artificial Intelligence, vol. 33, pp. 6949–6956 (2019)
11. Ahmadvand, A., Kallumadi, S., Javed, F., Agichtein, E.: Jointmap: joint query intent understanding for modeling intent hierarchies in e-commerce search. In: Proceedings of the 43rd International ACM SIGIR Conference on Research and Development in Information Retrievalm, pp. 1509–1512 (2020)
12. Gao, C., et al.: Neural multi-task recommendation from multi-behavior data. In: 2019 IEEE 35th International Conference on Data Engineering (ICDE), pp. 1554–1557. IEEE (2019)
13. Wang, J., et al.: A multi-task learning approach for improving product title compression with user search log data. In: Proceedings of the AAAI Conference on Artificial Intelligence, vol. 32 (2018)
14. Zhang, L., Wang, R., Zhou, J., Yu, J., Ling, Z., Xiong, H.: Joint intent detection and entity linking on spatial domain queries. In: Proceedings of the 2020 Conference on Empirical Methods in Natural Language Processing: Findings, pp. 4937–4947 (2020)
15. Babbar, R., Schölkopf, B.: Data scarcity, robustness and extreme multi-label classification. Mach. Learn. **108**(8), 1329–1351 (2019)
16. Babbar, R., Schölkopf, B.: DiSMEC distributed sparse machines for extreme multi-label classification. In: Proceedings of the Tenth ACM International Conference on Web Search and Data Mining, pp. 721–729 (2017)
17. Izmailov, P., Podoprikhin, D., Garipov, T., Vetrov, D., Wilson, A.G.: Averaging weights leads to wider optima and better generalization. arXiv preprint arXiv:1803.05407 (2018)
18. Prabhu, Y., Varma, M.: FastXML: a fast, accurate and stable tree-classifier for extreme multi-label learning. In: Proceedings of the 20th ACM SIGKDD International Conference on Knowledge Discovery and Data Mining, pp. 263–272 (2014)
19. Bhatia, K., Jain, H., Kar, P., Varma, M., Jain, P.: Sparse local embeddings for extreme multi-label classification. In: Advances in Neural Information Processing Systems 28: Annual Conference on Neural Information Processing Systems 2015, vol. 29, pp. 730–738 (2015)
20. Tagami, Y.: AnnexML: approximate nearest neighbor search for extreme multi-label classification. In: Proceedings of the 23rd ACM SIGKDD International Conference on Knowledge Discovery and Data Mining, pp. 455–464 (2017)
21. Huang, W., et al.: Hierarchical multi-label text classification: an attention-based recurrent network approach. In: Proceedings of the 28th ACM International Conference on Information and Knowledge Management, pp. 1051–1060 (2019)
22. Kendall, A., Gal, Y., Cipolla, R.: Multi-task learning using uncertainty to weigh losses for scene geometry and semantics. In: Proceedings of the IEEE Conference on Computer Vision and Pattern Recognition, pp. 7482–7491 (2018)
23. Van der Maaten, L., Hinton, G.: Visualizing data using t-SNE. J. Mach. Learn. Res. **9**(11), 1–28 (2008)

Exploiting Spatial Attention and Contextual Information for Document Image Segmentation

Yuman Sang[1], Yifeng Zeng[2(✉)], Ruiying Liu[1], Fan Yang[1], Zhangrui Yao[1], and Yinghui Pan[3]

[1] Department of Automation, Xiamen University, Xiamen, China
[2] Department of Computer and Information Sciences, Northumbria University, Newcastle upon Tyne, UK
`yifeng.zeng@northumbria.ac.uk`
[3] College of Computer Science and Software Engineering, Shenzhen University, Shenzhen, China

Abstract. We propose a new framework of combining an attention mechanism with a conditional random field to deal with a document image segmentation task. The framework aims to recognize homogeneous regions, e.g. text, figures, or tables, in document images through a pixel-wise spatial attention module. The attention module obtains essential global information and gathers long-distance pixel dependencies. To get extra knowledge around images, we use a conditional random field to model contextual information in the document. The new framework enables an effective combination of pixel features with their contextual information in the document image segmentation task. We conduct extensive experiments over multiple challenging datasets and demonstrate the performance of our new framework in comparison to a series of state-of-the-art segmentation methods.

Keywords: Document image segmentation · Conditional random field · Spatial attention mechanism

1 Introduction

Document image segmentation (DIS), or document layout analysis, plays an important role in many applications such as automatic document categorization, document content understanding, information retrieval, and so on. The goal of DIS is to identify homogeneous objects (e.g. background, text, figures, or tables) in a document image. A good DIS method can immediately boost the performance of some downstream tasks such as OCR and keyword spotting [6]. In addition, the accessibility of a document layout can add great value to the development of document digitization and massive educational applications.

Supported by the NSF in China (NSF: 62176225 and 61836005). Professor Yifeng Zeng is the corresponding author for this article.

J. Gama et al. (Eds.): PAKDD 2022, LNAI 13282, pp. 261–274, 2022.
https://doi.org/10.1007/978-3-031-05981-0_21

Existing DIS methods can roughly be clustered into two groups: (*i*) traditional image processing-based methods [13,15]; and (*ii*) omnipresent deep learning-based methods [4,23]. The former generally rely on hand-crafted features and complex heuristic rules, which are both experience-driven. In addition, some essential pre-processing, such as binarization, must be applied beforehand, which may result in the loss of color information of document images. Although traditional post-processing measures, like conditional random field (CRF) [11], are developed, they are badly used with a time-consuming process. Furthermore, heuristic rules are only suitable for some specific applications and have serious restrictions on the format of a document layout. Due to the amazing performance in various computer vision tasks, deep learning-based methods with attention mechanism [20] have been increasingly valued in document layout analysis. They exploit more representative features and show stronger generalization in complex layouts. However, most of the current work follows a general framework designed for natural scene images without fully considering the inherent characteristics of document images. We shall note that document images differ significantly from natural scene images in many aspects, such as richness of color, diversity of object aspect ratio, etc. The difference fundamentally compromises the performance of deep learning-based off-the-shelf methods.

Notice that local features extracted from images are not discriminative enough to analyze a document layout due to the absence of contextual information in the document image. As depicted in Fig. 1, characters contained in the figures and tables are likely to be mislabeled as text due to the similarity of their local features. Thus, contextual information in images, such as spatial dependencies among pixels and prior knowledge among their labels (e.g. the neighbors of a pixel labeled by *figure* are more likely to be labeled as *figure* rather than *table*), can play an important role in a document image segmentation task.

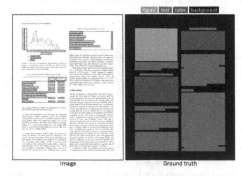

Image   Ground truth

Fig. 1. Samples that are difficult to identify if only based on local features. Characters (marked by red boxes) within the figures and tables might be mislabeled as *text* while their ground-truth is *figure* or *table*. (Color figure online)

Inspired by the findings above, in this paper, we consider document image segmentation as a pixel-wise classification problem and propose a new framework, namely DSAP, that integrates a spatial attention module to capture global information with a probabilistic graphical model, e.g. conditional random field (CRF), to learn contextual information. Our new framework can receive as input an original document image of arbitrary size. It does not need to extract restricted hand-crafted features, design complex heuristic rules, and implement laborious pre-processing.

The difficulty of developing such a new framework lies in the computational complexity of the CRF inference and optimization process. In traditional CRF formulation, message passing is implemented by applying entirely hand-crafted *Gaussian* filters for image features such as pixel locations and color values. Instead, we use neural networks deployed in a deep learning platform to train the CRF parameters so as to avoid all the time-consuming and memory-expensive operations. We conduct a experimental study over multiple datasets to analyze the performance of each component in DSAP and show expected performance by comparing DSAP with state-of-the-art methods.

2 Related Works

With the limited space, we will focus on the review of the approaches closely relevant to our method while referring readers to access a comprehensive survey about document layout analysis [4].

Document Image Segmentation. Many deep learning-based methods have achieved good performance on document image segmentation. For example, Oliveira et al. [17] proposed a single CNN-based pixel-wise predictor coupled with task-dependent post-processing blocks. Fink et al. [7] first applied the U-Net to a baseline extraction which aims to detect text lines of document images. Yi et al. [23] focuses on how to transfer and refine those object detection approaches from natural scene images to document images.

Attention Mechanism. With the powerful ability to capture long-range dependencies, an attention mechanism has been favored by a broad range of works. Wang et al. [20] presented non-local operations as a generic family of building blocks for capturing long-range dependencies. Fu et al. [8] proposed a dual attention network to integrate local features with their global dependencies. Niu et al. [16] further expanded the attention scope to channel and location attention.

Probabilistic Graphical Models. Probabilistic graphical models are widely used in graph analysis tasks since they allow direct label-level modeling of correlations and compatibility among neighboring nodes [9]. Conditional random field (CRF) [11] is one of the most popular probabilistic graphical models that can encode contextual information in a machine learning task. In the task of real-time stroke classification, Ye et al. [22] applied CRF to integrate contextual information, and optimized CRF parameters through joint training. Li et al. [12] used CRF to classify text and non-text blocks in test papers.

3 New Framework for Document Image Segmentation

In Fig. 2, we show the new framework (DSAP) for document image segmentation. It contains two parts: one feature extractor with a spatial attention mechanism and one CRF module. The workflow is described in Fig. 3. The framework

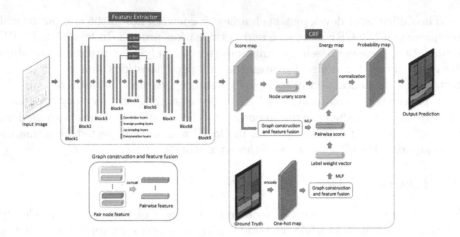

Fig. 2. Summary demonstration of DSAP for document image segmentation. It contains two components to achieve recognition of homogeneous regions. The Feature Extractor module is built on U-Net [18], including a contracting path and an expansion path. A novel spatial attention network (A-Net) is introduced in the skip connection to extract pixel dependencies. The CRF module constructs the score map and integrates contextual information into an energy map. The details are given in Sect. 3.

receives a document image of any size and the image first goes through a feature extractor. Subsequently, we construct a score map as a CRF model in which nodes represent pixels. We then obtain a label weight vector from corresponding ground truth data to represent the label prior and compute unary and pairwise scores. An energy map is the sum of the unary and pairwise scores and is normalized to get the probability map. Finally, the category with the highest probability is taken as the final segmentation result.

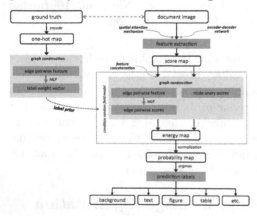

Fig. 3. The DSAP workflow implements the two components in Fig. 2.

Compared with existing methods, the DSAP framework differs in the following aspects. Firstly, DSAP does not require any pre-processing and an original image can be directly used as the input to the framework. Secondly, many methods ignore the fact that the raw fusion of deep and shallow features often results in huge redundancy of networks. Considering the dependencies among pixels at different locations, we introduce a spatial attention mechanism to refine the features. Thirdly, CRF is often used to extract contextual information in a segmentation task; however, many CRF-based techniques [10,14] only exploit either position or color information. In contrast, we leverage contextual information among labels in CRF. In addition, we introduce neural networks to fit a potential function so that CRF can be trained efficiently.

3.1 Feature Extraction Network

Encoder-decoder architectures, such as U-Net [18] and SegNet [3], have attracted much attention in image segmentation tasks. An encoder gets a feature map of an input image by down-sampling, while a decoder recovers the feature map to the original size by up-sampling to achieve the classification of each pixel. The network for our feature extraction is located in the left corner (Feature Extractor) within Fig. 2. It is a special extension of the U-Net [18] for document image segmentation. The encoder on the left consists of five blocks with the corresponding number of channels doubled after each down-sampling. The decoder on the right consists of four blocks and ends with a 1×1 convolution to turn the channel number of the final feature map into the desired number of categories. In addition, we introduce a novel attention module in the skip connection at each level. The skip connection structure can combine shallow, refined, and appearance features with deep, coarse, and semantic features.

3.2 Pixel-wise Spatial Attention Module

Following the formulation [20], we define a global attention mechanism below.

$$\mathbf{y}_i = \mathbf{x}_i + \frac{1}{\phi(\mathbf{x})} \sum_{\forall j} f(\mathbf{x}_i, \mathbf{x}_j) g(\mathbf{x}_j) \tag{1}$$

where $\mathbf{x}$ is the input feature map and $\mathbf{y}$ is the output feature map of the same size as $\mathbf{x}$. Note that i is the position index in the output and j is the index that enumerates all possible positions. $\phi(\mathbf{x})$ is the normalization coefficient. $f(\cdot)$ is a pairwise function that calculates a correlation value between $\mathbf{x}_i$ and $\mathbf{x}_j$. $g(\cdot)$ computes a certain mapping of the input feature $\mathbf{x}_j$.

In Eq. 1, we introduce a pixel-wise spatial attention network (A-Net) as shown in Fig. 4. The attention network is divided into three steps. To facilitate the implementation in a deep learning platform, we consider a combination of dot-product similarity with the *Gaussian* transformation to present the inter-correlation between features. Namely, we consider:

$$f(\mathbf{x}_i, \mathbf{x}_j) = e^{\mathbf{x}_i^T \mathbf{x}_j} \tag{2}$$

The normalization factor $\phi(\mathbf{x})$ is set in Eq. 3.

$$\phi(\mathbf{x}) = \sum_{\forall j} f(\mathbf{x}_i, \mathbf{x}_j) = \sum_{\forall j} e^{\mathbf{x}_i^T \mathbf{x}_j} \tag{3}$$

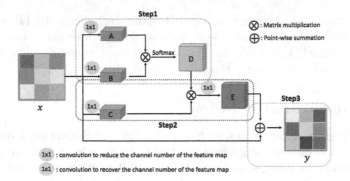

Fig. 4. A computational description for a pixel-wise spatial attention module which consists of three steps: a) two feature maps after convolution, A and B, are multiplied to obtain a weight map as the inter-correlation between features; b) the normalized weight map D is then multiplied with C; and c) a point-wise summation is performed on the feature map E with the input $\mathbf{x}$.

To ease the calculation, we reduce the channel number of the input feature map. In the first step, two 1×1 convolutions are performed on the input feature map $\mathbf{x}$ to convert it to A and B respectively, and then a multi-dimension matrix multiplication is executed on the transposed A and B. Finally, the weight map D is obtained by the *softmax* normalization in Eq. 4.

$$D(\mathbf{x}_i) = \frac{e^{A(\mathbf{x}_i)^T B(\mathbf{x}_j)}}{\sum_{\forall j} e^{A(\mathbf{x}_i)^T B(\mathbf{x}_j)}} \tag{4}$$

In the second step, the correlations obtained in the first step are taken as weights of the corresponding features. In particular, we first obtain the feature map C by performing 1×1 convolution on the input and then multiply it with the transposed weight map D.

Finally, in the third step, the channel number of the feature map E is recovered by 1×1 convolution, and then a point-wise summation is performed with the input feature map $\mathbf{x}$ as shown in Eq. 5.

$$\mathbf{y}_i = \alpha \sum_{\forall j} D(\mathbf{x}_i)^T C(\mathbf{x}_j) + \mathbf{x}_i \tag{5}$$

where α is a trainable factor with the initialization of zero and gradually gets the optimal value during the learning.

The output feature map $\mathbf{y}$ of our attention network has the same size as the input $\mathbf{x}$, and each pixel in the output is the weighted embedding of the corresponding pixel in the input.

3.3   Contextual Information Integration

To capture contextual information in a document image, we construct a probabilistic graphical model and consider each pixel in the feature map as a node.

Intuitively, correlations exist among the neighboring pixels, which could be modeled in a graph. In this paper, we use a pairwise potential function in CRF for modeling the correlations. As shown in Fig. 2, CRF (succeeding the feature extractor) integrates contextual information and refine the previous output.

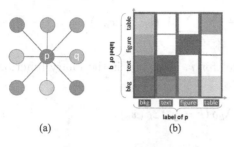

(a)   (b)

Fig. 5. A certain relationship exists among labels of neighboring pixels: (a) each node is connected with its eight nearest neighboring nodes (left, right, top, bottom, top left, top right, bottom left and bottom right); and (b) the relationship between labels of node p and its neighboring node q. A darker color indicates a stronger correlation.

In CRF, we define two key variables: pixels in a feature map as observed variables $\mathbf{X}$ and labels in the ground truth as implied variables $\mathbf{Y}$. The basic model can be formulated to the conditional probability distribution $P(\mathbf{X}|\mathbf{Y})$ of $\mathbf{X}$ given $\mathbf{Y}$. For one pixel (or one node), to get its prediction with high confidence, we consider not only the features of this pixel itself (such as color, position, texture, etc.), but also the features and label information of its neighbors.

Based on the aforementioned discussion, we exploit certain contextual relationships among the label categories in the CRF model. As shown in Fig. 5(b), the horizontal axis denotes the possible prediction of classes for a given node p, while the vertical axis denotes that for its neighboring node q. The color depth indicates the probability that node p and node q are predicted to be the corresponding classes simultaneously. The darker the color, the higher the probability is. For example, the probability of *(label of p, label of q) = (text, text)* is greater than the probability of *(label of p, label of q) = (text, background)*, and the probability of *(label of p, label of q) = (text, figure)* could be extremely small.

To tackle the challenge of optimizing the structured CRF component, we propose to use neural networks to reformulate a novel pairwise potential function in CRF. This allows us to implement the CRF through highly efficient GPU computation, in which back-propagation can be used to train all parameters fast and automatically. We define a second-order CRF model below.

$$P(y|x;\theta) = \frac{1}{\Phi(x;\theta)} \exp\left[-E(y,x;\theta)\right] \tag{6}$$

where $\Phi(x;\theta)$ is the partition function and $E(y,x;\theta)$ is the energy function.

We formulate the the energy function in Eq. 7.

$$E(y,x;\theta) = \sum_{p \in N_U} U\left(y_p, x_p; \theta_U\right) + \sum_{(p,q) \in S_V} V\left(y_p, y_q, x_{p,q}; \theta_V\right) \tag{7}$$

where U is the unary potential function which denotes the cost of node p being predicted as y_p, N_U the node set in CRF and θ_U parameters of U function.

Likewise, V is the pairwise potential function which denotes the cost of node p being predicted as y_p and node q being predicted as y_q simultaneously. S_V is the edge set and θ_V are parameters of the V function.

We compute the energy score for the given node p in Eq. 8.

$$E\left(y_p, x_p\right) = U\left(y_p, x_p\right) + \sum_{q \in \Omega_p} \sum_{\forall y_q} V\left(y_p, y_q, x_{p,q}\right) \tag{8}$$

where Ω_p is the set of neighboring nodes for p. $U\left(y_p, x_p\right)$ is the unary function for node p and is defined below.

$$U\left(y_p, x_p\right) = \sum_{\forall k} -\lambda_k \delta\left(k = y_p\right) z_k\left(x_p\right) \tag{9}$$

where $z_k\left(x_p\right)$ is the feature value corresponding to the node p belonging to the class k in the output of the feature extraction network. λ_k is the weight coefficient and can be represented by the presence frequency of pixels labeled as class k in the ground truth.

In addition, we model the pairwise function V through a neural network that is essentially a multi-layer perceptron (MLP). Firstly, we concatenate the feature vectors corresponding to the node p with the node q in the output of the feature extraction network. We show the specific process in Fig. 6. The concatenation vector as the initial pairwise feature is then fed into MLP for the training purpose, of which the output dimension is K^2.

Then we can get the pairwise potential function $V\left(y_p, x_{p,q}\right)$ in Eq. 10.

$$V\left(y_p, y_q, x_{p,q}\right) = \sum_{\forall k_p} \sum_{\forall k_q} \left(-\lambda_{k_p, k_q} \delta\left(k_p = y_p\right) \cdot \delta\left(k_q = y_q\right) z_{k_p, k_q}\left(x_{p,q}\right)\right) \tag{10}$$

where z_{k_p, k_q} is the feature value corresponding to the node pair (p, q) in the output of the MLP, and δ is the impulse function. λ_{k_p, k_q} is the weight coefficient and represents the correlation between the corresponding labels, which can be extracted from the ground truth. To get the λ_{k_p, k_q}, we use a similar measure, including a feature concatenation process and a MLP, for the easy fusion of label knowledge.

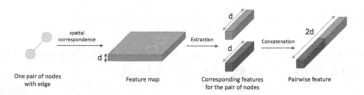

Fig. 6. The concatenation process of pairwise features: one pairwise edge in CRF goes through the feature concatenation of the corresponding two nodes to obtain a final pairwise feature.

Briefly speaking, we construct a CRF model to refine the coarse output from the feature extractor in DSAP. The CRF model leverages contextual information by integrating relationships of pixel labels and features into a pairwise potential function. We develop an efficient CRF training process by introducing two MLPs to fit the pairwise potential function. Towards this goal, we reformulate the exact and fast message-passing step in our CRF through the joint training on GPUs, which eliminates the need to compute the complex optimization function.

4   Experimental Results

We conducted a series of experiments to investigate: *a*) effectiveness and impact of the two modules on the baseline model (U-Net); *b*) the DSAP performance compared with other methods. We used four challenging datasets in the empirical evaluation - PubLayNet [25], RDCL2015 [2], DSSE-200 [21], and DIVA-HisDB [19]. For each dataset, we compare DSAP with state-of-the-art segmentation techniques. In order to provide an extensive evaluation, we consider five commonly used metrics [1]: Intersection-over-Union (IoU), precision, recall, F1-score and pixel accuracy (PA).

4.1   Implementation Details

We built our architecture based on *PyTorch* and used the pre-trained weights on ImageNet to initialize the baseline model. We trained the model for 200k iterations with a mini-batch size of 8 on a NVIDIA RTX 3090Ti GPU server of 16 GB memory. We used a Stochastic Gradient Descent (SGD) optimizer to update weights with an initial learning rate of 0.01 and then adopted a common decay strategy to reduce the learning rate by a factor of 10 at the 120k-160k iterations. Two 4-layer MLPs were used to implement our pairwise potential function in CRF with the *sigmoid* activation function after all hidden layers and the *softmax* function after the last layer. Notably, inference in our CRF can be performed in less than 10ms due to our new potential functions - an increase of two-orders magnitude compared to traditional CRF, which considerably speeds up the inference of the overall DSAP framework.

4.2   Ablation Study

We first conduct experiments on PubLayNet - the largest existing dataset for document layout analysis, and use its development set (containing about 11k pages) in the evaluation. In our experiments, each pixel in an image is labelled with four possible classes: *background* (*bkg*), *text*, *figure* (*fig*) and *table*. The mean values of five metrics over the classes are presented in Table 1. The results show that introducing the two modules (Attention and CRF) achieves a remarkable improvement on the baseline model (U-Net). Besides, we present the detailed results for each class in Table 2. The measurements of IoU and precision are significantly improved especially for the classes of *figure* and *table*.

4.3 Comparisons Against State-of-the Art Methods

Comparisons on PubLayNet-dev and DSSE-200 (Table 3). We compare
our method (DSAP) with the state-of-the-art methods on these two datasets.
The results of IoU are shown in Table 3 where − shows unavailable results from
the references. DSAP achieves better performance than all the other methods.
All the methods, including ours, are more accurate at detecting *background* and
text than *figure* and *table*. This is because the class *text* has more regular pixel
structures and monotonic features than other categories. We also evaluate the
comparison on the DSSE-200 dataset, which has six label categories. To focus on
appearance-based labels, we integrate *section, caption, list* and *paragraph* into a
generic category *text*. In Table 3, the MFCN (extract) represents the model using
extracted text by OCR engines and the MFCN (none) represents the vision-only
model [21]. Our method obviously outperforms other six methods. In addition,
we can see that the performance of DSAP on the DSSE-200 dataset is inferior to
that on the PubLayNet dataset. This is because PubLayNet is not so complicated
as DSSE-200 and the region recognition becomes easier.

Table 1. Ablation experiments on the PubLayNet-dev dataset. The architecture of
each model is characterized by the application of attention and CRF module. Mean
results of five metrics are presented (all values are in % and averaged over the classes).

Model			Mean scores				
Baseline	Atten	CRF	IoU	Precision	Recall	F1	PA
U-Net			66.67	62.62	94.21	75.23	91.20
U-Net	✓		91.51	87.67	97.11	92.15	96.96
U-Net		✓	90.19	91.64	95.32	93.44	95.74
U-Net	✓	✓	93.44	94.48	97.49	95.96	97.69

Table 2. Detailed class-wise scores (%) on the PubLayNet-dev dataset.

Model			IoU				Precision				Recall			
Baseline	Atten	CRF	bkg	Text	Fig	Table	bkg	Text	Fig	Table	bkg	Text	Fig	Table
U-Net			91.63	89.45	18.16	26.81	95.01	93.07	31.38	31.02	96.29	95.92	56.66	86.12
U-Net	✓		96.16	93.40	68.16	79.42	98.17	94.28	71.14	87.10	98.06	98.08	94.18	90.91
U-Net		✓	93.07	90.30	83.07	78.59	95.61	94.20	90.41	86.32	97.29	95.09	91.92	90.64
U-Net	✓	✓	96.50	93.89	77.80	84.76	97.70	95.46	92.17	92.60	98.49	97.83	95.89	92.26

Table 3. IoU (%) results of methods on the PubLayNet-dev and DSSE-200 datasets.

Dataset	PubLayNet-dev				DSSE-200			
Methods	bkg	Text	Fig	Table	bkg	Text	Fig	Table
FCN [14]	78.4	72.1	73.6	70.3	71.2	68.3	55.9	52.8
DeeplabV3 [5]	83	81.4	79.2	77.9	76.1	73.2	64.7	62.5
SegNet [3]	85.8	83.1	81.6	79.4	79.3	74.1	63.9	61.7
PSPNet [24]	90.4	88.7	78.2	77.1	78.1	71.2	73.5	69.6
MFCN (none) [21]	-	-	-	-	84.6	-	83.3	79.4
MFCN (extract) [21]	-	-	-	-	83.9	-	83.7	79.7
DSAP(Ours)	**96.5**	**93.89**	**77.8**	**84.76**	**87.9**	**81.3**	**84.5**	**81.2**

Comparisons on RDCL2015 (Table 4). We further test our method on the RDCL2015 dataset and extend our comparison to other five benchmarks in Table 4. The performances of the five benchmarks have been reported in [15]. For a fair comparison, we use the same classification criteria as those in the previous work [12], and only consider three classes (*background*, *figure* and *text*). DSAP performs competitively to DSPH, where the latter has shown superior performance to other methods. DSPH yields better performance under the classification criteria for *non-text* and *text* while its performance for *background* and *figure* is competitive to DSAP. This is partly due to the fact that the non-text classification in DSPH is only based on mathematical morphology, where valuable information may get lost when identifying figures. We show a visual demonstration of the segmentation results from three datasets in Fig. 7, which is pretty consistent with the actual document layouts.

Table 4. IoU (%) results of methods on the RDCL2015 dataset

Methods	Background	Figure	Non-Text	Text
Leptonica	-	-	84.7	86.8
Bukhari	-	-	90.6	90.3
Fernández	-	70.1	-	85.8
MFCN	-	77.1	94.5	91
DSPH	92.2	77.8	96.3	94.7
DSAP (Ours)	**95.6**	**86.4**	**95.7**	**93.3**

Table 5. Performances (in % averaged over the classes) on DIVA-HisDB

Methods	IoU	Precision	Recall	F1	PA
KFUPM	65.36	81.32	75.07	75.50	94.70
IAIS	74.07	84.47	84.77	83.32	95.12
DSAP (Ours)	**81.03**	**86.92**	**87.9**	**87.41**	**96.59**
MindGarage-2	89.58	95.15	93.92	93.86	98.34
BYU	90.18	93.96	94.87	93.57	99.01
Demokritos	92.27	95.65	96.09	95.32	99.17
MindGarage-1	93.95	96.55	97.10	96.04	99.60
NLPR	94.90	97.58	97.20	96.81	99.57

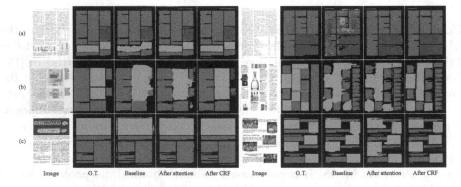

Fig. 7. A visual demonstration of the segmentation results on images using DSAP from (a) PubLayNet, (b) RDCL2015 and (c) DSSE-200 datasets. Red, blue, green represent text, table, and figure respectively. (Color figure online)

Comparisons on DIVA-HisDB (Table 5). Lastly, we turn to a history manuscript dataset (DIVA-HisDB) which is completely different from the former three datasets. We repeat the task [19]: each pixel in the image is assigned a label among four classes: *background*, *comment*, *decoration* and *text*. We compare DSAP with seven methods from the competition where performances have been reported [17]. The results averaged over classes are shown in Table 5. Since there is a great difference between history and ordinary document layouts as well as the existence of handwritten texts in the DIVA-HisDB dataset, the not-so-good results are expected. Although slightly inferior to the optimal results, DSAP still outperforms KFUPM and IAIS - indicating its good generalization on the history datasets. Further improvement will be explored in our future work.

5   Conclusion

Deep neural network that is popular in computer vision has yet to be fully explored in document layout analysis. In this paper, we propose a novel and effective framework (DSAP) for document image segmentation. In particular, we exploit a pixel-wise spatial attention mechanism to capture global information with pixel dependencies. We further integrate contextual information, which we conceptualize as the label relationships modeled by conditional random field (CRF). This allows the effective integration of pixel features with label knowledge in the homogeneous regions extraction process, which is currently neglected by existing document segmentation techniques. Our new framework demonstrates superior performance and good generalization on various challenging datasets in terms of multiple metrics. In addition, the ablation experiments show that both the two proposed modules (Attention and CRF) can boost the segmentation performance with a measurable margin upon the baseline model.

With the growing needs of document digitalization in a world-wide, the layout analysis of complicated documents, such as historical manuscripts and test papers, will be an important topic in our future work. Notably, from the results on the challenging DIVA-HisDB dataset, we could perceive potential improvement for the new framework. For example, we could extend the CRF capability in learning contextual information even in the pre-processing phase so as to extract more informative connected components in DSAP.

References

1. Alberti, M., Bouillon, M., Ingold, R., Liwicki, M.: Open evaluation tool for layout analysis of document images. In: ICDAR, vol. 4, pp. 43–47. IEEE (2017)
2. Antonacopoulos, A., Clausner, C., Papadopoulos, C., Pletschacher, S.: ICDAR2015 competition on recognition of documents with complex layouts-RDCL2015. In: ICDAR, pp. 1151–1155. IEEE (2015)
3. Badrinarayanan, V., Kendall, A., Cipolla, R.: SegNet: a deep convolutional encoder-decoder architecture for image segmentation. IEEE Trans. Pattern Anal. Mach. Intell. **39**(12), 2481–2495 (2017)
4. Binmakhashen, G.M., Mahmoud, S.A.: Document layout analysis: a comprehensive survey. ACM Comput. Surv. **52**(6), 1–36 (2019)
5. Chen, L.C., Papandreou, G., Schroff, F., Adam, H.: Rethinking atrous convolution for semantic image segmentation. arXiv preprint arXiv:1706.05587 (2017)
6. Eskenazi, S., Gomez-Krämer, P., Ogier, J.M.: A comprehensive survey of mostly textual document segmentation algorithms since 2008. Pattern Recogn. **64**, 1–14 (2017)
7. Fink, M., Layer, T., Mackenbrock, G., Sprinzl, M.: Baseline detection in historical documents using convolutional U-Nets. In: IAPR International Workshop on Document Analysis Systems, pp. 37–42. IEEE (2018)
8. Fu, J., et al.: Dual attention network for scene segmentation. In: CVPR, pp. 3146–3154 (2019)
9. Koller, D., Friedman, N.: Probabilistic Graphical Models: Principles and Techniques. MIT Press, Cambridge (2009)
10. Krähenbühl, P., Koltun, V.: Efficient inference in fully connected CRFs with Gaussian edge potentials. Adv. Neural Inf. Process. Syst. **24**, 109–117 (2011)
11. Lafferty, J., McCallum, A., Pereira, F.C.: Conditional random fields: probabilistic models for segmenting and labeling sequence data (2001)

12. Li, X.-H., Yin, F., Liu, C.-L.: Page segmentation using convolutional neural network and graphical model. In: Bai, X., Karatzas, D., Lopresti, D. (eds.) DAS 2020. LNCS, vol. 12116, pp. 231–245. Springer, Cham (2020). https://doi.org/10.1007/978-3-030-57058-3_17

13. Li, X., Yin, F., Xue, T., Liu, L., Ogier, J.M., Liu, C.: Instance aware document image segmentation using label pyramid networks and deep watershed transformation. In: ICDAR, pp. 514–519. IEEE (2019)

14. Long, J., Shelhamer, E., Darrell, T.: Fully convolutional networks for semantic segmentation. In: CVPR, pp. 3431–3440 (2015)

15. Lu, T., Dooms, A.: Probabilistic homogeneity for document image segmentation. Pattern Recogn. **109**, 107591 (2021)

16. Niu, R., Sun, X., Tian, Y., Diao, W., Chen, K., Fu, K.: Hybrid multiple attention network for semantic segmentation in aerial images. IEEE Trans. Geosci. Remote Sensing **60**, 1–18 (2021)

17. Oliveira, S.A., Seguin, B., Kaplan, F.: dhSegment: a generic deep-learning approach for document segmentation. In: ICFHR, pp. 7–12. IEEE (2018)

18. Ronneberger, O., Fischer, P., Brox, T.: U-Net: convolutional networks for biomedical image segmentation. In: Navab, N., Hornegger, J., Wells, W.M., Frangi, A.F. (eds.) MICCAI 2015. LNCS, vol. 9351, pp. 234–241. Springer, Cham (2015). https://doi.org/10.1007/978-3-319-24574-4_28

19. Simistira, F., et al.: ICDAR2017 competition on layout analysis for challenging medieval manuscripts. In: ICDAR, vol. 1, pp. 1361–1370. IEEE (2017)

20. Wang, X., Girshick, R., Gupta, A., He, K.: Non-local neural networks. In: CVPR, pp. 7794–7803 (2018)

21. Yang, X., Yumer, E., Asente, P., Kraley, M., Kifer, D., Lee Giles, C.: Learning to extract semantic structure from documents using multimodal fully convolutional neural networks. In: CVPR, pp. 5315–5324 (2017)

22. Ye, J., Zhang, Y., Liu, C.: Joint training of conditional random fields and neural networks for stroke classification in online handwritten documents. In: ICPR, pp. 3264–3269. IEEE (2016)

23. Yi, X., Gao, L., Liao, Y., Zhang, X., Liu, R., Jiang, Z.: CNN based page object detection in document images. In: ICDAR, vol. 1, pp. 230–235. IEEE (2017)

24. Zhao, H., Shi, J., Qi, X., Wang, X., Jia, J.: Pyramid scene parsing network. In: CVPR, pp. 2881–2890 (2017)

25. Zhong, X., Tang, J., Yepes, A.J.: PubLayNet: largest dataset ever for document layout analysis. In: ICDAR, pp. 1015–1022. IEEE (2019)

DHA: Product Title Generation with Discriminative Hierarchical Attention for E-commerce

Wenya Zhu[1]([✉]), Yinghua Zhang[2], Yu Zhang[3], Yuhang Zhou[1], Yinfu Feng[1], Yuxiang Wu[4], Qing Da[1], and Anxiang Zeng[5]

[1] Alibaba, Hangzhou, China
zhuwenya1991@gmail.com, {zyh174606,yinfu.fyf,daqing.dq}@alibaba-inc.com
[2] Hong Kong University of Science and Technology, Kowloon, Hong Kong
yzhangdx@cse.ust.hk
[3] Southern University of Science and Technology, Shenzhen, China
[4] University College London, London, UK
[5] Nanyang Technological University, Singapore, Singapore
zeng0118@ntu.edu.sg

Abstract. Product titles play an important role in E-Commerce sites. However, manually crafting product titles needs tremendous time and human effort. It is expected that product titles can be automatically generated, but existing generation methods usually require densely labeled data that are unavailable in the real world. We formulate a novel product title generation task that generates the title from the image and auxiliary information (e.g., category) to address the gap. To generate titles that are consistent with search queries, we construct the first large-scale dataset (AEPro) and propose a Discriminative Hierarchical Attention (DHA) model. The DHA model first identifies the image regions related to the product of interest (POI) with a *POI* attention module. Then, based on the title context, the identified image regions are further revised by a *generation* attention module. Finally, the titles are generated by dynamically attending to these image regions. Experiments on the AEPro dataset demonstrate the effectiveness of the DHA model. Besides, online A/B testing results show that 61.8% of the titles generated by the DHA model are accepted directly or with minor modifications. The exposure rate of the products with machine-generated titles is improved by 40%.

Keywords: Natural language generation · Discriminative attention · Hierarchical attention

1 Introduction

Nowadays, E-Commerce websites have gained a growing popularity all over the world. When shopping online, a customer expresses his/her shopping intention by a search query, and the search engine returns a list of products that match the query. The product title, which is usually provided by the seller, plays an

J. Gama et al. (Eds.): PAKDD 2022, LNAI 13282, pp. 275–287, 2022.
https://doi.org/10.1007/978-3-031-05981-0_22

important role in the above process. Yet, it is challenging for sellers to compose product titles. Firstly, customers might express the same shopping intention with different queries, for example, "pullover" and "hoodies" might be used interchangeably. Thus, a seller need to consider various expression styles. Secondly, the title composition cost grows due to daily new products. Therefore, it is expected that product titles can be automatically generated.

Automatic product title generation has been previously studied in [3,21]. However, these methods require either structured product information in terms of slot-value pairs (e.g. watch type: wrist watch) [3], or product descriptions provided by sellers [21], which are unavailable in most E-commerce websites. Instead, basic product information such as product images and its category is readily available, containing rich information about the product. Hence, we propose a product title generation task which aims to automatically generates the product title from its image and category.

Although image captioning also aims to transform the visual information into the textual sequence, the proposed task is different from image captioning [22]. Figure 1 demonstrates the differences between the two tasks. Specifically, image captioning involves the interactions among multiple objects, and hence most regions of the image need to be considered. On the other hand, there is only one product of interest for title generation while other objects are irrelevant. For example, the product of interest is the top while the midi skirt are irrelevant. With the product of interest as the main focus, the title describes it from various fine-grained aspects (e.g., color and functionality).

To generate titles that are consistent with search queries, we construct the first large-scale dataset (AEPro) from the AliExpress where there are $720K$ (product image, title) pairs and the titles with low search frequency are filtered out. Moreover, we develop a Discriminative Hierarchical Attention (DHA) model. To leverage the visual information at a fine-grained scale, two novel attention mechanisms are designed. A *product-of-interest (POI) attention* mechanism is first utilized to highlight the image areas that contain the product of interest. Then a *generation* attention mechanism attentively reads the image areas emphasized by the POI attention mechanism, and uses the attended visual information for title generation. To facilitate the learning of the attention modules, we utilize the Grad-CAM [20] as the weakly supervised information for the POI attention, which can coarsely locate the image areas of the product category by flowing the gradients in a CNN-based image classifier. The tie to the product category is critical, as it can avoid generating incorrect descriptions due to unrelated image areas. We validate the effectiveness of the proposed DHA model on the AEPro dataset. Besides, the proposed method has been employed in a large online E-commerce system to help sellers to craft titles for new products.

2   Related Works

2.1   Product Title Generation

Automatic title generation plays an important role in the E-commerce domain. Many methods have been proposed to save time and increase revenues, including

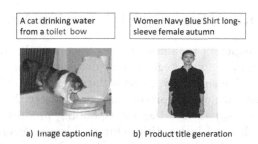

a) Image captioning   b) Product title generation

Fig. 1. The differences between image captioning and product title generation. a) The caption models the interaction between the cat and toilet bow with the verb *drink*; b) the long-sleeve woman shirt is the product of interest and the title describes the shirt from various aspects: the style (long-sleeve), color (navy blue), gender (female), etc. The product image also includes a midi skirt, but it is not the main focus and might mislead title generation. (Color figure online)

rule-based models [3], statistical models [21], and hybrid methods [5]. Structured information about the product in terms of slot/value pairs (e.g., "watch type": "wrist watch") is leveraged in [3]. In [21], a title hypothesis is first generated by reorganizing frequent n-grams of listing titles provided by sellers, and then evaluated by a supervised model trained from a large volume of human-annotated data. In [5], short natural-language descriptions are generated by taking structured DBpedia[1] data as the input. Recently, there is a trend towards developing generative neural models. A recurrent neural network is adopted to generate texts from a knowledge base in [16]. Unlike previous work, the proposed DHA model uses product images as the information source to generate product titles.

2.2 Image Captioning

The goal of image captioning is to generate one sentence to describe the image. Previous works on image captioning can be mainly categorized into three approaches, namely template-based methods, retrieval-based methods and generative neural network models. Template-based methods rely on predefined templates, and fill the slots in the template with object detection and attribute discover [6,11,17]. Retrieval-based methods encode representations for images and sentences, and retrieve most relevant sentences based on the similarity computed with hand-crafted features [7,8]. These two methods are not flexible since the generated sentences are restricted by templates. With the success of the encoder-decoder framework in machine translation, generative neural networks are more popular nowadays. Kiros et al. [10] combine a feed-forward neural network with a multi-modal log-bilinear model to predict the next word based on the image and the previous word. Other methods adopt the encoder-decoder framework [23]. Google NIC [22] first generates the caption with an RNN-based decoder based on the image information encoded by a convolutional network.

---

[1] http://dbpedia.org.

3   Discriminative Hierarchical Attention Model

The task of this paper is to automatically generate the title of a product based on the its image and category. The dataset is composed of m tuples $\mathcal{D} = \{(\mathbf{I}_i, c_i, \mathbf{S}_i)\}_{i=1}^m$. For simplicity, we omit the superscript i of related notations in the following. Let $\mathbf{I}$ denote the product image, c denote the product category, and $\mathbf{S} = \{s_1, s_2, \cdots, s_n\}$ denote the corresponding product title of length n, where s_j is the j-th word in the title. Our problem is formulated as follows: Given a product image $\mathbf{I}$ and its category c, the goal is to generate product title S that is coherent with $\mathbf{I}$ and c. Essentially, the model developed in this paper estimates the generative probability of $\mathbf{S}$ conditioned on $\mathbf{I}$ and $c : P(\mathbf{S}|\mathbf{I}, c)$.

In this section, we present the proposed DHA model for the product title generation problem. An overview of the DHA model is shown in Fig. 2. The whole model consists of five components: an image encoder, two attention modules, a language decoder, and a category classifier. The image encoder converts the input image into visual features. The hierarchical attention modules first highlight the product of interest, and then dynamically attends to the highlighted areas given the title generation context. With the attended visual features as input, the language decoder generates the product titles. The category classifier is introduced to provide auxiliary supervision for learning the attention module.

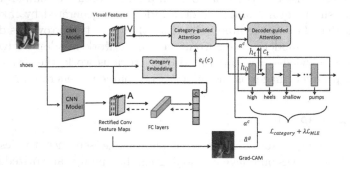

Fig. 2. Detailed depiction of Discriminative Hierarchical Attention (DHA) model for product title generation.

3.1   Background: Encoder-Decoder Framework

In this section, as a basis of DHA, a general encoder-decoder framework for image captioning is introduced.

The image encoder that is usually a convolutional neural network converts the input image $\mathbf{I}$ into a set of hidden representations, denoted by $\mathbf{V}$, and there is $\mathbf{V} = \{\mathbf{v}_1, \mathbf{v}_2, \cdots, \mathbf{v}_K\}$, where K denotes the number of the image regions and $\mathbf{v}_k \in \mathbb{R}^d$ is a d-dimensional hidden representation for the k-th image region. Specifically, in a ResNet-50 [9], the dimension of feature maps output by the last convolutional layer is $1024 \times 7 \times 7$, which can be flattened into 1024×49. In this case, there are $d = 1024$ and $K = 49$.

The language decoder outputs the product title as a sequence of words. It is usually a recurrent neural network (RNN). The decoder maintains a hidden state and uses previous generated words as inputs. To generate the j-th word, the decoder takes the visual context vector $\mathbf{c}_j$ and the embedding of the previously decoded word $\mathbf{e}(s_{j-1})$ as input, and updates its hidden state $\mathbf{h}_j$ as

$$\mathbf{h}_j = \text{RNN} \left(\mathbf{h}_{j-1}, \mathbf{c}_{j-1}, \mathbf{e}(s_{j-1}) \right), \tag{1}$$

where the visual context vector $\mathbf{c}_{t-1}$ is a weighted sum of the hidden representations over all regions. It is defined as

$$\mathbf{c}_{j-1} = \sum_k a_{k,j-1} \mathbf{v}_k,$$

where the attention map $a_{k,j-1}$ denotes the relevance of the image region k for s_j. The attention map can be obtained from various attention models. Generally, similar to [24], the attention map is determined by a multi-layer perceptron (MLP) as

$$\mathbf{a}_{j-1} = \text{softmax}[\mathbf{w}_c^T (\tanh (\mathbf{W}_v \mathbf{V}) + \tanh (\mathbf{W}_h \mathbf{h}_{j-2}))].$$

The decoder generates a word by sampling the generation distribution $\mathbf{o}_j$ derived from the decoder's hidden state $\mathbf{h}_j$ as

$$s_j \sim \mathbf{o}_j = P(s_j | s_{1:j-1}, \mathbf{I}) = \text{softmax} (\mathbf{W}_o \mathbf{h}_j),$$

where $s_{1:j-1} = s_1 s_2 \cdots s_{j-1}$ denotes the words previously generated.

We denote the parameters in the above model by θ. The loss function for learning θ is the sum of the negative log likelihood of the correct word at all steps as

$$\mathcal{L}_{MLE} = - \sum_j \log P(s_j | s_{1:j-1}, \mathbf{I}; \theta). \tag{2}$$

3.2 Hierarchical Attention Mechanism

The base attention mechanism introduced in Sect. 3.1 is designed for image captioning benchmarks, such as MSCOCO [14]. In those settings, there are multiple objects in an image and it is necessary to model the interactions among the objects. Hence, most regions of the image need to be considered. On the other hand, there is only one product of interest for title generation while other objects are irrelevant, making the base attention mechanism inappropriate for product title generation. To address the gap, we propose a hierarchical attention mechanism, including a *POI attention* and a *generation* attention. Firstly, the POI attention localizes the image regions relevant to the product. Secondly, the generation attention dynamically attends to these regions depending on the title context. The two attention modules are described consequently.

POI Attention. The product image usually contains areas that are irrelevant to the product, which may introduce noise to the title generation. Therefore, it is necessary to locate relevant areas of the product in the image first. Image segmentation is a straightforward method to identify regions of target objects. For instance, in the region-level attention mechanism, the segmented regions enhance the image representation [1,25]. However, image segmentation requires densely annotated objects, attributes, and relationships. To avoid the time-consuming annotation, we devise the POI attention mechanism by computing attention map based on the product category. In detail, given the embedding of the product category, denoted by $\mathbf{e}_c(c)$, and the image hidden representations $\mathbf{V}$, the POI attention map, denoted by $\mathbf{a}^c$, measures the relevance between the visual features and the product category:

$$\mathbf{a}^c = \text{softmax}(\mathbf{W}_c^c(\tanh(\mathbf{W}_v^c\mathbf{V}) + \mathbf{W}_h^c\mathbf{e}_c(c))), \tag{3}$$

where $\mathbf{W}_c^c$, $\mathbf{W}_v^c$, $\mathbf{W}_h^c$, and the category embedding matrix $\mathbf{e}_c$ are the parameters in the POI attention module. The POI attention map acts as a "soft" bounding box of the target product. The highlighted image representation, which is the weighted average of visual features, serves as the initial state of the language decoder:

$$\mathbf{h}_0 = \sum_{k=1}^n a_k^c \mathbf{v}_k. \tag{4}$$

Generation Attention. The POI attention defined in Eq. (3) is static, it cannot dynamically select the image regions during the title generation. [24] showed that dynamically selecting image regions can improve the performance of a language decoder. Hence, we further propose the generation attention, which revises the POI attention map based on the context information of the language decoder. Firstly, we generate the attention map $\mathbf{a}_j^r$ based on the hidden state of the language decoder $\mathbf{h}_j$ as

$$a_{j,k}^r = \mathbf{W}_c^r(\tanh(\mathbf{W}_v^r\mathbf{v}_k) + \tanh(\mathbf{W}_h^r\mathbf{h}_j)), \tag{5}$$

where $\mathbf{W}_c^r$, $\mathbf{W}_v^r$ and $\mathbf{W}_h^r$ are the transformation matrices to be learned. Secondly, the generation attention map $\mathbf{a}^d$ is modeled as a mixture of the $\mathbf{a}^r$ and POI attention $\mathbf{a}^c$ and it is formulated as

$$\mathbf{a}^d = \text{softmax}(\mathbf{a}^r \odot \mathbf{a}^c), \tag{6}$$

where $\odot$ represents the element-wise product. The generation attention selects the image regions that are emphasized by both the POI attention and title context. The visual context vector in our model is computed by aggregating the region-level features weighted with decoder-guided attention map $\mathbf{a}^d$:

$$\mathbf{c}_j = \sum_{k=1}^K a_k^d \mathbf{v}_k. \tag{7}$$

3.3 Discriminative Attention Mechanism

The POI attention plays a crucial role in the DHA model. If it fails to identify the product of interest, the visual features are incorrect for downstream model components. However, our empirical experiments show that the distant $(\mathbf{I}, \mathbf{S})$ pairs do not provide sufficient supervision for accurate POI localization.

Additional supervision signals are necessary for learning the POI attention module. The Gradient-weighted Class Activation Mapping (Grad-CAM) proposed in [20] highlight the dedicated image areas given a category. For instance, for image a) in Fig. 3, the Grad-CAM highlights both the top and the bottom for product category "Women Set", while ignoring the top region for "Pants-Capris". This indicates that Grad-CAM is a candidate supervision for the POI attention. The Grad-CAM, denoted by $\mathbf{a}^g$, is obtained by back-propagating the gradients of the predicted score to the last convolutional layer. It is defined as

$$\alpha_k^c = \frac{1}{Z} \sum_i \sum_j \frac{\hat{y}^c}{A_{ij}^k},\tag{8}$$

$$\mathbf{a}^g = \text{ReLU}(\sum_k \alpha_k^c \mathbf{A}^k),\tag{9}$$

where $\hat{y}_c$ is the predicted score before the softmax layer for category c, $\mathbf{A}^k$ is the k-th convolutional layer output. The ReLU function neglects pixels with negative weights, because they do not have positive influence on predicting image class. The Grad-CAM has the same spatial size as the feature maps of the convolutional layer i. Notice that the Grad-CAM $\mathbf{a}^g$ is sensitive to the predicted score. To remove the bias introduced by $\hat{y}_c$, the Grad-CAM should be normalized before supervising the POI attention. Due to the ReLU function in Eq. (9), the Grad-CAM is a non-negative matrix, where only positive values should be normalized. Hence, the normalization function is defined as

$$\bar{\mathbf{a}}^g = \sum_k \alpha_k^c \mathbf{A}^k,\tag{10}$$

$$\hat{\mathbf{a}}^g = \text{softmax}(\text{ReLU}(\bar{\mathbf{a}}^g) - P * \text{ReLU}(-\bar{\mathbf{a}}^g)),\tag{11}$$

where P is a large positive number, and $\hat{\mathbf{a}}^g$ is the normalized Grad-CAM. Unlike the ideal bounding box annotation selecting all relevant regions, the Grad-CAM tends to highlight image regions that are sufficient to determine the product category. For example, in Fig. 3, the Grad-CAM only highlights the skirt part for a one-piece dress. However, the top is also informative for product title generation, since it provides essential information about the sleeve style and neck type. Thus, the POI attention loss $\mathcal{L}_{POI}$ is defined as the cross-entropy loss between the normalized Grad-CAM and the POI attention map

$$\mathcal{L}_{POI} = - \sum_i \hat{a}_i^g \log a_i^c.$$

In $\mathcal{L}_{POI}$, the normalized Grad-CAM can be interpreted as the confidence when penalizing the POI attention map. To obtain the Grad-CAM, we learn a category classification model . The product classification model is trained with the cross-entropy loss between the predicted class and the true class.

3.4   Training Algorithm

We need to learn an image classifier for obtaining Grad-CAM the DHA for product title generation. Instead of using the pre-trained CNN model from ImageNet [19], we learn our own CNN model on the product image set as the image classifier, whose training loss is the cross entropy loss between the predicted category $\hat{y}_i$ and ground-truth category y_i.

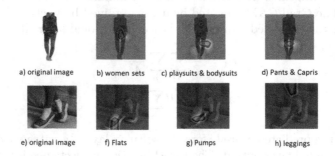

Fig. 3. The visualization of Grad-Cam for top three predicted category: (b-d) the Grad-CAM visualizations for product category "Women Sets", "Playsuits", "Pants-Capris" for image a); (f–h) the Grad-CAM visualizations for "Flats", "Pumps" and "Leggings" for image b).

For DHA, we learn the parameters for the image encoder, the language decoder, and the hierarchical attention networks. The DHA is trained with the loss presented in Eq. 12, which is the weighted sum of word-likelihood $\mathcal{L}_{MLE}$ defined in Eq. 2 and the POI attention loss:

$$\mathcal{L}_{DHA} = \mathcal{L}_{MLE} + \lambda \cdot \mathcal{L}_{POI}, \tag{12}$$

where the λ is the hyper-parameter to balance the two losses. Notice that we do not share the CNN parameters between the category classifier for the Grad-CAM and the image encoder in the DHA, due to the huge gap between two tasks. The CNN model for Grad-CAM is not fine-tuned during the DHA model learning.

4   The AEPro Dataset

We collect a dataset from AliExpress which is an international E-commerce site serving 243 countries. The collected dataset, named AEPro, provides basic information about the product such as the product image and product title. We sample products from 170 categories including "Clothes", "Shoes", etc. We conduct the pre-processing for the product titles. Firstly, we discard the products whose titles have less than 8 terms, since our goal aims to generate the title describing the product detailed. Secondly, to generate titles that semantically align with search queries, we filter out the terms with low search frequency. Finally, we remove the terms that denote the product type and brand, since it is hard to generate them from the image information. Specifically, each record in the AEPro dataset is represented as a quaternion (*product ID*, *product image*, *product category ID*, *product title*), where the *product ID* and *product category ID* is hashed. The dataset is randomly split into a training set and a test set with the proportion as 90% and 10%, respectively. The training dataset contains $3,909,811$ products and the test dataset contains $434,423$ products. After lowercasing the terms and discarding the terms occurring fewer than 20 times, we get a vocabulary of $4,390$ unique words. The max sequence length of the title is 12. Although the past dataset like eBay SIGIR 2019 eCom challenge[2] contains the products' information, the eBay dataset do not consider the search preferences of users, and thus cannot make the product title consistent with search queries.

Table 1. Automatic performances on the test split of our dataset.

Method	BLEU1	BLEU2	ROUGE-L	METEOR
Google NIC [22]	47.8	32.0	48.0	23.3
Google NIC (category)	49.6	32.7	49.1	23.7
Soft attention [24]	49.2	32.4	48.4	23.5
Soft attention (category)	51.2	34.3	50.4	24.5
HA (category)	51.1	34.0	50.3	24.4
HA (hire)	52.8	35.9	51.9	25.3
DHA (category)	51.3	34.6	50.2	24.5
DHA	**52.9**	**36.1**	**52.1**	**25.4**

5 Experiments

5.1 Experimental Setup

The image encoders for product classification and product title generation both use the ResNet-50 without sharing any parameter, due to the task gap. The ResNet-50 has been pre-trained on ImageNet. A GRU [2] with 1024 hidden units is used as the language decoder. The word embedding size is 1024. λ in the training loss referred to Eq. 12 is set to 0.01, which leads to the best performance. We first train the product classification model with batch size 64 for 4 epochs and then train the proposed DHA model for title generation with batch size 32. Both models are learned by the Adam optimizer with $\beta_1 = 0.9$, $\beta_2 = 0.999$, and L2 weight decay of 0.01. The initial learning rate is 10^{-3} for the product classification model and 10^{-4} for the DHA model.

---

[2] https://github.com/eBay/sigir-2019-ecom-challenge.

5.2   Baselines and Ablated Models

As aforementioned, this work is the first work to address the product title generation task based on the product image and category. So far as we are concerned, there is no closely related baselines in the literature. Since our task is similar to image captioning, we compare our model with two popular image captioning models: **Google NIC** [22] and **Soft Attention** [24]. Since these models do not consider the product category, we improve them by concatenating the product category embedding into the hidden state of the language decoder, and denote them by **Google NIC (category)** and **Soft Attention (category)**. Moreover, we ablate the full DHA model and demonstrate the contribution of each component. The ablated models are denoted as follows: **HA (category)** only has the category-guided attention; **HA (hire)** has the hierarchical attention but without Grad-CAM as the penalty; **DHA (category)** and **DHA** augment **HA (category)** and **HA (hire)** with the guidance from Grad-CAM, respectively.

Table 2. Case study for the product title generation. Incorrect words are marked in red while informative ones are in blue.

Image	Category	Model	Generated title
		Soft Attention	women top t-shirt sleeve tee summer cotton shirt short neck
	Pants	Soft Attention (category)	pants women trousers high waist wide leg summer loose casual
		HA(hire)	pants women waist black ladies casual pencil sexy lace short
		DHA	jeans women denim high pants waist size plus skinny trousers
		Soft Attention	women shirt tops long sleeve plus size summer chiffon print
	Dress	Soft Attention (category)	dress women long summer size plus maxi party sexy sleeve
		HA(hire)	dress women long party sleeve lace casual print ladies neck
		DHA	dress dresses long summer size plus party sleeve lace floral

5.3 Experiment Results

Table 1 shows the performance of both the baselines and the ablated models, using a set of standard metrics, namely BLEU [18], ROUGE [13], and METEOR [4]. Compared with the baselines, DHA achieves the best performance in terms of all metrics. Specifically, it outperforms Soft Attention (category), the most competitive baseline, by 3.37% (ROUGE) and 3.67% (METEOR). The overall performance of the models relying on product category is much better than that of category-independent models, indicating that the product category contributes to the title generation. For the category-based models, DHA (category) exhibits better performance than Google NIC (category) and Soft Attention (category). This demonstrates the advantage of the POI attention mechanism over directly adding the category embedding into the hidden state of the language decoder. In addition, the performance of the generation attention will degrade significantly without limiting from POI attention (Soft Attention). For the ablated models, HA (hire) performs better than HA (category), which illustrates that dynamically attended image areas can help the title generation. Moreover, Grad-CAM improves the performance by a large margin. This result is consistent with previous studies [12,15]. In summary, the hierarchical attention is the prime contributor for the improvements on all metrics and the Grad-CAM further improves the performance.

5.4 Online A/B Testing

This section presents the results of online evaluation in a real-world E-commerce application with a standard A/B testing configuration. The product titles are provided by sellers in the basic bucket, while those in the experiment bucket are generated by DHA. We adopt the exposure rate (the number of the products browsed by the users within one day) to measure the performance. With A/B testing of a whole week, DHA improves the exposure rate by 40%. The increase of exposure rate indicates that the generated titles by DHA are more informative. Moreover, we provide the generated titles for sellers to update the profiles of products. Among the generated titles, 39.3% are accepted directly, whereas 28.2% are rejected. The rest 32.5% are partially adopted with minor modifications. This demonstrates that automatic product title generation model such as DHA can significantly ease the burden of sellers while achieving decent performance.

5.5 Case Study

A case study on different methods is shown in Table 2. The improper words are highlighted in red, and informative ones are in blue. Without the product category, Soft Attention is unable to detect the product of interest, thereby generating the wrong nouns to present the product. For example, Soft Attention generates *t-shirts* while the product is a pair of pants, as shown in the first example in Table 2. The Soft Attention (category) and HA (hire) can generate the words related to the product of interest, but it still outputs wrong attributes such as *loose* for the skinny pants. With the proposed hierarchical attention and Grad-CAM supervision, DHA returns more accurate words for product attributes. These examples show that DHA can generate more appropriate and informative titles than the baselines.

6 Conclusion and Future Work

Inspired by the importance of product title in E-commerce, we propose a generative neural model to automatically generate product titles. Currently, we only utilize the Grad-CAM to guide the learning of the POI attention. In the future, we will study what penalty can be applied to the generation attention. Besides, the task studied in this paper is to generate product title which can describe product accurately, without considering whether the generated title is good for customers and sellers. We are going to generate titles which can improve the likelihood of purchase.

References

1. Anderson, P., et al.: Bottom-up and top-down attention for image captioning and visual question answering. In: Proceedings of the IEEE Conference on Computer Vision and Pattern Recognition, pp. 6077–6086 (2018)
2. Cho, K., et al.: Learning phrase representations using RNN encoder-decoder for statistical machine translation. arXiv preprint arXiv:1406.1078 (2014)
3. Dale, R., Green, S.J., Milosavljevic, M., Paris, C., Verspoor, C., Williams, S.: The realities of generating natural language from databases. In: Proceedings of the 11th Australian Joint Conference on Artificial Intelligence, pp. 13–17. Citeseer (1998)
4. Denkowski, M., Lavie, A.: Meteor universal: language specific translation evaluation for any target language. In: Proceedings of the Ninth Workshop on Statistical Machine Translation, pp. 376–380 (2014)
5. Duma, D., Klein, E.: Generating natural language from linked data: unsupervised template extraction. In: Proceedings of the 10th International Conference on Computational Semantics (IWCS 2013)-Long Papers, pp. 83–94 (2013)
6. Farhadi, A., et al.: Every picture tells a story: generating sentences from images. In: Daniilidis, K., Maragos, P., Paragios, N. (eds.) ECCV 2010. LNCS, vol. 6314, pp. 15–29. Springer, Heidelberg (2010). https://doi.org/10.1007/978-3-642-15561-1_2
7. Gong, Y., Wang, L., Hodosh, M., Hockenmaier, J., Lazebnik, S.: Improving image-sentence embeddings using large weakly annotated photo collections. In: Fleet, D., Pajdla, T., Schiele, B., Tuytelaars, T. (eds.) ECCV 2014. LNCS, vol. 8692, pp. 529–545. Springer, Cham (2014). https://doi.org/10.1007/978-3-319-10593-2_35
8. Gupta, A., Verma, Y., Jawahar, C.: Choosing linguistics over vision to describe images. In: Twenty-Sixth AAAI Conference on Artificial Intelligence (2012)
9. He, K., Zhang, X., Ren, S., Sun, J.: Deep residual learning for image recognition. In: Proceedings of the IEEE Conference on Computer Vision and Pattern Recognition, pp. 770–778 (2016)

10. Kiros, R., Salakhutdinov, R., Zemel, R.: Multimodal neural language models. In: International Conference on Machine Learning, pp. 595–603 (2014)

11. Kulkarni, G., et al.: Babytalk: understanding and generating simple image descriptions. IEEE Trans. Pattern Anal. Mach. Intell. **35**(12), 2891–2903 (2013)

12. Li, J., Ebrahimpour, M.K., Moghtaderi, A., Yu, Y.Y.: Image captioning with weakly-supervised attention penalty. arXiv preprint arXiv:1903.02507 (2019)

13. Lin, C.Y.: ROUGE: a package for automatic evaluation of summaries. In: Text Summarization Branches Out, pp. 74–81 (2004)

14. Lin, T.-Y., et al.: Microsoft COCO: common objects in context. In: Fleet, D., Pajdla, T., Schiele, B., Tuytelaars, T. (eds.) ECCV 2014. LNCS, vol. 8693, pp. 740–755. Springer, Cham (2014). https://doi.org/10.1007/978-3-319-10602-1_48

15. Liu, C., Mao, J., Sha, F., Yuille, A.: Attention correctness in neural image captioning. In: Thirty-First AAAI Conference on Artificial Intelligence (2017)

16. Mei, H., Bansal, M., Walter, M.R.: What to talk about and how? Selective generation using LSTMs with coarse-to-fine alignment. arXiv preprint arXiv:1509.00838 (2015)

17. Mitchell, M., et al.: Midge: generating image descriptions from computer vision detections. In: Proceedings of the 13th Conference of the European Chapter of the Association for Computational Linguistics, pp. 747–756. Association for Computational Linguistics (2012)

18. Papineni, K., Roukos, S., Ward, T., Zhu, W.J.: BLEU: a method for automatic evaluation of machine translation. In: Proceedings of the 40th Annual Meeting on Association for Computational Linguistics. pp. 311–318. Association for Computational Linguistics (2002)

19. Russakovsky, O., Bernstein, M., et al.: ImageNet large scale visual recognition challenge. Int. J. Comput. Vision **115**(3), 211–252 (2015)

20. Selvaraju, R.R., Cogswell, M., Das, A., Vedantam, R., Parikh, D., Batra, D.: Grad-CAM: visual explanations from deep networks via gradient-based localization. In: Proceedings of the IEEE International Conference on Computer Vision, pp. 618–626 (2017)

21. de Souza, J.G.C., et al.: Generating e-commerce product titles and predicting their quality. In: Proceedings of the 11th International Conference on Natural Language Generation, pp. 233–243 (2018)

22. Vinyals, O., Toshev, A., Bengio, S., Erhan, D.: Show and tell: a neural image caption generator. In: Proceedings of the IEEE Conference on Computer Vision and Pattern Recognition, pp. 3156–3164 (2015)

23. Wang, H., Zhang, Y., Yu, X.: An overview of image caption generation methods. Comput. Intell. Neurosci. **2020** (2020)

24. Xu, K., et al.: Show, attend and tell: neural image caption generation with visual attention. In: International Conference on Machine Learning, pp. 2048–2057 (2015)

25. Yao, T., Pan, Y., Li, Y., Mei, T.: Exploring visual relationship for image captioning. In: Ferrari, V., Hebert, M., Sminchisescu, C., Weiss, Y. (eds.) Computer Vision – ECCV 2018. LNCS, vol. 11218, pp. 711–727. Springer, Cham (2018). https://doi.org/10.1007/978-3-030-01264-9_42

Multi-channel Orthogonal Decomposition Attention Network for Sequential Recommendation

Jia Guo, Wendi Ji, Jiahao Yuan, and Xiaoling Wang[✉]

School of Computer Science and Technology, East China Normal University,
Shanghai, China
{jiaguo,jhyuang}@stu.ecnu.edu.cn, xlwang@cs.ecnu.edu.cn

Abstract. Sequential recommender systems aim to model users' evolving interests from historical behaviors and make customized recommendations. Except for items, the feature carried by the interaction also contains a wealth of information (e.g., item category and user rating). Therefore, many researches tried to leverage features, which directly fuse various types of features into the item vector. However, items and features are in different vector spaces, so the direct fusion destroys the consistency of the item vector space. Furthermore, the direct fusion of multiple features leads to mutual interference, making it hard to capture the transfer patterns of feature sequences. In this paper, we propose a novel **M**ulti-channel **O**rthogonal **D**ecomposition **A**ttention **N**etwork (**MODAN**) for the sequential recommendation. Specifically, we apply two kinds of channels. One is the item channel, which only focuses on the pure dependency among items. The other is the feature channel, which captures the feature transfer patterns. In the feature channels, we adopt orthogonal decomposition and reverse orthogonal decomposition to maintain the consistency of both the item and feature vector space. Experimental results on three datasets demonstrate that MODAN achieves substantial improvement over state-of-the-art methods.

Keywords: Sequential recommendation · Orthogonal decomposition · Feature information · Attention network

1 Introduction

With the explosive growth of information on the Internet, the recommender system has become essential in various applications, such as e-commerce, advertising and information retrieval to alleviate information overload. The sequential recommendation is one of the fundamental tasks in the recommender systems, which aims to recommend the next item that a user will likely interact with by capturing useful sequential patterns from users' interaction history [3].

Increasing research interests have been put in the sequential recommendation with various models proposed. Early methods usually utilize the Markov assumption that the current behavior is tightly related to the previous ones [4,12].

J. Gama et al. (Eds.): PAKDD 2022, LNAI 13282, pp. 288–300, 2022.
https://doi.org/10.1007/978-3-031-05981-0_23

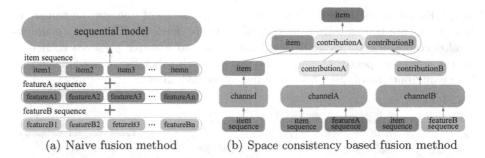

(a) Naive fusion method   (b) Space consistency based fusion method

Fig. 1. An illustration of two kinds of methods. (a) Naive fusion methods fuse all kinds of information, and feed them into sequential models. (b) In space consistency based fusion method, different channels process different features.

Recently, sequential neural networks such as recurrent neural networks [2,6], convolutional neural network [14] and Transformer [7], have been applied to recommendation tasks. These networks can encode sequential user-item interactions and learn effective representations of user preferences. However, these methods only consider the sequential patterns between items, ignoring the sequential patterns between features (e.g., user rating and item category), which are beneficial for capturing the user's fine-grained preferences [10]. Some researchers try to utilize feature information through auxiliary modules. For example, FDSA [16] applies an additional self-attention block to process all feature information to obtain the feature transfer pattern. BERT-NOVA [11] only uses feature information to calculate the attention matrix of BERT [13] model.

Although the previous methods have achieved satisfactory results, there are still some challenges: (1) Vector space consistency problem. Naive fusion methods directly and irreversibly fuse item and feature vectors, including adding or concatenating, and then input the fusion result into the sequence model, as shown in Fig. 1(a). Item and feature are in different vector spaces, and the naive fusion method destroys the consistency of item and feature vector space, respectively. (2) Noise interference problem. Features usually contain two parts of information, valid information and invalid information, which are the features that users care about and the noise they don't care about. The fusion method of direct concatenation among feature vectors causes unrelated noises to interfere with normal features, and ultimately unable to capture the user's accurate preferences.

To this end, we propose a Multi-Channel Orthogonal Decomposition Attention Network (MODAN), as shown in Fig. 1(b). For challenge (2), we adopt multiple channels. Every feature channel (FC) processes one kind of feature to avoid the mutual interference between different features, and the item channel (IC) only focuses on the pure dependency among items. For challenge (1), we adopt orthogonal decomposition (OD) and reverse orthogonal decomposition (ROD) in the feature channels. Based on the item, these two kinds of decomposition extract the fine-grained contribution of features to the item in the item vector space. Based on the feature, they extract the influence of item on the feature to better capture the transition pattern of users' preferences. These two symmetrical decomposition operations make the item and features preserve their own vector spaces respectively, and contribute to each other for reasonable fusion. The contributions of our work are summarized as follows:

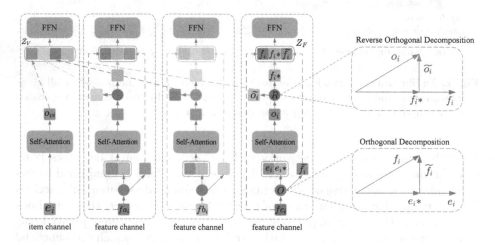

Fig. 2. The overall workflow of the MODAN model.

- This paper emphasizes the importance of independent modeling of different feature information. At the same time, the relationship transfer mode between item and feature needs to be modeled at a fine-grained level to maintain the consistency of the item and feature vector space.
- We use a multi-channel attention network to process different feature information, which eliminates the noise interference between features.
- We first introduce orthogonal decomposition and reverse orthogonal decomposition to extract the contribution of items and features on each other to maintain the consistency of the item and feature vector space, respectively.
- We conduct extensive experiments on three real-world datasets to verify the effectiveness of our proposed model.

2 Related Work

2.1 Sequential Recommendation

Early works on the sequential recommendation are mainly based on Markov Chain assumption. For instance, FPMC [12] fuse the matrix factorization and first-order Markov Chain for modeling global user preference and short-term intentions, respectively. In recent years, due to the powerful representation capabilities of deep learning, neural network-based methods have made significant progress in sequence recommendation. GRU4Rec [6] firstly introduces the gated recurrent unit (GRU) to model the short-term intention based on the user's action sequences. SASRec [7] successfully employed self-attention mechanism to aggregate relevant items for user modeling selectively. BERT4Rec [13] further improves the performance by bidirectional self-attention mechanism. However, these approaches ignore the sequential patterns between features that are beneficial for capturing the user's fine-grained preferences.

2.2 Feature Fusion Sequential Recommendation

Some methods try to make better use of feature information [1,9]. FDSA [16] applies separated self-attention networks on item sequences and feature sequences, respectively, to capture the item-item and feature-feature relationships while supplementing with vanilla attention to select essential features adaptively. BERT-NOVA [11] consider feature as the input of query and key, and item as the input of value. Since the feature only participates in calculating the attention matrix, the consistency of item vector space can be maintained to a certain extent. However, these methods cannot accurately capture user preferences because features interfere with each other, resulting in limited utilization of feature information. The consistency of the vector space cannot be fully guaranteed.

3   Proposed Method

3.1   Problem Formulation

Given a user's interaction history, the sequential recommendation task asks the next item to interact with. Let $\mathcal{U}, \mathcal{V}$ and $\mathcal{F}\lambda$ denote a set of users, items and features relatively, where $\mathcal{U} = \{u_1, u_2, ...u_{|\mathcal{U}|}\}, \mathcal{V} = \{v_1, v_2, ...v_{|\mathcal{V}|}\}$ and $\mathcal{F}\lambda = \{f\lambda_1, f\lambda_2, ...f\lambda_{|\mathcal{F}\lambda|}\}$. We uniformly denote the item-related feature (e.g., category, price, producer) and behavior-related feature (e.g., purchase, rating, time of execution) as $\mathcal{F}\lambda, \lambda \in \{a, b, c, ...\}$. The user's interaction history $u \in \mathcal{U}$ can be represented as a chronological sequence $\mathcal{S}_u = \{b_1^u, b_2^u, ..., b_{|\mathcal{S}_u|}^u\}$. We denote the term $b_i^u = (v_i^u, fa_i^u, fb_i^u, ..., fk_i^u)$ as the i-th interaction the user has made, where $v_i^u \in \mathcal{V}$ is the item that user u interacts with and $f\lambda_i^u \in \mathcal{F}\lambda$ presents one kind of feature. Our proposed model M learns to output a score $y_v = M(\mathcal{S}_u)$ for every item v in the item set $\mathcal{V}$, then the top-K items with the highest scores will be recommended to the user.

3.2   Multi-channel Orthogonal Decomposition Attention Network

In Multi-channel Orthogonal Decomposition Attention Network (MODAN), we first transform the user-item interaction sequences into item and feature vectors, namely embeddings. Next, we apply item channel and feature channel to process different embeddings, respectively. Orthogonal decomposition and reverse orthogonal decomposition methods are applied in the feature channel. Then, we integrate the feature contribution extracted by different feature channels into the item channel. Embedding is constantly updated and optimized through stacking modules. Finally, we only use the representation of the last position in the item sequence through a prediction layer to predict the next item. The overall overview and details of MODAN are shown in Fig. 1b and Fig. 2.

Embedding Layer. We create embedding matrix $M^V \in \mathbb{R}^{|\mathcal{V}| \times d}$ and $M^F \in \mathbb{R}^{|\mathcal{F}| \times d}$ for items and features relatively, where d is the latent dimension. Following [7,13], we use learnable positional embedding matrices $P \in \mathbb{R}^{n \times d}$ to increase the position information, where n is the sequence length. Formally, the representation for each item $v_i \in \mathcal{S}_u$ can be formulated as:

$$e_i = m_i^v + p_i, \tag{1}$$

For the input feature sequence, we process in the same way. Specifically,

$$f_i = m_i^f + p_i, \tag{2}$$

where $m_i^v, m_i^f, p_i \in \mathbb{R}^d$ denotes the item embedding, feature embedding and the positional embedding, respectively. Hence, we get the item and feature sequence representation, namely $E = \{e_1, e_2, ..., e_n\}, F = \{f_1, f_2, ..., f_n\}$. Here we only use one feature as an example, so $f\lambda$ is abbreviated as f. If there are multiple features, the same processing method is adopted.

IC: Item Channel. For the item channel (IC), we focus on the pure item representation and maintain the consistency of the item vector space, that is, only the item embedding is used as input, and no feature information is integrated. We adopt the self-attention structure to capture the correlation between items, which was proposed by [15], and defined as follows:

$$Attention(Q, K, V) = softmax(\frac{QK^T}{\sqrt{d}})V, \tag{3}$$

where Q, K, V denote query, key, and value, respectively, and d represents the dimension of each item embedding. The scale factor $\sqrt{d}$ is to avoid overly large values of the inner product. We first convert item sequence E through linear transformation, and then feed them into the self-attention block as follows:

$$O_V = Attention(EW_Q^I, EW_K^I, EW_V^I), \tag{4}$$

where $W_Q^I, W_K^I, W_V^I \in \mathbb{R}^{d \times d}$ are the projection matrices. We get the item sequence representation O_V.

FC: Feature Channel. The naive fusion method with self-attention integrates item and all feature embeddings as input. In fact, it cannot accurately capture the user preferences due to the implicit use of features and mutual interference between embeddings. We propose feature channel. We adopt a corresponding feature channel (FC) for each feature type, eliminating the mutual interference between features and effectively capturing the transfer patterns in the feature sequence. To predict the next item the user will interact with, the final result should be based on item embedding, that is, the result of the IC. We apply orthogonal decomposition (OD) and reverse orthogonal decomposition (ROD) in the FC to extract the contribution of features on the item and integrate it into item embedding, avoiding damage to the consistency of the vector space.

Figure 2 illustrates the idea of OD using a two-dimensional space example. For the module input embeddings $e_i \in E$ and $f_i \in F$, we decompose the f_i in parallel and orthogonal directions in the basic direction of e_i. We first project the input feature f_i onto the input item e_i to obtain the parallel vector of the feature in the item vector space:

$$Proj(x, y) = \frac{x \cdot y}{|y|} \frac{y}{|y|}, \tag{5}$$

$$e_i* = Proj(f_i, e_i), \tag{6}$$

where $Proj$ is a projection function, x, y are vectors, and e_i* represents the degree of contribution of the feature on the item. We then make the projection in the orthogonal direction of the projected feature e_i* to get the purer feature vector:

$$\widetilde{f}_i = Proj(f_i, (f_i - e_i*)), \tag{7}$$

where $\widetilde{f}_i$ is the feature independent of the item, and can be regarded as interference information in the item space. The representation e_i* contains the contribution of the feature on the item, which is still in the item vector space, so we can directly concatenate the item representation e_i and the mapped item representation e_i* without destroying the consistency of vector space:

$$h_i = [e_i : e_i*], \tag{8}$$

where h_i denote the feature enhanced item representation. By stacking it, we get the sequence representation $H = \{h_1, h_2, ..., h_n\}$. Same as IC, we also apply self-attention structure to process H:

$$O_F = Attention(HW_Q^F, HW_K^F, HW_V^F), \tag{9}$$

where $W_Q^F, W_K^F, W_V^F \in \mathbb{R}^{d \times d}$ are the weight matrices. Since the self-attention operation is only performed in the same space, so the output O_F is still in item vector space.

After the self-attention operation, the feature interference information hidden in e_i* may be stimulated, so we use reverse orthogonal decomposition (ROD) to refine the item representation more purely. At the same time, ROD can also capture the contribution of item on feature for later update and optimization of feature representation. We decompose the item representation $o_i \in O_F$ in parallel and orthogonal directions in the basic direction of $f_i \in F$:

$$f_i* = Proj(o_i, f_i), \tag{10}$$

$$\widetilde{o_i} = Proj(o_i, (o_i - f_i*)), \tag{11}$$

Where f_i* represents the degree of contribution of the item on the feature, which is applied for subsequent feature optimization, and $\widetilde{o_i}$ represents the purer item vector, which further eliminates the interference information hidden in e_i*.

Feature Update and Item Update. In order to more accurately capture the user preference hidden in the feature information, we integrate different parts of the feature information to update and optimize the feature representation:

$$z_i = [f_i : \widetilde{f_i} : f_i*]W_z + b_z, \tag{12}$$

where : denote concatenation operation, f_i is the original input of feature channel, $\widetilde{f_i}$ is item irrelevant feature obtained by OD, f_i* is the enlightenment information of the direction in which the feature will be updated after ROD, $W_z \in \mathbb{R}^{3d \times d}$ is the weight matrice, and $b_z \in \mathbb{R}^d$ is the bias term. We get the new feature sequence $Z_F = \{z_1, z_2, ..., z_n\}$.

Different channel contain different item information. For IC, it focus on the pure item representation. For FC, due to OD operation and the ROD operation, the contribution of the feature on the item is extracted, which is still in the item vector space. We integrate the item representation of all channels as the final output of the IC. The item continuously absorbs the contribution of the feature, while maintaining its purity, and finally reach the optimum after continuous iteration:

$$Z_V = [O_V : \widetilde{O_F}]W_V + b_V, \tag{13}$$

where $W_V \in \mathbb{R}^{2d \times d}$ is the weight matrice, b_V is the bias term, O_V is the output sequence of self-attention in the IC, and $\widetilde{O_F} = \{\widetilde{o_1}, \widetilde{o_2}, ..., \widetilde{o_n}\}$ is the sequence obtained by this certain FC.

To endow the model with nonlinearity and interactions between different dimensions, we apply Position-wise Feed-Forward Network [15]. Also, layer normalization and residual network are employed to improve the model stability. Specifically,

$$E^{(1)} = LayerNorm(Z_V + \sigma(Z_V W_1 + b_1)W_2 + b_2), \tag{14}$$

$$F^{(1)} = LayerNorm(Z_F + \sigma(Z_F W_3 + b_3)W_4 + b_4), \tag{15}$$

where $W_1, W_2, W_3, W_4 \in \mathbb{R}^{d \times d}$ is the weight matrix, $b_1, b_2, b_3, b_4 \in \mathbb{R}^d$ is the bias term, and $\sigma(\cdot)$ is the gelu activation function. We stack multiple transformer layers to learn higher-order sequential dependencies and more informative item embeddings:

$$E^{(1)}, F^{(1)} = MODAN(E, F), \tag{16}$$

The l-th ($l >= 1$) layer is defined as:

$$E^{(l)}, F^{(l)} = MODAN(E^{(l-1)}, F^{(l-1)}), \tag{17}$$

where $E^{(0)} = E, F^{(0)} = F$, $E^{(l)}$ is the item sequence output, and $F^{(l)}$ is the feature sequence output.

Prediction Layer. After l layer self-attention blocks, we get the final item and feature sequence output $E^{(l)}, F^{(l)}$. Since the set of candidate items to be predicted for the user's next interaction is still in the item vector space, we only use the item representation $E^{(l)}$, which already contains the contribution of different features After multiple processing channels. We select the representation of last position $e_n^{(l)}$ as user preference given the previous n times interactions. The score of each candidate item $v_i \in \mathcal{V}$ is calculated by

$$\hat{y}_i = softmax(e_n^{(l)} v_i^{\mathrm{T}}), \tag{18}$$

where $\hat{y}_i$ denotes the recommendation probability of the item v_i to be the next user interacts. Finally, we adopt the cross-entropy loss as the objective function:

$$L(y, \hat{y}) = -\sum_{i=1}^{|\mathcal{V}|} y_i log(\hat{y}_i) \tag{19}$$

where y is the one-hot encoding vector exclusively activated by the ground truth.

Table 1. Statistics of datasets

Dataset	Users	Items	Actions	Feature
Amazon Apps	87271	13209	0.75M	Category, rating, timestamp
MovieLens-1M	6040	3416	1.0M	Genre, rating, timestamp
Diginetica	48157	20612	0.35M	Category, price, timestamp

4 Experiments

4.1 Experimental Setup

Datasets. We evaluate our proposed model on three real-world datasets, i.e., *Amazon Apps*[1], *MovieLens-1M*[2] and *Diginetica*[3]. *Amazon* is a series of product review datasets crawled from Amazon.com, and we adopt the *Apps* category. *MovieLens-1M* is a widely used benchmark dataset, including users' ratings on movies. *Diginetica* is collected from the CIKM Cup platform 2016, which records the user's clicks from six months, and we used the released click data.

Following [7], we treat the presence of a review or a rating as an interaction between the user and the item. Furthermore, the interaction orders are determined by timestamps. To ensure the dataset's quality, we filter out all items with fewer than five occurrences and users with fewer than five interactions. For all datasets, we follow [7] to split user's interaction sequences $\mathcal{S}_u$ into three parts: (1) the most recent action $v^u_{|\mathcal{S}_u|}$ for testing, (2) the second most recent action $v^u_{|\mathcal{S}_u|-1}$ for validation, and (3) all remaining actions for training. The statistics of the processed datasets are summarized in Table 1.

---

[1] http://jmcauley.ucsd.edu/data/amazon/.
[2] https://grouplens.org/datasets/movielens/1m/.
[3] https://competitions.codalab.org/competitions/11161.

Baseline Methods. To evaluate the effectiveness of the model we proposed, we compare it with the following baseline methods, including the general sequential recommendation methods [5,7,12,13] and the feature fusion methods [11,16]: **FPMC** [12] used a two-order Markov chain to model the transition in behavior sequences. **GRU4Rec+** [5] is a session-based recommendation model based on GRU cells. We treat each user's interaction sequence as a session. GRU4Rec+ [5] is an improved version of GRU4Rec [6], which takes up an advanced loss function. We adopt the latter in our experiments. **SASRec** [7] utilizes self-attention blocks to recommend the next item, only considering items in the historical interaction sequence. **BERT4Rec** [13] utilizes a bi-directional self-attention module to capture the context information in the historical interaction sequences. **SASRecF** is our extension of the SASRec method, which concatenates item representations and various features together as the input of the self-attention module. **FDSA** [16] constructs a self-attention block to model item transition patterns and another one to model feature transition patterns. **BERT-NOVA** [11] integrates the feature information into the query and key of the bi-directional self-attention module and input the item into the value of the bi-directional self-attention module.

Evaluation Metrics. To evaluate the performance of all models, we adopt three metrics, including *Hit Ratio* (HR@K), *Mean Reciprocal Rank* (MRR@K) and *Normalized Discounted Cumulative Gain* (NDCG@K). In this work, due to space limitations, we report HR, MRR, and NDCG with $K = 10$.

Implementation Details. In this work, all hyper-parameters and initialization strategies follow the original paper's suggestion and are fine-tuned on the validation sets via grid search on each dataset. For FPMC, GRU4Rec, SASRec and FDSA, we use code provided by *RecBole* [17], an open source framework for research purpose in recommendation. For BERT4Rec[4], SASRecF[5] and BERT-NOVA, we refer to the open source code and implement ourselves. It is worth noting that we don't adopt negative sampling strategies because sampled metrics do not persist relative statements [8]. We implement our proposed model with Pytorch. The sequence length equals 50, and the latent dimension d equals 128 for all models. According to the experimental results, the optimal hyper-parameters are $\{\eta : 0.0001, \varepsilon : 0.25, l : 3\}$ on *Amazon Apps* dataset, $\{\eta : 0.0002, \varepsilon : 0.25, l : 3\}$ on *MovieLens-1M* dataset, and $\{\eta : 0.0001, \varepsilon : 0.25, l : 3\}$ on *Diginetica* dataset, where η, ε, l denotes learning rate, dropout rate and layer number. All source code is available at GitHub[6].

---

[4] https://github.com/FeiSun/BERT4Rec.
[5] https://github.com/kang205/SASRec.
[6] https://github.com/kuangwushijian/MODAN.

Table 2. Performance comparison on three datasets. The best results overall methods are boldfaced, while the second-best results are underlined. Improvements over baseline methods are statistically significant with $p < 0.001$.

Models	Amazon Apps			MovieLens-1M			Diginetica		
	H@10	M@10	N@10	H@10	M@10	N@10	H@10	M@10	N@10
FPMC	0.0922	0.0349	0.0482	0.1636	0.059	0.0833	0.1797	0.0565	0.0848
GRU4Rec+	0.1283	0.0537	0.071	0.2599	0.1121	0.1466	0.2027	0.0705	0.101
SASRec	0.1342	0.0469	0.0672	0.2642	0.1086	0.1448	0.2267	0.0622	0.1001
BERT4Rec	0.133	0.0535	0.072	0.2685	0.1132	0.1494	0.201	0.0707	0.1008
SASRecF	0.1301	0.0561	0.0733	0.2725	0.119	0.1548	0.2223	0.0808	0.1136
FDSA	0.1369	0.0564	0.0751	0.2798	0.1218	0.1587	0.2272	0.0821	0.1156
BERT-NOVA	0.1364	0.0565	0.0751	0.2724	0.1201	0.1556	0.2117	0.0766	0.1079
MODAN (ours)	0.1441	0.0623	0.0813	0.2881	0.1284	0.1657	0.2389	0.0877	0.1226

4.2 Comparison with Baseline Methods

To illustrate the effectiveness of our proposed model, we compare it with baseline methods as shown in Table 2. GRU4Rec is better than FPMC, proving that the RNN-based model can more accurately capture the user preference from the user's interaction sequence than the traditional Markov chain-based model. Attention-based models, SASRec and BERT4Rec, capture the user's long-term preferences through a self-attention network. BERT4Rec achieves the best result in the general sequential model through the deep bidirectional structure. Compared with the general sequential recommendation models, feature fusion methods show slight improvement. SASRecF is slightly better than SASRec, indicating that the feature information is practical for item prediction. FDSA achieves the best results on two datasets, and one dataset is approximately optimal, thanks to its use of an independent self-attention to better capture feature relationships. The result of BERT-NOVA is slightly better than BERT, which shows the importance of vector space consistency. According to the results, our model MODAN outperforms all baseline methods consistently. It is attributed to multiple channels to model the feature sequence relationship and orthogonal decomposition to ensure the consistency of the vector space.

4.3 Ablation Study

In order to verify the effect of each module, we design three contrast models for comparison: MODAN-No directly concatenates item embedding and all feature embedding as the input of self-attention. MODAN-MC uses multiple channels to process items and features independently. Pure item embedding is used as the input of IC, and the concatenation of the item and one kind of feature is used as the input of one of the FC, and the other FC are the same. On the basis of MODAN-MC, MODAN-OD only uses orthogonal decomposition in FC to process feature embedding without reverse orthogonal decomposition.

Table 3. Performances of different variants of MODAN.

Models	Amazon Apps			MovieLens-1M			Diginetica		
	H@10	M@10	N@10	H@10	M@10	N@10	H@10	M@10	N@10
MODAN-No	0.1301	0.0561	0.0733	0.2725	0.119	0.1548	0.2223	0.0808	0.1136
MODAN-MC	0.1353	0.0564	0.0748	0.2773	0.1228	0.159	0.23	0.0809	0.1153
MODAN-OD	0.1385	0.0584	0.0771	0.28	0.1257	0.1618	0.2354	0.0857	0.1203
MODAN	**0.1441**	**0.0623**	**0.0813**	**0.2881**	**0.1284**	**0.1657**	**0.2389**	**0.0877**	**0.1226**

Table 3 illustrates the performances of all variants and our proposed method on the three datasets. MODAN-MC outperforms MODAN-No, proving that using different channels to model features can better capture user's intentions. The weak performance improvement may be due to the direct concatenation of features that destroys the consistency of the item vector space. The improvement of MODAN-OD result proves the effectiveness of the orthogonal decomposition scheme we proposed. It can extract the contribution of features on the item and ensure the consistency of item vector space. MODAN is significantly better than MODAN-OD, which shows that the combination of orthogonal decomposition and reverse orthogonal decomposition in FC can more completely extract the contribution of different features, thereby more completely ensuring the consistency of the item vector space in the IC.

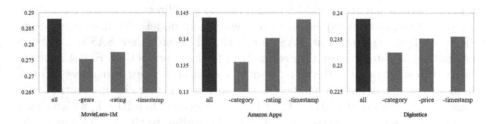

Fig. 3. Effect of different kinds of feature information in our framework.

4.4   Effect of Different Feature

In this part, we study how each feature contributes to the final performance for the sequential recommendation. We first prepare the complete MODAN model for all features and then remove one kind of feature information at each time. Figure 3 presents the HR@10 results. As we can see, all the features help to improve the performance of our MODAN. The most useful feature is the category (or genre) on the three datasets, directly reflecting the user preference type and scope. On *MovieLens-1M* and *Amazon Apps* dataset, the rating result indicates that the user pays attention to the quality of the item, such as the quality of the movie and the user experience on the applications. On *Diginetica* dataset, item price may explain the user's consumption concept. The timestamp is useful to improve the performance on *MovieLens-1M* and *Diginetica* dataset, which may imply the periodicity of the user's interaction. However, it does not help the model on the *Amazon Apps* dataset. Intuitively, Applications on the user's mobile phone rarely change, so the timestamp feature is considered invalid noise to a certain extent. However, Our model can eliminate the interference of feature noise and only extract its effective contribution on the item, especially there is almost no effective contribution here.

5 Conclusion

In this paper, we propose a Multi-Channel Orthogonal Decomposition Attention Network (MODAN) for the sequential recommendation. First, we use multiple channels to process items and different features separately to avoid mutual interference in the vector space between features. Subsequently, we use orthogonal decomposition and reverse orthogonal decomposition to extract the contribution of the feature on the item, which avoids the interference of the feature on the item vector space. Finally, the contribution of different features on the item and the item representation is fused to ensure the consistency of the item vector space. Experimental results on three benchmark datasets show that our model outperforms all the other baselines.

Acknowledgements. This work was supported by NSFC grants (No. 61972155), the Science and Technology Commission of Shanghai Municipality (20DZ1100300) and the Open Project Fund from Shenzhen Institute of Artificial Intelligence and Robotics for Society, under Grant No. AC01202005020, Shanghai Knowledge Service Platform Project (No. ZF1213), Shanghai Trusted Industry Internet Software Collaborative Innovation Center.

References

1. Cai, R., Wu, J., San, A., Wang, C., Wang, H.: Category-aware collaborative sequential recommendation. In: SIGIR, pp. 388–397 (2021)
2. Chung, J., Gulcehre, C., Cho, K., Bengio, Y.: Empirical evaluation of gated recurrent neural networks on sequence modeling. arXiv preprint arXiv:1412.3555 (2014)
3. Fang, H., Guo, G., Zhang, D., Shu, Y.: Deep learning-based sequential recommender systems: concepts, algorithms, and evaluations. In: Bakaev, M., Frasincar, F., Ko, I.-Y. (eds.) ICWE 2019. LNCS, vol. 11496, pp. 574–577. Springer, Cham (2019). https://doi.org/10.1007/978-3-030-19274-7_47
4. He, Q., et al.: Web query recommendation via sequential query prediction. In: ICDE, pp. 1443–1454. IEEE (2009)
5. Hidasi, B., Karatzoglou, A.: Recurrent neural networks with top-k gains for session-based recommendations. In: CIKM, pp. 843–852 (2018)
6. Hidasi, B., Karatzoglou, A., Baltrunas, L., Tikk, D.: Session-based recommendations with recurrent neural networks. arXiv preprint arXiv:1511.06939 (2015)
7. Kang, W.C., McAuley, J.: Self-attentive sequential recommendation. In: ICDM, pp. 197–206. IEEE (2018)

8. Krichene, W., Rendle, S.: On sampled metrics for item recommendation. In: KDD, pp. 1748–1757 (2020)
9. Lei, C., Ji, S., Li, Z.: TiSSA: a time slice self-attention approach for modeling sequential user behaviors. In: WWW, pp. 2964–2970 (2019)
10. Li, C., Niu, X., Luo, X., Chen, Z., Quan, C.: A review-driven neural model for sequential recommendation. arXiv preprint arXiv:1907.00590 (2019)
11. Liu, C., Li, X., Cai, G., Dong, Z., Zhu, H., Shang, L.: Noninvasive self-attention for side information fusion in sequential recommendation. In: AAAI, pp. 4249–4256 (2021)
12. Rendle, S., Freudenthaler, C., Schmidt-Thieme, L.: Factorizing personalized Markov chains for next-basket recommendation. In: WWW, pp. 811–820 (2010)
13. Sun, F., et al.: BERT4Rec: sequential recommendation with bidirectional encoder representations from transformer. In: CIKM, pp. 1441–1450 (2019)
14. Tang, J., Wang, K.: Personalized top-n sequential recommendation via convolutional sequence embedding. In: WSDM, pp. 565–573 (2018)
15. Vaswani, A., et al.: Attention is all you need. In: NIPS. pp. 5998–6008 (2017)
16. Zhang, T., et al.: Feature-level deeper self-attention network for sequential recommendation. In: IJCAI, pp. 4320–4326 (2019)
17. Zhao, W.X., et al.: RecBole: towards a unified, comprehensive and efficient framework for recommendation algorithms. arXiv preprint arXiv:2011.01731 (2020)

Rethinking Adjacent Dependency
in Session-Based Recommendations

Qian Zhang[1], Shoujin Wang[2], Wenpeng Lu[1(✉)], Chong Feng[3], Xueping Peng[4],
and Qingxiang Wang[1]

[1] School of Computer Science and Technology, Qilu University of Technology
(Shandong Academy of Sciences), Jinan, China
{wenpeng.lu,wangqx}@qlu.edu.cn
[2] School of Computing, Macquarie University, Sydney, Australia
shoujin.wang@mq.edu.au
[3] School of Computer Science and Technology, Beijing Institute of Technology,
Beijing, China
fengchong@bit.edu.cn
[4] Australian Artificial Intelligence Institute, University of Technology Sydney,
Sydney, Australia
xueping.peng@uts.edu.au

Abstract. Session-based recommendations (SBRs) recommend the next item for an anonymous user by modeling the dependencies between items in a session. Benefiting from the superiority of graph neural networks (GNN) in learning complex dependencies, GNN-based SBRs have become the main stream of SBRs in recent years. Most GNN-based SBRs are based on a strong assumption of *adjacent dependency*, which means any two adjacent items in a session are necessarily dependent here. However, based on our observation, the adjacency does not necessarily indicate dependency due to the uncertainty and complexity of user behaviours. Therefore, the aforementioned assumption does not always hold in the real-world cases and thus easily leads to two deficiencies: (1) the introduction of *false dependencies* between items which are adjacent in a session but are not really dependent, and (2) the missing of *true dependencies* between items which are not adjacent but are actually dependent. Such deficiencies significantly downgrade accurate dependency learning and thus reduce the recommendation performance. Aiming to address these deficiencies, we propose a novel review-refined inter-item graph neural network (RI-GNN), which utilizes the topic information extracted from items' reviews to refine dependencies between items. Experiments on two public real-world datasets demonstrate that RI-GNN outperforms the state-of-the-art methods (The implementation is available at https://github.com/Nishikata97/RI-GNN.).

Keywords: Recommender system · Session-based recommendation · Graph neural network · Adjacent dependency

J. Gama et al. (Eds.): PAKDD 2022, LNAI 13282, pp. 301–313, 2022.
https://doi.org/10.1007/978-3-031-05981-0_24

1   Introduction

In recent years, session-based recommendations (SBRs) have attracted extensive attention [4,14] for its strong capability to capture users' dynamic and short term preference. SBRs recommend the next item to a user by modeling the sequential dependencies over items within sessions.

Driven by the development of deep learning, many neural network based SBRs have been developed. Among them, recurrent neural network (RNN) and graph neural network (GNN) [17] based approaches have shown good performance. RNN-based methods attempt to capture sequential dependencies between items within the sessions, which is based on the assumption that there is a strict chronological order inside the session [4]. However, this assumption does not always hold in the real-world scenarios since users' behaviours are usually uncertain and dynamic and thus not all interacted items in one session are sequentially dependent. Benefiting from the capability of GNN in learning complex dependencies, many GNN-based SBRs [3,9,18,19] have been proposed. Leveraging the flexibility of graph structure used in GNN, the problem of strict chronological order confusing RNN-based methods is thus alleviated.

However, GNN-based SBRs often rely heavily on the strong assumption of adjacent dependency, namely, the adjacent items within one session are necessarily dependent. This is determined by its particular work mechanism. Specifically, most GNN-based SBRs first convert a given session consisting of a sequence of interacted items into a session graph by mapping each item to a node and the adjacency relation between any two items to an edge [19] to indicate the dependency between them, as shown in Fig. 1(b). This common practice of constructing session graphs often leads to two significant deficiencies: (1) the introduction of **false dependencies between adjacent but actually independent items** in a session, e.g., the item v_1 (i.e., a bird cage) and item v_2 (i.e., cat food) in the session described in Fig. 1(a), and (2) the missing of **true dependencies between items which are non-adjacent but actually dependent** in a session, e.g., the item v_1 and item v_3 (i.e., a bird) in the session described in Fig. 1(a). In practice, both false dependencies and true dependencies mentioned above are not uncommon in the real-world cases [16]. Obviously, these two deficiencies significantly downgrade the accurate learning of inter-item dependencies embedded in session data and thus reduce the performance of the downstream next-item recommendations. Therefore, it is critical to refine the dependencies between items by identifying and keeping all true dependencies while removing the false ones.

In practice, in addition to the session information, the review information associated with items can reveal dependencies between them to some degree. For example, the reviews associated with item v_1 and v_3 shown in Fig. 1(a) are closely related and fall into the same topic. This actually provides extra information to enable the possibility to refine dependencies between items in sessions. To this end, we propose review-refined inter-item graph neural network (RI-GNN) to address the two deficiencies mentioned above in this paper. By leveraging the topic information from reviews written for items, RI-GNN can not only reduces

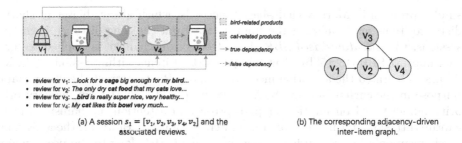

(a) A session $s_1 = [v_1, v_2, v_3, v_4, v_2]$ and the associated reviews.

(b) The corresponding adjacency-driven inter-item graph.

Fig. 1. A toy example for the construction of adjacency-driven inter-item graph.

the false dependencies between adjacent but actually independent items, but also well captures true dependencies between non-adjacent but dependent items which are usually ignored or weaken by existing GNN-based SBRs.

The main contributions of this work are summarized below:

- We propose and discuss a novel and important research question: *does adjacency necessarily indicate dependency between items in sessions?* We perform a preliminary exploration with the hope of shedding some light in this area.
- We propose a novel review-refined inter-item graph neural network, called RI-GNN, for session-based recommendations. To the best of our knowledge, this is the first work for leveraging reviews to enhance the dependency learning for SBR on anonymous sessions.
- We propose a novel method for constructing a novel review-refined inter-item graph for each session. In the graph, reviews for items are employed to filter out the false dependencies between some adjacent items, and recall the true dependencies between non-adjacent items missed by existing methods.

2   Related Work

2.1   Session-Based Recommendation

Existing methods for SBR can be summarized into: (1) Markov chain-based SBR; (2) RNN-based SBR; (3) Attention-based SBR; and (4) GNN-based SBR.

(1) *Markov chain-based SBR.* Early researches on SBR rely Markov chain to model the short-term dependencies to predict the next item. For example, FPMC [10] combined Markov chain and matrix factorization to model sequential behavior between two adjacent items and recommend next item. However, Markov chain-based methods only focus on first-order dependencies between adjacent items, while neglecting the high-order dependencies between long-distance items. (2) *RNN-based SBR.* Due to the powerful ability in modeling sequential data, RNN-based methods are applied widely to SBR [4,7,8]. GRU4Rec [4] first applied RNN to SBR, which adopted gated recurrent unit (GRU) to model the dependencies within sessions. However, RNN-based SBR also suffers the

similar problem in Markov chain-based methods, which always biases to short-distance items while missing the information from the long-distance items in sessions. (3) *Attention-based SBR*. The attention mechanism is applied to further improve SBR [13,15] by identifying important items within sessions. NARM [7] first integrated the attention mechanism into SBR to extract the user's main purpose in the current session. STAMP [8] proposed a short-term memory priority model based on multi-layer perceptron and attention mechanism, which captured the long-term and short-term interests of users. However, the attention mechanism only focuses on few important items that belong to the user's main purpose, while neglecting the other purposes indicated by few inferior items. (4) *GNN-based SBR*. Due to the superiority of GNN on modeling transition dependencies between items, GNN-based methods have been applied widely to SBR. SR-GNN [19] modeled all sessions via directed graphs, utilized GNN to capture the dependencies between items with sessions, and extracted the long-term and short-term interest of users to suggest the next item. FGNN [9] proposed to capture the sequence order and latent order in session graph, which devised the weighted attention graph layer to learn item embeddings and session embeddings for more accurate next item recommendation. GCE-GNN [18] proposed to learn the transitions between items from local and global perspectives simultaneously, so as to make better recommendations by leveraging the information from other sessions. Although GNN-based methods have achieved great success on SBR, all of these approaches construct the session graph according to the adjacent items, which ignore the dependencies from the non-adjacent items.

2.2 Review-Based Recommendation

Considering the great value of user reviews on items, some works strive to model reviews to improve the performance of SR [6,21]. DeepCoNN [21] employed two parallel cooperative neural networks to learn user behaviors by exploiting reviews written by the user and learn item properties from the reviews written for the item, then utilized factorization machine to predict item ratings. RNS [6] proposed a review-driven neural sequential recommendation, which learned user's long-term preference according to her historical reviews. Although the existing methods improve the recommendation performance, most of them are devised for the task of rating prediction instead of session-based recommendation. Although RNS is proposed for sequential recommendation, it requires to collect all reviews written by a user according to the user's explicit ID. This means that it is unable to work well on the anonymous session-based recommendation. Neither of the existing works really solve the problem of session-based recommendations based on review information.

3 Preliminary

3.1 Problem Statement

Let $V = \{v_1, v_2, \ldots, v_m\}$ represent the whole item set. Each anonymous session $s = [v_1, v_2, \ldots, v_n](v_i \in V)$ is an ordered list of items, where all the items in s

are interacted by an anonymous user in a chronological order. We embed each item into the same embedding space and let $\mathbf{h}_{v_i} \in \mathbb{R}^d$ denote the embedding of item v_i, where d is the dimensionality. To accurately identify item dependencies within the session, we utilize review information to enhance item representations. Given an item v_i, all of its reviews are collected to form the review document D_i, where each word is represented by the corresponding embedding with the dimension d_w. For the session-based recommendation problem, the goal is to predict the top-N items that the user is most likely to click in the next step.

3.2   Graph Construction

In this subsection, we first introduce the *adjaceny-driven inter-item graph* (AIG), which is widely adopted by existing GNN-based approaches [18,19], and then we present a novel graph, i.e., *review-refined inter-item graph* (RIG). RIG is used as an additional graph to complement AIG rather than to replace it by enhancing the learning of true dependencies.

Adjaceny-Driven Inter-item Graph (AIG). AIG captures important sequential patterns based on pair-wise adjacent items within the current session, which is first proposed by SR-GNN [19]. AIG converts each session s into a directed graph $\mathcal{G}_s^{adj} = (\mathcal{V}_s, \mathcal{E}_s^{adj})$, where $\mathcal{V}_s \subseteq V$ denotes the node set, $\mathcal{E}_s^{adj}$ denotes the edge set. The weight of each edge is set as the value of the occurrences of the edge divided by the outdegree of the edge's start node. This means that the more frequent occurrences of the edge, the stronger dependency between the items connected by it. The connection matrix $\mathbf{A}_s \in \mathbb{R}^{n \times 2n}$ describes how nodes are connected with each other in the graph, and $\mathbf{A}_{s,i:} \in \mathbb{R}^{1 \times 2n}$ are the two columns of blocks in $\mathbf{A}_s$, corresponding to node v_i. $\mathbf{A}_s = \mathbf{A}_s^{(out)} \| \mathbf{A}_s^{(in)}$, where $\mathbf{A}_s^{(out)}$ and $\mathbf{A}_s^{(in)}$ are the outgoing and incoming adjacency matrix respectively. $\|$ indicates the concatenation operation, and n is the length of session s.

Review-Refined Inter-item Graph (RIG). The aforementioned AIG faces two deficiencies caused by the adjacent but independent items and the nonadjacent but dependent items. To address them, we utilize reviews to refine the session graph, and thus devise the review-refined inter-item graph (RIG).

We convert the session s into a directed graph $\mathcal{G}_s^{re} = (\mathcal{V}_s, \mathcal{E}_s^{re})$, where $\mathcal{V}_s \subseteq V$ indicates the node set, $\mathcal{E}_s^{re}$ denotes the edge set. As shown in Fig. 2, there are two steps for RIG to recognize the *cross-item dependencies* (i.e., the dependencies

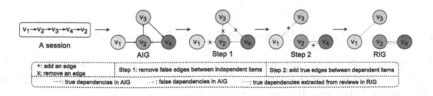

Fig. 2. Construction of AIG.

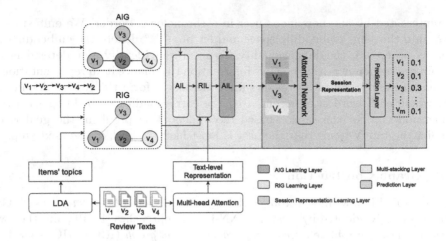

Fig. 3. Architecture of our proposed RI-GNN model.

between non-adjacent items) to obtain the correct edges. *Firstly*, once AIG is generated, in order to filter out the noise edges and refine the session graph, we only reserve the edges between items sharing the same topic in the AIG, i.e., the same user purpose, and remove the other edges. For recognizing the topic information conveniently, we collect all reviews written for the item into the document set D together, and utilize LDA [1] to extract their topics. Once obtaining the topics of each item, we filter out the edges between items that do not belong to the same topic. *Secondly*, for each ordered pair of nodes (v_{t_1}, v_{t_2}) in the session sequence, we add the directed edge $(v_{t_1} \rightarrow v_{t_2})$ if item v_{t_1} shares the same topic with item v_{t_2} and $t_1 < t_2$. This makes the current item directly connect to all the following dependent items in the session for capturing the cross-item dependencies. Finally, we obtain the refined edge set $\mathcal{E}_s^{re}$. Similar to AIG, we calculate the connection matrix $\mathbf{B}_s \in \mathbb{R}^{n \times 2n}$ for RIG, where $\mathbf{B}_s = \mathbf{B}_s^{(\text{out})} \| \mathbf{B}_s^{(\text{in})}$.

4   Architecture of RI-GNN Model

The architecture of our proposed RI-GNN is shown in Fig. 3, which mainly consists of five components, i.e., *adjacency-driven inter-item graph (AIG) learning layer*, *review-refined inter-item graph (RIG) learning layer*, *multi-stacking layer*, *session representation learning layer* and *prediction layer*.

4.1   Adjaceny-Driven Inter-item Graph Learning Layer (AIL)

AIL aims to capture the sequential dependencies between items based on AIG within the current session. Next, we will present how to learn the sequential dependencies between adjacent pair-wise items, as follows:

$$\mathbf{a}_{s,i}^t = \mathbf{A}_{s,i:}^s \left[\mathbf{h}_{v_1}^{t-1}, \ldots, \mathbf{h}_{v_n}^{t-1}\right]^\top \mathbf{H} + \mathbf{b_1},$$
$$\mathbf{z}_{s,i}^t = \sigma\left(\mathbf{W}_z \mathbf{a}_{s,i}^t + \mathbf{U}_z \mathbf{h}_{v_i}^{t-1}\right),$$
$$\mathbf{r}_{s,i}^t = \sigma\left(\mathbf{W}_r \mathbf{a}_{s,i}^t + \mathbf{U}_r \mathbf{h}_{v_i}^{t-1}\right), \tag{1}$$
$$\tilde{\mathbf{h}}_{v_i}^t = \tanh\left(\mathbf{W}_o \mathbf{a}_{s,i}^t + \mathbf{U}_o \left(\mathbf{r}_{s,i}^t \odot \mathbf{h}_{v_i}^{t-1}\right)\right),$$
$$\mathbf{h}_{v_i}^t = \left(1 - \mathbf{z}_{s,i}^t\right) \odot \mathbf{h}_{v_i}^{t-1} + \mathbf{z}_{s,i}^t \odot \tilde{\mathbf{h}}_{v_i}^t,$$

where $\mathbf{a}_{s,i}^t \in \mathbb{R}^{2d}$ is the current state at time step t, which aggregates the adjacent items' embeddings for item v_i in AIG. $[\mathbf{h}_{v_1}^{t-1}, \ldots, \mathbf{h}_{v_n}^{t-1}]$ is the list of item embeddings in session s at previous time step $t-1$, $\mathbf{A}_{s,i:} \in \mathbb{R}^{1 \times 2n}$ are the two columns of blocks in $\mathbf{A}_s$ corresponding to item v_i^s, $\mathbf{H} \in \mathbb{R}^{d \times 2d}$, $\mathbf{W}_z, \mathbf{W}_r, \mathbf{W}_o \in \mathbb{R}^{d \times 2d}$, $\mathbf{U}_z, \mathbf{U}_r, \mathbf{U}_o \in \mathbb{R}^{d \times d}$, $\mathbf{b_1} \in \mathbb{R}^{2d}$ are trainable parameters, $\sigma(\cdot)$ is the sigmoid function, and $\odot$ is the element-wise multiplication operator, $\mathbf{z}_{s,i}^t \in \mathbb{R}^d$ and $\mathbf{r}_{s,i}^t \in \mathbb{R}^d$ are the update and reset gates respectively, $\mathbf{h}_{v_i}^t$ is the final state at the time step t. We mark the final representation of item v_i in AIL as $\mathbf{h}_{v_i}^{adj}$.

4.2   Review-Refined Inter-item Graph Learning Layer (RIL)

In order to filter out the noise information and improve the representations of items, we next present how to propagate features on RIG to encode item dependencies from reviews. This layer is built based on the architecture of graph neural network, and we generate attention weights based on the similarity of reviews between items by exploiting the idea of graph attention network.

For each item's review document D_i obtained from Sect. 3.2, we first convert it into a representation vector $\mathbf{E}_i \in \mathbb{R}^{l \times d_w}$ through word embeddings. In order to better extract item features from the review representation $\mathbf{E}_i$, we utilize the self-attention method proposed by Transformer:

$$\text{Attention}(\mathbf{Q}, \mathbf{K}, \mathbf{V}) = \text{softmax}\left(\frac{\mathbf{Q}\mathbf{K}^\top}{\sqrt{d}}\right)\mathbf{V}, \tag{2}$$

where $\mathbf{Q}$ is the queries, $\mathbf{K}$ is the keys, $\mathbf{V}$ is the values and $\sqrt{d}$ is the scale factor. We adopt multi-head attention to enable the model to jointly focus on information from different representation subspaces from different positions. For item v_i, the detailed operations are described as below:

$$\text{head}_k = \text{Attention}\left(\mathbf{E}_i \mathbf{W}_k^{\mathbf{Q}}, \mathbf{E}_i \mathbf{W}_k^{\mathbf{K}}, \mathbf{E}_i \mathbf{W}_k^{\mathbf{V}}\right),$$
$$\mathbf{r}_i = \text{MultiHead}(\mathbf{E_i}) = \text{Concat}(\text{head}_1, \ldots, \text{head}_h)\mathbf{W}_1, \tag{3}$$

where $\mathbf{r}_i$ is the review representation of item v_i extracted by the multi-head attention, $\mathbf{W}^{\mathbf{Q}} \in \mathbb{R}^{d_w \times d_q}$, $\mathbf{W}^{\mathbf{K}} \in \mathbb{R}^{d_w \times d_k}$, $\mathbf{W}^{\mathbf{V}} \in \mathbb{R}^{d_w \times d_v}$, and $\mathbf{W}_1 \in \mathbb{R}^{hd_v \times d_w}$ are the learnable parameters. In our experiments, we set the number of parallel attention heads h to 3, set the dimensions of d_q, d_k, d_v, d_w to 100, 100, 100, 300.

In order to distinguish the importance of neighbor items for obtaining the representation of current item, we adopt attention mechanism and calculate the attention weight by cosine similarity:

$$\pi\left(v_i, v_j\right) = \begin{cases} \sin\left(\mathbf{r}_i, \mathbf{r}_j\right), \text{ if } TP_i = TP_j \\ \qquad 0, \text{ if } TP_i \neq TP_j \end{cases}, \tag{4}$$

where $\pi\left(v_i, v_j\right)$ estimates the importance weight of different neighbor items, $\sin()$ is the cosine similarity function, $\mathbf{r}_i$ and $\mathbf{r}_j$ are multi-head review representation of item v_i and item v_j respectively, TP_i and TP_j are the topics of item v_i and item v_j respectively. Next, we can obtain the final item representation by the linear combination of neighbor items:

$$\hat{\pi}(v_i, v_j) = \frac{\exp\left(\pi\left(v_i, v_j\right)\right)}{\sum_{v_k \in \mathcal{N}_{v_i}^{re}} \exp\left(\pi\left(v_i, v_k\right)\right)}, \quad \mathbf{h}_{v_i}^{re} = \sum_{v_j \in \mathcal{N}_{v_i}^{re}} \hat{\pi}\left(v_i, v_j\right) \mathbf{h}_{v_j}, \tag{5}$$

where $\hat{\pi}(v_i, v_j)$ is attention coefficient normalized by softmax, which means the different contribution of neighbor item v_j to the current item v_i. $\mathcal{N}_{v_i}^{re}$ is the neighbor set of item v_i in the RIG, $\mathbf{h}_{v_j}$ is the representation of the neighbor item v_j of item v_i, $\mathbf{h}_{v_i}^{re}$ is the final representation of item v_i in the RIL.

4.3 Multi-stacking Layer

In order to fully capture the deep dependencies between items, inspired by the work of Chen et al. [2], we stack multiple AIL and RIL layers, which can capture the complex dependencies (i.e., both adjacent-item and cross-item) within the session, described as below:

$$\mathbf{h}_{0,v_i}^{adj} \rightarrow \mathbf{h}_{1,v_i}^{re} \rightarrow \ldots \rightarrow \mathbf{h}_{l,v_i}^{*} \rightarrow \ldots \rightarrow \mathbf{h}_{k,v_i}^{*}, \tag{6}$$

where $\mathbf{h}_{l,v_i}^{*}$ denotes the representation of item v_i which is the output of layer l, $l \in (2, k)$, k is the hyper-parameter, and $*$ indicates either adj or re.

To fully utilize all features captured by all layers, we apply dense connections in our work. The input of each layer consists of the output representations of all previous layers. More specifically, the input of the l-th layer is $[\mathbf{h}_{0,v_i}^{adj}\|\mathbf{h}_{1,v_i}^{re}\| \cdots \|\mathbf{h}_{l-1,v_i}^{*}]$. This allows the higher layers to utilize not only the features through their previous layer, but also the low-level features at lower layers. For each item v_i, we obtain its representations $\mathbf{h}_{v_i}' \in \mathbb{R}^d$ by stacking multiple AIL and RIL.

4.4 Session Representation Learning Layer

Through the three layers mentioned above, given session $s = [v_1, v_2, \ldots, v_n]$, we can obtain all item representations in it, i.e., $\mathbf{H} = [\mathbf{h}_{v_1}', \mathbf{h}_{v_2}', \ldots, \mathbf{h}_{v_n}']$. Then, we can generate session representation with the representations of items in it.

To reflect the different importance of different positions in the session sequence to the target item, we utilize a learnable position embedding matrix $\mathbf{P} = [\mathbf{p}_1, \mathbf{p}_2, \ldots, \mathbf{p}_n]$, where $\mathbf{p}_i \in \mathbb{R}^d$ is a position vector for specific position i and n is the length of the session. We combine the position information with item representations through concatenation and non-linear transformation:

$$z_i = \tanh\left(\mathbf{W}_2\left[\mathbf{h}'_{v_i}\|\mathbf{p}_{n-i+1}\right] + \mathbf{b}_2\right), \quad \mathbf{s}' = \frac{1}{n}\sum_{i=1}^{n}\mathbf{h}'_{v_i}, \tag{7}$$

where parameters $\mathbf{W}_2 \in \mathbb{R}^{d\times 2d}$ and $\mathbf{b}_2 \in \mathbb{R}^d$ are trainable parameters, $\mathbf{s}'$ is the session information computed as the average of representations of items in the session.

Next, we adopt a soft-attention mechanism to learn the contribution of item v_i to the next prediction, and then we can obtain the session representation by linearly combining the item representations:

$$\beta_i = \mathbf{q}^\top\sigma\left(\mathbf{W}_3\mathbf{z}_i + \mathbf{W}_4\mathbf{s}' + \mathbf{b}_3\right), \quad \mathbf{s} = \sum_{i=1}^{l}\beta_i\mathbf{h}'_{v_i}, \tag{8}$$

where $\mathbf{W}_3, \mathbf{W}_4 \in \mathbb{R}^{d\times d}$ and $\mathbf{q}, \mathbf{b}_3 \in \mathbb{R}^d$ are learnable parameters.

4.5 Prediction Layer

We first utilize dot product and then apply softmax function to predict the click probability $\hat{\mathbf{y}}$ for the item v_i:

$$\hat{\mathbf{y}}_i = \text{Softmax}\left(\mathbf{s}^\top\mathbf{h}_{v_i}\right), \tag{9}$$

where $\hat{\mathbf{y}}_i \in \hat{\mathbf{y}}$ denotes the probability of item v_i to be the true next item. The loss function is defined as the cross-entropy of the prediction results.

5 Experiments and Analysis

5.1 Experimental Settings

Datasets. We select two datasets from the Amazon dataset[1] for our experiments: *Pet Supplies* and *Movies and TV*. The datasets contain purchase history of users and users' reviews for products. Following a common manner, we remove items appearing less than 5 times. Following [11], we split user's purchase behaviors into week-long sessions. To evaluate our method more comprehensively, we prepare two versions for each dataset. The first version (Case 1) keeps all sessions with more than 1 item [7,18,19] while the second version

---

[1] https://nijianmo.github.io/amazon/index.html.

(Case 2) keeps sessions with more than 5 items [19]. Obviously, Case 2 is a subset of Case 1 which keeps long sessions only. We set the sessions of last year as the test data, and the remaining sessions for training. Then, we adopt sequence splitting preprocessing method which is commonly adopted in SBR. For an input session $[v_1^s, v_2^s, \ldots, v_n^s]$, we generate multiple input sequence-label pairs, i.e., $([v_1^s], v_2^s), ([v_1^s, v_2^s], v_3^s), \ldots, ([v_1^s, v_2^s, \ldots, v_{n-1}^s], v_n^s)$.

Evaluation Metrics and Baselines. Following [18,19], we adopt P@K (Precision) and MRR@K (Mean Reciprocal Rank) as evaluation metrics. We compare RI-GNN with the following representative methods, including S-POP [5], S-KNN [5], GRU4Rec [4], NARM [7], STAMP [8], BERT4Rec [12], SR-GNN [19], GCE-GNN [18] and DHCN [20].

Hyper-parameters Settings. Following previous studies [7,18,19], the dimension of node embedding is 100, the size of mini-batch is 100, and the L_2 regularization is 10^{-5} for all models. For RI-GNN, we use the Adam optimizer with the initial learning rate 0.001. The dropout ratio of session graph is 0.2. Moreover, the number of topics are empirically set to 24 and 20 on *Pet Supplies* and *Movies and TV* dataset, respectively.

Table 1. Experimental results on sessions with more than 1 item.

Dataset	Pet supplies				Movies and TV			
Metrics	P@10	P@20	MRR@10	MRR@20	P@10	P@20	MRR@10	MRR@20
S-POP	6.52	6.52	5.25	5.26	1.83	1.84	1.59	1.59
S-KNN	21.86	24.30	12.22	12.39	20.33	23.46	8.40	8.62
GRU4Rec	8.79	9.84	6.26	6.34	8.32	9.47	5.64	5.72
NARM	24.48	27.68	19.48	19.70	22.40	25.40	17.05	17.26
STAMP	21.50	24.66	16.41	16.60	20.59	23.96	14.69	14.92
BERT4Rec	24.18	27.40	19.29	19.51	23.21	26.46	17.75	17.97
SR-GNN	24.64	27.81	<u>19.53</u>	<u>19.75</u>	23.69	26.74	<u>18.24</u>	<u>18.45</u>
GCE-GNN	<u>24.73</u>	<u>28.28</u>	18.16	18.41	<u>24.25</u>	<u>27.64</u>	17.02	17.25
DHCN	24.23	27.45	17.80	18.02	23.33	26.49	15.94	16.16
RI-GNN	**25.43***	**28.88***	**19.93***	**20.15***	**24.69***	**27.94***	**18.93***	**19.14***
Improv. (%)	2.83	2.12	2.05	2.02	1.81	1.09	3.78	3.74

[a][1] The best results of each column are highlighted in boldface, the suboptimal one is underlined, the improvements are calculated by using the difference between the performance of our proposed RI-GNN and the best baseline, and * denotes the significant difference for t-test.

Table 2. Experimental results on sessions with more than 5 items.

Dataset	Pet supplies				Movies and TV			
Metrics	P@10	P@20	MRR@10	MRR@20	P@10	P@20	MRR@10	MRR@20
NARM	26.07	30.34	19.33	19.62	20.98	24.73	14.27	14.53
STAMP	24.44	28.06	18.28	18.53	21.27	25.40	14.49	14.77
BERT4Rec	25.85	30.04	<u>19.49</u>	<u>19.77</u>	22.24	26.43	15.87	16.15
SR-GNN	<u>26.29</u>	<u>30.49</u>	19.31	19.60	22.73	26.57	<u>16.11</u>	<u>16.37</u>
GCE-GNN	25.95	30.36	17.51	17.81	<u>23.18</u>	<u>27.51</u>	14.39	14.71
DHCN	24.96	29.20	17.25	17.54	21.83	25.71	13.01	13.27
RI-GNN	**27.72***	**32.11***	**20.26***	**20.57***	**24.15***	**28.30***	**16.99***	**17.27***
Improv. (%)	5.44	5.31	3.95	4.05	4.19	2.87	5.46	5.50

Overall Performance. The experimental results of overall performance on sessions of different lengths are reported in Table 1 and Table 2 respectively, according to the tables, we can draw the following conclusions:

Case 1 (Performance on sessions with more than 1 item): (1) Traditional methods (i.e., S-POP, S-KNN) show a significant inferiority to neural methods, except for GRU4Rec. This demonstrates that neural networks can learn more sophisticated features than the traditional methods. (2) Neural network based methods (i.e., GRU4Rec, NARM, STAMP, BERT4Rec) usually have better performance for SBR. GRU4Rec shows worse performance, which is probably because it strictly defines a session as a sequence. The other neural methods (i.e., NARM, STAMP, BERT4Rec) outperform GRU4Rec significantly. Among them, NARM combines RNN and attention mechanism, STAMP and BERT4Rec are completely based on attention mechanism. This result demonstrates that attention-based methods are also an effective way besides the RNN-based methods. (3) Among all the baseline methods, the GNN-based methods (i.e. SR-GNN, GCE-GNN and DHCN) demonstrate the superiority over the others. This indicates that GNN-based models would be more effective than RNN-based and attention-based models when capturing the complex dependencies between items in SBR. (4) It is obvious that our proposed RI-GNN model outperforms all the baselines on all datasets. Compared with the GNN-based methods, the superiority of RI-GNN may due to that it is equipped with RIL and review information, which can model item dependencies more accurately.

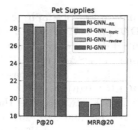 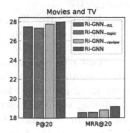

Fig. 4. Comparison of Ablation variants.

Case 2 (Performance on sessions with more than 5 items): Table 2 shows that our proposed RI-GNN model can outperform the state-of-the-art SBR methods by 2.87% to 5.50% on different metrics. Compared with the result in Table 1, RI-GNN outperforms state-of-the-art SBR methods with a larger margin on long sessions. This demonstrates the superiority of RI-GNN is more outstanding for the long sessions. The reason may be that there exist more dependencies between non-adjacent items within the long sessions. This further verifies our hypothesis.

Ablation Study. To investigate the effectiveness of RIL component and review information for RI-GNN, we implement three variants of RI-GNN, denoted as RI-GNN$_{-RIL}$, RI-GNN$_{-topic}$, and RI-GNN$_{-review}$, by removing RIL, topic and review respectively. The comparison of them is shown in Fig. 4. It is clear that all variants are inferior to the standard RI-GNN, which demonstrates that RIL, topics and review information are critical and necessary for the success of RI-GNN model.

6 Conclusion

In this paper, we rethink adjacent dependencies in SBR, argue that adjacent items in a session are not always dependent and non-adjacent items are not necessarily independent. Accordingly, we propose a novel review-refined inter-item graph neural network (RI-GNN), which leverages reviews to reduce the false dependencies between adjacent but actually independent items and capture true dependencies between non-adjacent but dependent items.

Acknowledgment. Wenpeng Lu is the corresponding author. The research work is partly supported by National Natural Science Foundation of China under Grant No. 11901325 and No. 61502259, National Key R&D Program of China under Grant No. 2018YFC0831700, Key Program of Science and Technology of Shandong Province under Grant No. 2020CXGC010901 and No. 2019JZZY020124, and Natural Science Foundation of Shandong Province under Grant ZR2021MF079.

References

1. Blei, D.M., Ng, A.Y., Jordan, M.I.: Latent Dirichlet allocation. J. Mach. Learn. Res **3**, 993–1022 (2003)
2. Chen, T., Wong, R.C.W.: Handling information loss of graph neural networks for session-based recommendation. In: SIGKDD, pp. 1172–1180 (2020)
3. Guo, W., Wang, S., Lu, W., et al.: Sequential dependency enhanced graph neural networks for session-based recommendations. In: DSAA, pp. 1–10 (2021)
4. Hidasi, B., Karatzoglou, A., Baltrunas, L., Tikk, D.: Session-based recommendations with recurrent neural networks. In: ICLR (2016)
5. Jannach, D., Ludewig, M.: When recurrent neural networks meet the neighborhood for session-based recommendation. In: RecSys. pp. 306–310 (2017)
6. Li, C., Niu, X., Luo, X., et al.: A review-driven neural model for sequential recommendation. In: IJCAI, pp. 2866–2872 (2019)

7. Li, J., Ren, P., Chen, Z., et al.: Neural attentive session-based recommendation. In: CIKM, pp. 1419–1428 (2017)
8. Liu, Q., Zeng, Y., Mokhosi, R., Zhang, H.: STAMP: short-term attention/memory priority model for session-based recommendation. In: KDD, pp. 1831–1839 (2018)
9. Qiu, R., Li, J., Huang, Z., Yin, H.: Rethinking the item order in session-based recommendation with graph neural networks. In: CIKM, pp. 579–588 (2019)
10. Rendle, S., Freudenthaler, C., Schmidt-Thieme, L.: Factorizing personalized markov chains for next-basket recommendation. In: WWW, pp. 811–820 (2010)
11. Song, W., Xiao, Z., Wang, Y., et al.: Session-based social recommendation via dynamic graph attention networks. In: WSDM, pp. 555–563 (2019)
12. Sun, F., Liu, J., et al.: BERT4Rec: sequential recommendation with bidirectional encoder representations from transformer. In: CIKM, pp. 1441–1450 (2019)
13. Wang, S., Cao, L., Hu, L., et al.: Hierarchical attentive transaction embedding with intra-and inter-transaction dependencies for next-item recommendation. IEEE Intell. Syst. **36**(04), 56–64 (2021)
14. Wang, S., Cao, L., Wang, Y., et al.: A survey on session-based recommender systems. ACM Comput. Surv. **54**(7), 1–38 (2021)
15. Wang, S., Hu, L., Cao, L., et al.: Attention-based transactional context embedding for next-item recommendation. In: AAAI, pp. 2532–2539 (2018)
16. Wang, S., Hu, L., Wang, Y., et al.: Sequential recommender systems: challenges, progress and prospects. In: IJCAI, pp. 6332–6338 (2019)
17. Wang, S., Hu, L., Wang, Y., et al.: Graph learning based recommender systems: a review. In: IJCAI, pp. 4644–4652 (2021)
18. Wang, Z., Wei, W., Cong, G., et al.: Global context enhanced graph neural networks for session-based recommendation. In: SIGIR, pp. 169–178 (2020)
19. Wu, S., Tang, Y., Zhu, Y., et al.: Session-based recommendation with graph neural networks. In: AAAI. pp. 346–353 (2019)
20. Xia, X., Yin, H., Yu, J., et al.: Self-supervised hypergraph convolutional networks for session-based recommendation. In: AAAI, pp. 4503–4511 (2021)
21. Zheng, L., Noroozi, V., Yu, P.S.: Joint deep modeling of users and items using reviews for recommendation. In: WSDM, pp. 425–434 (2017)

Insomnia Disorder Detection Using EEG Sleep Trajectories

Stephen McCloskey[1]($\boxtimes$), Bryn Jeffries[1], Irena Koprinska[1],
Christopher Gordon[2], and Ronald R. Grunstein[2]

[1] School of Computer Science, University of Sydney, Sydney, NSW, Australia
smcc7913@sydney.edu.au
[2] Woolcock Institute of Medical Research, University of Sydney, Sydney, NSW,
Australia

Abstract. In this paper, we present a novel objective method for insomnia disorder detection based on sleep trajectories, without sleep stage scoring. The sleep trajectories are computed from EEG recordings, generated by BrainTrak, which estimates the physiological parameters of the brain activity during sleep using a neural-field brain model. This method also allows combining multiple different EEG datasets, as sleep trajectories are not affected by systemic differences in EEG collection. We then propose a data-driven semi-supervised approach based on multi-class conditional deep convolutional GAN (CDCGAN) to distinguish between people with insomnia and normal sleepers. Our method uses CDCGAN as a semi-supervised classifier on 20-min subtrajectories of sleep to learn and identify the characteristics of insomnia disorder compared to normal sleepers. We conducted an evaluation using two datasets: Insomnia-100 and MASS. CDCGAN achieved an accuracy of 74.5%, substantially outperforming CNN, kNN and SVM approaches used for comparison. More generally, our work demonstrates the potential of GANs for medical informatics applications.

Keywords: Insomnia · Sleep trajectories · GAN

1 Introduction

Insomnia disorder is a widespread sleep disorder impacting 10.8% of the general population. It is characterised with daytime sleepiness, cognitive impairment, irritability, reduced energy and motivation, and also strongly contributes to anxiety and depression [2]. At present, insomnia disorder is diagnosed subjectively, by interviewing the patient and applying the criteria from the Diagnostic and Statistical Manual of Mental Disorders (DSM-5) which defines it as a dissatisfaction with sleep quality or quantity. However, patients often exhibit sleep misperception, underestimating sleep and awake times. This motivates the development of *objective* diagnostic criteria for insomnia disorder detection.

Clinical studies investigating sleep disorders regard polysomnography (PSG) data collected in a sleep study as the gold standard for measuring sleep objectively. Polysomnography is typically performed in a sleep laboratory and consists

J. Gama et al. (Eds.): PAKDD 2022, LNAI 13282, pp. 314–325, 2022.
https://doi.org/10.1007/978-3-031-05981-0_25

of electroencephalography (EEG) and other measures such as electrocardiogram, electromyography pulse oximetry, nasal airflow and thoracic impedance. Previous studies have found that people with insomnia disorder have differences in EEG activity [9]. Sleep misperception is another symptom of insomnia disorder and has been linked to abnormal beta EEG activity during sleep [7].

Sleep is well known to consist of several stages including non rapid-eye movement (NREM) and rapid-eye movement (REM). Traditionally, sleep is manually scored in 30-s epochs by visual inspection of several EEG channels. This method originated from cutting the EEG recording, rather than based on physiological evidence. The 30-s epoch is currently used since it is economical for hand scoring; although it follows the macro-structure of normal sleep, it is less reliable for people with sleep disorders that have fragmented sleep.

In this paper we propose to use *sleep trajectories* to objectively detect insomnia disorder. The brain state during sleep, observed through EEG, can be represented in terms of the underlying physiology, in a three-dimensional space with X,Y,Z coordinates relating to cortical, corticothalamic and intrathalamic stability, respectively [1] as shown in Fig. 1. We employ BrainTrak [1], a method that analyses sleep EEG samples and estimates the corresponding corticothalamic parameters using a neural-field brain model. An important advantage of BrainTrak is that it seeks to determine physiological-based parameters from the electroencephalography (EEG) power spectrum and thus is less affected by systematic differences in the EEG collection, including sample rate.

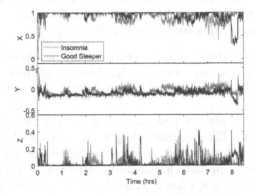

Fig. 1. Sleep trajectories of a person with insomnia and a normal sleeper. The brain state changes during sleep within a 3-dimensional space of physiological parameters: cortical (X), corticothalamic (Y) and intrathalamic stability (Z).

It is challenging to source medical PSG datasets in the large quantities required to train robust machine learning models. Although some institutions curate their data for re-use, it is often for small number of patients and has little representation from target sleep disorder populations. One method that can be used to alleviate this problem is to generate synthetic data with similar properties of real data, for example by using a generative adversarial network

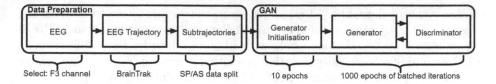

Fig. 2. Overview of the proposed approach.

(GAN). GANs are deep adversarial generative models consisting of two neural networks: a generator to create synthetic data to fool the discriminator and a discriminator that tries to identify real and fake data. The generator transforms simple random uncorrelated input into a complex random variable that is shaped to the targeted distribution after training. It requires a loss function to directly or indirectly compare the generated distribution to the true distribution. The discriminator's goal is to identify the class of the real and fake data.

Various types of GANs have been developed and used for different purposes. Multi-class GANs can be used as a semi-supervised classifier by adjusting the loss function [8]. Conditional GANs train the generator to produce data from a particular class and use the discriminator to determine whether the data is real or fake, and which class it belongs to. A deep convolutional GAN (DCGAN) uses transposed convolutional layers in the generator to increase the spatial dimension of the data and convolutional layers in the discriminator for locally connected filters. In this paper we used a conditional deep convolutional GAN (CDCGAN) that combines the properties of these three types of GANs.

The contributions of this study are as follows:

1. We propose a novel objective method to identify people with insomnia disorder. It uses sleep subtrajectories computed from EEG recordings and doesn't require sleep staging. Sleep trajectories haven't been used before for insomnia detection.
2. We show how multiple different EEG datasets can be combined using the BrainTrak method. The generated sleep trajectories are based on physiological signals and are not affected by systemic differences in EEG collection.
3. We propose a novel approach for insomnia detection and sleep analysis based on GANs. It generates synthetic examples of sleep subtrajectories using an exploratory multi-class CDCGAN and then uses the discriminator from the trained CDCGAN as a semi-supervised classifier to distinguish subtrajectories of normal sleepers and people with insomnia. Our results are promising, demonstrating the potential of GANs for medical informatics applications.

2   Approach

An overview of our approach is presented in Fig. 2. We seek to distinguish between people with insomnia and normal sleepers, based upon sleep trajectories—a 3-dimensional representations of sleep EEG derived from Brain-Trak. We formulate the task as a semi-supervised classification problem, where a classifier is trained using the labelled data from each sleeper type (normal and insomnia), with augmented synthetic data from a generative model. Both the classifier and generator are part of our CDCGAN model.

The data preparation procedure is outlined in Fig. 2. EEG data is typically collected simultaneously from multiple electrode sites. Previous work has shown the significance of particular EEG channels for insomnia disorder - F3, C3 and O1 [4]. We decided to use only the F3 channel at this stage to investigate differences in the prefrontal cortex. We used BrainTrak to process the EEG and determine physiological-based parameters based on a neural-field brain model, in order to reduce the impact of systematic differences from the EEG collection. In the BrainTrak processing we analysed the EEG in 1-s segments as a reasonable trade-off between having sufficient resolution in the power spectrum whilst restricting the trajectory to a localised region of the XYZ space.

Rather than attempting to classify entire sleep trajectories, we instead restrict our approach to working with *subtrajectories* of 20 min' duration. This was motivated by several reasons including reducing the complexity of CDC-GAN, increasing the number of training data points for the relatively small dataset and training with translation invariance. The original trajectories of the full sleep recordings are long ($\sim$8 h) with a sample every second, which was not feasible for networks like GANs that require large sample sizes. This design choice was also supported by the evidence in Sect. 1 that insomnia disorder is associated with differences in EEG activity. The classification part includes the initialisation of the generator and then the adversarial back and forth of the generator and discriminator as shown in Fig. 2. Using this approach, we explored a range of hyperparameters for the generator and discriminator architecture which was complicated by the adversarial nature of GANs. After the CDCGAN was trained, we evaluated its performance on the hold-out data set, assessing its accuracy for classifying people with insomnia disorder and normal sleepers.

3 Data

We combined two different datasets: the Insomnia-100 dataset [5] and the the Mon-treal Archive of Sleep Studies (MASS) dataset [6]. The Insomnia-100 dataset was collected from a single night of PSG data and consists of 89 people with clinically-diagnosed insomnia disorder. The MASS dataset consists of four subsets of healthy sleepers with common EEG channels with no diagnosed disorders, except for 15 in the first subset with mild cognitive impairment. We used a total of 160 subjects from the MASS dataset: 53 subjects from subset 1, 19 from subset 2, 62 from subset 3 and 26 from subset 5. Both datasets used EEG electrodes in the common 10–20 system. The Insomnia-100 dataset was recorded at a sample rate 512 Hz, while the MASS subsets were recorded at a rate 256 Hz.

The F3 EEG channel for each PSG recording was processed in 1 s segments using BrainTrak[1] to obtain a trajectory for each subject's full sleep. It took approximately one week to process each recording on a high-performance cluster

---

[1] https://github.com/BrainDynamicsUSYD/braintrak.

node. The sleep trajectory of each subject was then split into 1200 s (20 min) *subtrajectory* segments with 50% overlap.

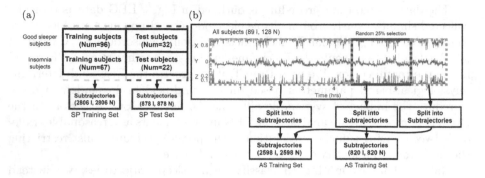

Fig. 3. (a) Subject-partitioned (SP) dataset - 75% of the subjects from the two cohorts were selected for the training set and the rest for the test set. A balanced set of subtrajectories were randomly chosen from the two cohorts of subtrajectories. (b) All-subject (AS) dataset - a random 25% section of the full trajectory was selected for the test set, with the rest used in the AS training set.

We defined two subtrajectory classes: I if the recording is from a person with insomnia disorder, or N if it is from a normal sleeper.

For GAN hyperparameter selection, training and initial evaluation we separated the data by subject, which we refer to as the subject-partitioned (SP) dataset. As shown in Fig. 3a, 25% of the subjects (22 insomnia, 32 normal) were used as a test set and the remaining 75% (67 insomnia, 96 normal) as a training set. From the training set a 10% validation set (6 insomnia; 9 normal) was reserved for parameter tuning. The subtrajectories for both classes were then balanced by randomly selecting the same number of subtrajectories for the test (878 insomnia; 878 normal) and training (2806 insomnia; 2806 normal) sets.

If the model was to be trained on a larger population, we would expect the generator and discriminator to cover a wider selection of subjects' characteristics. To evaluate the classification of subtrajectories in this situation, we created a new test and training set, sampling both from each subject. Each subject's trajectory was divided at random into two disjoint groups of subtrajectories, with 25% in the first group (going to the test set) and 75% in the second (for the training set) as demonstrated in Fig. 3b. This resulted in 820 insomnia and 820 normal subtrajectories in the test set, and 2598 insomnia and 2598 normal in the training set for this all-subject approach (AS).

Unlabeled training data is commonly used to improve the performance of GANs as it focuses on distinguishing real data from generated data; it is included in the loss function of the discriminator as shown in Sect. 4.1. To form the unlabeled dataset, we used the full training set without labels.

Each subtrajectory consists of a time series of the X, Y and Z coordinates. In the neural field model there are restricted ranges for the X, Y and Z dimensions with $0 \leq X \leq 1$, $-0.4 \lessapprox Y \lessapprox 0.6$ and $0 \leq Z \leq 1$. To restrict the range of the generated subtrajectories to valid X, Y and Z coordinates, the training coordinates were normalised with those respective ranges to $[-1, 1]$ to fit with the tanh function. The Y dimension had some outliers ($\approx 0.04\%$) outside that range which were set to the minimum or maximum. This was done with the rescaling as a cubic dimension to not bias the Euclidean distances used in the loss functions of the CDCGAN described in Sect. 4.1.

4   CDCGAN Design

To tune the CDCGAN, we went through a number of different CDCGAN architectures described in Sect. 4.2, assessing the performance on the validation set. Tuning GANs is complicated due to the long training time required and the large number of hyperparameters of both the generator and discriminator. Since it is an adversarial network, it is difficult to assess the performance of the individual architecture as it is balancing the strength of the generator and discriminator. If the generator learns much faster than the discriminator, the discriminator will not be able to learn to identify the generated samples. Conversely, if the discriminator keeps outperforming the generator, the discriminator will not learn much from the generator. Therefore, it is difficult to assess the quality of CDC-GAN as high accuracy from the discriminator does not necessarily mean that the generated samples represent good examples.

4.1   Loss Functions

Since the objectives of the generator and discriminator are adversarial, each of them has its own loss function to quantify their performance.

The generator's loss is defined as $L_G = L_{M_G} + L_{G_D}$, where L_{M_G} is the feature matching loss and L_{G_D} is and loss for not fooling the discriminator.

Feature matching [8] was used to help the generated data match the distribution of real data. This was done by comparing the real and fake data at an intermediate layer of the discriminator:

$$L_{M_G} = \sqrt{\sum_{k \in (I,N)} \sum_{i=1}^{n_k} \left(\overline{D_{r_3}(x_i)} - \overline{D_{r_3}(G_{z_i}(k))} \right)^2} \qquad (1)$$

where I are people with insomnia disorder, N are normal sleepers and n_k is the number of subtrajectories for each class; x_i is the ith real subtrajectory, G_{z_i} is the ith fake subtrajectory created by the generator; $\overline{D_{r_3}(x_i)}$ is the mean of the output from the r_3 leaky ReLU layer shown in Fig. 4.

L_{G_D}, the loss from the discriminator for not classifying the generated data as the intended real class, is defined as:

$$L_{G_D} = \mathbb{E}_{k \in (I,N), j \in [1, i_k]} \log \left(\frac{e^{D_{G_j}(k)}_F}{\sum e^{D_j}} \right) \qquad (2)$$

where $D_{G_j(k)}$ is the output from the discriminator D for the fake class F for the jth generated subtrajectory $G_j(k)$, S_D is the output from the discriminator for all classes, i_k is the ith trajectory of class k.

The discriminator's loss is defined as $L_D = L_{D_G} + L_{D_R} + L_{D_U}$ and considers the loss of incorrectly classifying (1) the fake data as real (L_{D_G}), (2) the real supervised data (L_{D_R}), and (3) the real unlabeled data as generated data (L_{D_U}). The three loss components are defined as:

$$L_{D_G} = \mathbb{E}_{k \in (I,N), j \in [1,i_k]} log \left(\frac{e^{D_{G_j(k)}_R}}{\sum e^{D_j}} \right) \tag{3}$$

$$L_{D_R} = \mathbb{E}_{k \in (I,N), j \in [1,i_k]} log \left(\frac{e^{D_{j_k}}}{\sum e^{D_j}} \right) \tag{4}$$

$$L_{D_U} = \sum_{j=1}^{i} \overline{log \left(1 - \frac{e^{D_{j,I_F}} - e^{D_{j,S_F}}}{S_D} \right)} \tag{5}$$

where $D_{G_j(k)}$ is the output from the discriminator D for the real class R for the fake subtrajectory $G_j(k)$, and I_F and N_F are the fake insomnia disorder and fake normal sleeper class respectively.

4.2 Tuning

The CDCGAN architecture was selected by evaluating the performance on the validation set. Since the generator was producing fake data of either the insomnia disorder or normal sleeper classes that sought to fool the discriminator, we calculated the accuracy by combining the real and fake of each class.

We evaluated architectures with different number of convolutional layers (from 1 to 3) in both the generator and discriminator, with the size of the convolutional layers varying from 5 to 240 neurons depending on the layer. Different activation functions were also tested: tanh, sigmoid, ReLu and leaky ReLU. The final layer in the discriminator was also investigated by testing fully connected layer of different sizes.

The generator in the CDCGAN was initialised by fitting the output to real data for 10 epochs with iterations of batch size 128:

$$G_{init} = \sqrt{\sum_{k \in (I,N)} \sum_{i=1}^{n_k} \left(\overline{(x_i)_k} - \overline{D(G(z_i))_k} \right)^2} \tag{6}$$

The CDCGAN was then trained for 1000 epochs with multiple iterations of subtrajectories with batches of size 128 on the generated, unlabeled and labeled data. During training, 10% of the class labels of the real labeled data were swapped with the fake class. This is a common technique used to create noise for the discriminator to balance the learning of the generator and discriminator. The learning rate for the generator and discriminator was set to 0.0001.

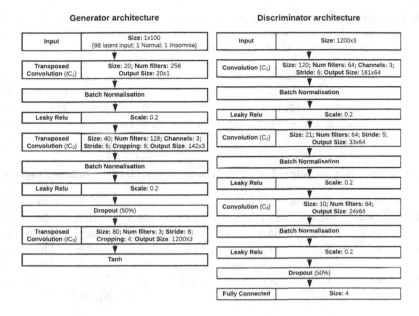

Fig. 4. Architecture of the generator and discriminator of CDCGAN

4.3 Final Architecture

The architecture of the generator and discriminator chosen after parameter tuning is shown in Fig. 4.

In the generator, the input layer consists of 100 neurons: 98 random latent variables and 2 variables to hot encode either people with insomnia disorder or normal sleepers. The generator has 3 transposed convolutional layers (tC_1, tC_2 and tC_3) that upscale the inputs into the synthetic subtrajectory. The first two have batch normalisation and leaky ReLU layers (gr_1 and gr_2), following techniques and recommendations from previous GANs [8]

For the discriminator, the size of the input layer is, that corresponds to 1200 s seconds and 3 channels (X, Y, Z). The discriminator has 3 convolutional layers (C_1, C_2 and C_3) each with batch normalisation and a leaky ReLU layer, a 50% dropout layer and a fully connected layer of size 4. Each output from the fully connected layer relates to a class: Real Insomnia (RI), Real Normal sleeper (RN), Fake Insomnia (FI) and Fake Normal sleeper (FN).

5 Results and Discussion

5.1 Subject-Partitioned (SP) CDCGAN

After the architecture was selected, the whole 75% set was used to retrain the CDCGAN with the selected architecture and its performance was evaluated on the test set of subjects whose data had not been used in the training of the model or hyperparameter selection.

Table 1. Confusion matrix of the subject-partitioned (SP) and all-subject (AS) method on the test data. The discriminator classified each subtrajectory as real or fake, and coming from an insomnia patient (I) or normal sleeper (N).

True class	Classified as real			Classified as fake			True class	Classified as real			Classified as fake		
	I	N	Total	I	N	Total		I	N	Total	I	N	Total
I	**506**	139	645	**166**	67	233	I	**430**	88	518	**218**	84	302
N	159	**480**	639	82	**157**	239	N	124	**389**	513	90	**217**	307

(a) SP   (b) AS

Table 2. Insomnia vs. Normal, with the classified real and fake summed together.

	Accuracy	Precision	Recall	F1		Accuracy	Precision	Recall	F1
CDCGAN	**74.5**	**76.5**	73.6	**75.0**	CDCGAN	**76.5**	**79.0**	75.2	**77.1**
kNN	62.8	58.3	**90.2**	70.8	kNN	64.6	60.2	**86.6**	71.0
SVM	61.9	60.3	69.7	64.7	SVM	68.1	67.2	70.5	68.8
NL-SVM	68.1	65.9	74.8	70.1	NL-SVM	71.4	68.7	78.7	73.3
CNN	65.6	63.9	66.1	65.0	CNN	67.3	66.2	67.7	67.0

(a) SP   (b) AS

The accuracy of the Insomnia vs. Normal classifier was 74.5% (Table 2a), which is approaching the accuracy that a human scorer achieves from a full night's sleep data. The performance on the SP test set demonstrates the generalisation ability of CDCGAN on new subjects, not present in the training set. The confusion matrix of the classification results is shown in Table 1a; we can see that the two classes are similarly missclassified - 23% of the Insomnia examples are missclassified as Normal and 27% of the Normal are missclassified as Insomnia. 73% of subtrajectories are correctly classified as real (1284/1756).

During the training process, due to the adversarial nature of the CDCGAN, the proportion of Real v. Fake subtrajectories fluctuates significantly, ≈20–80%. The accuracy on the validation set also fluctuates during training, as CDCGAN appears to not reach saturation after 1000 epochs because of its adversarial nature. This means the generator has been able to produce good enough examples to trick the discriminator, and then the discriminator adjusts to identify them and the generator adjusts to trick the discriminator again and so on.

5.2 Sampling from All-Subjects (AS) CDCGAN

To compare the AS and SP approaches, we used the same CDCGAN architecture as for SP. The 75% training set of the AS dataset was used to train the CDCGAN and its performance was evaluated on the 25% test set.

The accuracy of the CDCGAN was 76.5% (Table 2b), showing a small improvement over the SP dataset above. The confusion matrix in Table 1b also shows that the two classes (Normal and Insomnia) were similarly missclassified - 21% Insomnia as Normal and 26% Normal as Insomnia, and that the majority

(63%) of the test data is correctly identified as real. The AS approach has a similar instability as described in Sect. 5.1. We expected the performance of the AS dataset would be improved over the SP dataset, since the GAN would have learned from subtrajectory samples that belong to the same subjects from the training and test set. This demonstrates that different subtrajectories from the same individual have more similar characteristics even when being taken from different times during sleep.

5.3   Comparison with kNN and SVM Classifiers

To provide comparison with alternate methods for classifying subtrajectories, we evaluated two other common approaches for trajectory classification: k-Nearest Neighbour (kNN) and support vector machine - a linear (L-SVM) and a non-linear SVM (NL-SVM) with a Gaussian kernel.

For (kNN) we used the Euclidean distance due to computational constraints. Other distance measurements (such as the Fretchet distance and DTW) have complexities of $O(nm)$, which requires n subtrajectories and m as the length of the subtrajectories. We compared kNN with k from 1–50 using stratified 10-fold cross-validation on subtrajectories from all subjects (as for Sect. 5.2) and chose the k with the highest accuracy.

The kNN classifier achieved an accuracy of 62.8% on the hold-out test set with $k = 23$ when using the SP dataset (Table 2a) and 64.6% when using the AS dataset with $k = 11$ (Tables 2a and 2b). Although this is above the base-line of 50% for classifying the two-class balanced dataset, it was significantly outperformed by our CDCGAN approach. As shown in Tables 2a and 2b, kNN outperforms CDCGAN significantly on recall, but underperforms significantly on precision. This indicates that kN has a low false negative rate but a high false positive rate, which means a bias towards classifying data as insomnia disorder.

On the other hand, the L-SVM and NL-SVM classifiers achieved an accuracy of 61.9% and 68.1% respectively when the data was partitioned by subject and an accuracy of 68.1% and 71.4% respectively when data is sampled from all subjects (Tables 2a and 2b). NL-SVM performed better than L-SVM and kNN, but was outperformed by CDCGAN.

5.4   Comparison with CNN Classifier

In order to evaluate the benefit of the CDCGAN approach over a convolutional neural network (CNN) classifier, we took the discriminator architecture from our CDCGAN model and used this in isolation, testing and training with just the real data from Sect. 5.2. The accuracy of the CNN classifier was 65.6% and 67.3% for SP and AS respectively as shown in Tables 2a and 2b, still outperforming the kNN model, but significantly worse than the full CDCGAN approach. This suggests that the extra training provided from the generated data provides an important contribution to the performance of the classifier.

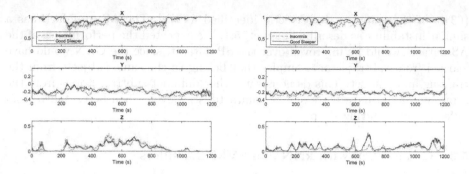

Fig. 5. Comparing the outcome from the same latent input variables of the trained generator between insomnia and normal sleepers for 2 subtrajectories.

5.5 CDCGAN Generated Subtrajectories

In Fig. 5, the same random latent values for the generator input were used, with only the hot-encoded inputs for insomnia disorder and the normal sleepers being different. Therefore, the generator should produce two variations: either with insomnia disorder or as a normal sleeper. As stated in Sect. 3, the generated subtrajectories were rescaled to the X, Y and Z range, so that was inversed in Fig. 5. This rescaling method was found to improve the accuracy of the CDCGAN $\approx$2–3%, since otherwise the generator would produce invalid subtrajectories.

When comparing Fig. 5 with Fig. 1, we can see that the generated subtrajectories have some similar characteristics such as irregular spikes and periodicity. However, due to the noisy nature of this data it is difficult to assess the generated subtrajectories visually, as would typically be done with GANs producing synthetic images that can be inspected visually or with image quality indices [3]. Another important characteristic of the generated subtrajectories that can be seen in Fig. 5 is that the normal sleepers typically have greater activity spikes in X, Y and Z and often has lower values of X, Y and Z on average. This has some interesting implications for potential differences in brain activity during sleep between good sleepers and people with insomnia.

6   Conclusion

In this paper, we present a novel objective method to identify people with insomnia disorder using EEG sleep recordings, without sleep stage scoring. The traditional methods involve visual inspection of EEG recordings and manual sleep stage scoring, which is subjective and time-consuming. We propose to use subtrajectories computed from EEG recordings using the BrainTrak method; these trajectories represent the brain state during sleep in terms of physiological parameters based on a neural-field brain model. In particular, we propose a data-driven semi-supervised approach based on multi-class CDCGAN to distinguish between people with insomnia disorder and normal sleepers using 20-min samples of sleep trajectories generated by BrainTrak. We conducted an evaluation using data from two datasets: Insomnia-100 and MASS. Our results showed that CDCGAN achieved an accuracy of 74.5% on the hold-out test set, considerably outperforming kNN (61.6%), CNN (65.6%), linear SVM (61.9%) and non-linear SVM (68.1%) classifiers. This work also demonstrated that by using data from a short (20 min) period of sleep, it was possible to distinguish people with insomnia and normal sleepers with good accuracy. The generated subtrajectories for each class provided insight into the different physiological characteristics attributed to insomnia disorder and good sleeping. The use of BrainTrak trajectories is also beneficial for combining multiple different EEG datasets as they are not affected by systemic differences in EEG collection. More generally, our work shows the potential of GANs for analysing biosignals and for medical informatics, in addition to their success for image and computer vision applications.

References

1. Abeysuriya, R., Rennie, C., Robinson, P.: Physiologically based arousal state estimation and dynamics. J. Neurosci. Methods **253**, 55–69 (2015)
2. Edinger, J.D., et al.: Derivation of research diagnostic criteria for insomnia. Sleep **27**(8), 1567–1596 (2004)
3. Kancharla, P., Channappayya, S.S.: Improving the visual quality of generative adversarial network (GAN)-generated images using the multi-scale structural similarity index. In: 25th International Conference on Image Processing (ICIP), pp. 3908–3912. IEEE (2018)
4. McCloskey, S., Jeffries, B., Koprinska, I., Miller, C.B., Grunstein, R.R.: Data-driven cluster analysis of insomnia disorder with physiology-based qEEG variables. Knowl.-Based Syst. **183**, 104863 (2019)
5. Miller, C.B., et al.: Clusters of insomnia disorder: an exploratory cluster analysis of objective sleep parameters reveals differences in neurocognitive functioning, quantitative EEG, and heart rate variability. Sleep **39**(11), 1993–2004 (2016)
6. O'Reilly, C., Gosselin, N., Carrier, J., Nielsen, T.: Montreal archive of sleep studies: an open-access resource for instrument benchmarking and exploratory research. J. Sleep Res. **23**(6), 628–635 (2014)
7. Perlis, M.L., Smith, M.T., Andrews, P.J., Orff, H., Giles, D.E.: Beta/Gamma EEG activity in patients with primary and secondary insomnia and good sleeper controls. Sleep **24**(1), 110–117 (2001)
8. Salimans, T., Goodfellow, I., Zaremba, W., Cheung, V., Radford, A., Chen, X.: Improved techniques for training GANs. In: Advances in Neural Information Processing Systems, pp. 2234–2242 (2016)
9. Schwabedal, J.T., Riedl, M., Penzel, T., Wessel, N.: Alpha-wave frequency characteristics in health and insomnia during sleep. J. Sleep Res. **25**(3), 278–286 (2016)

Parametric Bandits for Search Engine Marketing Optimisation

Marco Gigli[✉] and Fabio Stella

Dipartimento di Informatica, Sistemistica e Comunicazione,
University of Milano-Bicocca, 20126 Milan, Italy
m.gigli2@campus.unimib.it, fabio.stella@unimib.it

Abstract. Expense optimisation for online marketing is a relevant and challenging task. In particular, the problem of splitting daily budget among campaigns, together with the problem of setting bids for the auctions that regulate ad appearance, have been recently cast as a multi-armed bandit problem. However, at the current state of the art several shortcomings limit practical applications. Indeed, campaigns are routinely divided by practitioners into sub-entities called ad groups, while current approaches take into account only the case of single ad groups: in this paper, we extend the state of the art to multiple ad groups. Moreover, we propose a contextual bandit model which achieves high data efficiency, especially important for campaigns with few clicks and/or small conversion rate. Our model exploits domain knowledge to greatly reduce the exploration space by using parametric Bayesian regression. Elicitation of prior distributions from domain experts is simplified by interpretability, while action selection is carried out by Thompson sampling and local optimisation methods. A simulation environment was built to compare the proposed approach to current state-of-the-art methods. Effectiveness of the proposed approach is confirmed by a rich set of numerical experiments, especially in the early days of marketing expense optimisation.

Keywords: Marketing expense optimization · Multi armed bandit · Bayesian regression

1 Introduction

The digital advertising market tallied a total expense of 140 billions USD in 2020 in the US alone, showing a 12% year-over-year growth despite the impact of COVID-19 [14]. This shows the importance of developing principled and reliable methods to maximize return on investment.

Research interest stems from the fact that this can be seen as an instance of the *exploration-exploitation dilemma*; trying to maximise the number and value of *conversions* (whether they are clicks, contacts or sales), the advertiser can tweak various parameters (detailed below): a balance must be found between gathering new information on under-explored settings and avoiding *exploring* too much, i.e. *exploiting* the settings that proved most fruitful so far.

© The Author(s), under exclusive license to Springer Nature Switzerland AG 2022
J. Gama et al. (Eds.): PAKDD 2022, LNAI 13282, pp. 326–337, 2022.
https://doi.org/10.1007/978-3-031-05981-0_26

Among digital advertising formats, the top two by market share are *search ads* and *display ads*, which account for three quarters of the total spend. Every time a user performs a search or visits a website, an automated auction takes place (typically a *Generalized Second Price* or a *VCG* auction [21]), and the user is shown the winning ads. For every ad, the advertiser is thus called to choose wisely a *target* (i.e. search keywords and user profiles), a *bid* and a maximum *daily budget*. The bid, in particular, represents the maximum cost the advertiser is willing to pay if the user clicks on the ad.

Search engine marketing keywords and ads sharing a common theme are usually gathered by practitioners into *ad groups* [10], which are in turn grouped in *campaigns*, which can be used to specify location targeting. The budget is set at the campaign level; we consider here the common case in which a total daily budget is given and must be split among several campaigns. The bid is instead set on a per-ad group basis.

Two recent works [12,13] have cast the daily bid/budget optimization as a multi-armed bandit problem. In particular, the goal is to maximise the long term revenue, choosing daily the combination of bids and budgets for the whole *portfolio* (the collection of campaigns over which the total budget is set).[1] To this end, the whole combination of bids and budgets is seen as a *super arm* of a *combinatorial bandit* [5], playing which one observes not only the total reward, but also the full set of single-campaign rewards. Campaigns are in turn seen as *static contextual bandits* [1]: similar bid/budget combinations for a campaign share information. The fixed context (feature) vector is indeed given by the budget and bid choice. The application of these results in practical settings is, however, limited by the following shortcomings:

- The literature concentrates on campaigns without substructure (single ad group campaigns). This means that, besides the budget, only one bid must be chosen per campaign. As we have seen, however, the typical campaign is divided in ad groups, thus requiring generalisation. Given the regression model introduced in [12] this generalisation is non-trivial.
- Agnostic nature of Gaussian Processes (GPs) with respect to the functional form which links bid to observed clicks. While this method is very flexible, it also reflects in the need for more data to converge to a sensible posterior, if compared to a more informed model (in particular, a parametric model).
- The need to extrapolate, from the observed clicks, the so-called *saturation clicks*, i.e. the number of clicks we would have observed if we did not have budget constraints. The authors of [12] suggest using the time of the day when the daily budget finished and an estimate of the distribution of the number of clicks during the day to perform the extrapolation (see also [13] where the example of constant distribution is given). However, this information is not available, and ad frequency is explicitly reduced throughout the day to make

---

[1] We specify that the term "portfolio" is less established in the industry with respect to terms such as "ad group" and "campaign": we use it here just to refer to the collection of campaigns involved in the optimisation. If these make up the totality of an advertiser's campaigns, one could use the term *account* interchangeably.

sure the budget lasts until the end of the day [10]. Other metrics could be used to this end, but they are inherently noisy. Perhaps more importantly, this extrapolation method needs a certain share of the total budget to be *always* reserved for exploration (akin to ε-greedy strategies). The need for this missing data imputation stems from the use of vanilla GPs (i.e. with a Gaussian likelihood [22]). On the other hand, *censored regression* is a principled way to avoid the need for missing data imputation. While GPs can accomodate non-Gaussian likelihoods, this requires giving up exact update formulas, and switching to approximate methods [8].

These reasons make parametric regression appealing, exploiting a functional form suggested by domain knowledge. Bayesian regression, needed for Thompson sampling, can be conducted with Markov Chain Monte Carlo (MCMC) [18].

This paper overcomes these shortcomings making the following contributions:

- A multi-ad group generalisation of the relation between bid/budget and clicks (Sect. 2), suitable regardless of the regression model one employs.
- An informed alternative to GP regression, which accounts for censoring in a principled way and with interpretable parameters (Sect. 3).
- The use of such a model in the context of Thompson sampling is explored in Sect. 4. In particular, bid selection is recast as a local constrained optimisation problem, which can be tackled with off-the-shelf methods. The algorithm for budget selection is given in detail.
- A simulation environment, built on what is known about the inner workings of the auctions, is developed to test and compare performances of the proposed approach (Sect. 5). Numerical results are reported in Sect. 6.

2 Model and Notation

We follow [12] and assume for now that we have a portfolio of N campaigns, each with just one ad group. Let's call $n_j(b_j, B_j)$ the average number of clicks obtained by the j-th campaign with budget B_j and bid b_j. Let v_j be the average value of one click from the j-th campaign. The task of maximising the revenue can then be formulated as the following constrained optimisation problem:

$$\max \sum_{j=1}^{N} v_j n_j(b_j, B_j) \quad \text{s.t.} \sum_{j=1}^{N} B_j \leq B, \quad b_j \in [\underline{b}_j, \overline{b}_j] \, \forall j \tag{1}$$

where the maximum is taken over the budgets $B_1, \ldots, B_N$ and the bids $b_1, \ldots, b_N$, B is the total budget and $\underline{b}_j$ and $\overline{b}_j$ are the (possibly campaign-dependent) minimum and maximum allowed bids.

Both the functions n_j and the values v_j are unknown, and must be estimated from the collected data, hence the need to balance exploration and exploitation. Early estimates could be inaccurate and lead to sub-optimal decisions, but one does not want to spend too many resources on data gathering either, since this comes at the expense of exploiting acquired knowledge.

Note that, in accordance with [12], we are here assuming *stationarity*, i.e. that the probability distributions of click value and number of clicks given a bid and budget don't change with time. This is a realistic approximation only on short time spans: as detailed in Sect. 6, this motivates the need for fast-learning models, as the one we present in Sect. 3.

In order to reduce the burden of exploration, in [12] an ansatz for the form of n_j is proposed, reducing the complexity of a two-variable regression to two one-variable functions:

$$n_j(b_j, B_j) \approx n_j^{\text{sat}}(b_j) \min\left(1, \frac{B_j}{c_j^{\text{sat}}(b_j)}\right). \tag{2}$$

Here the function n^{sat} denotes the *saturation clicks*, i.e. the number of clicks a campaign would obtain if there were no budget limits. Likewise, c^{sat} denotes the *saturation cost*, i.e. the cost faced in the same situation. Since the right hand side depends non-linearly on n^{sat} and c^{sat}, and given that we are speaking about averages, the equality strictly holds only in the deterministic case.

The issue in generalising the problem (1) to the multi-ad group setting is that the budget is shared by all the ad groups of the same campaign, as stated in Sect. 1. While we can let an index k run over ad groups, and define v_{jk} as the value of a click from the ad group k of campaign j and do similarly for the bid b_{jk}, we cannot define a corresponding click function $n_{jk}(b_{jk}, B_j)$: the number of clicks gathered by an ad group depends also on the bids of all the other ad groups belonging to the same campaign. Intuitively, raising the bid b_{jk} will bring more clicks for the corresponding ad group, but it will also erode the budget B_j more quickly, thus lowering the clicks received by the other ad groups. This difficulty can be circumvented introducing the *total value* function $V_j(\boldsymbol{b}_j, B_j)$ of a campaign, which depends on the whole vector of bids $\boldsymbol{b}_j$. Therefore, the optimisation problem (1) generalises to

$$\max \sum_{j=1}^{N} V_j(\boldsymbol{b}_j, B_j) \quad \text{s.t.} \quad \sum_{j=1}^{N} B_j \leq B, \quad b_{jk} \in [\underline{b}_{jk}, \overline{b}_{jk}] \, \forall j, k. \tag{3}$$

In order to preserve data efficiency, the ansatz (2) must be generalised too, linking the total value function to the corresponding saturation quantities. Note that the aforementioned interdependence among different ad groups is a consequence of a limited budget, while the dependence of saturation quantities n_{jk}^{sat} and c_{jk}^{sat} on the single bid b_{jk} is well defined. If the j-th campaign contains m_j ad groups, and we let

$$V_j^{\text{sat}}(\boldsymbol{b}_j) = \sum_{k=1}^{m_j} v_{jk} n_{jk}^{\text{sat}}(b_{jk}), \qquad c_j^{\text{sat}}(\boldsymbol{b}_j) = \sum_{k=1}^{m_j} c_{jk}^{\text{sat}}(b_{jk}), \tag{4}$$

then ansatz (2) generalises to

$$V_j(\boldsymbol{b}_j, B_j) \approx V_j^{\text{sat}}(\boldsymbol{b}_j) \min\left(1, \frac{B_j}{c_j^{\text{sat}}(\boldsymbol{b}_j)}\right). \tag{5}$$

Since the right hand side is not a sum over single-ad group contributions, this formula captures the interaction among different ad groups.

As a way to intuitively justify (5) for fixed bids and budget, we can think of the single n_{jk}^{sat} as the sizes of "reservoirs", one for each ad group, from which clicks are randomly drawn, up until the moment when the total cost paid matches the assigned budget. If the clicks pertaining to different ad groups are well mixed, each will bring approximately the same fraction of its *saturation value* $V_{jk}^{\text{sat}}(b_{jk}) = v_{jk} n_{jk}^{\text{sat}}(b_{jk})$ and saturation cost. The value of this fraction is found equating the total cost paid and the assigned budget B_j, hence (5).

3   Parametric Regression Model

Among the strategies to face the exploration-exploitation dilemma, Thompson sampling [19] is of particular interest to practitioners, due both to its performance [4], generality and conceptual simplicity [16].

Viewing the j-th campaign as a contextual bandit, Thompson sampling requires performing Bayesian regression on the correspondence between $(\boldsymbol{b}_j, B_j)$ and the reward r_j.

We can restrict the search space by placing few, sensible hypotheses on the shape of the functions n^{sat} and c^{sat}, introduced in equations (5) and (4) (we are here dropping indices for simplicity). Clicks and cost paid are of course highly correlated, so to be able to perform separately the two regressions it is convenient to introduce the cost-per-click (CPC) function $\varphi(b) = \frac{c^{\text{sat}}(b)}{n^{\text{sat}}(b)}$. Both functions n^{sat} and φ must be positive, be monotonic increasing with the bid, saturate for high enough bid and vanish for vanishing bid. Moreover, φ was empirically found to be linear for small bids (in accordance with the law of diminishing returns), and must be strictly smaller than the identity (because of the meaning of bid as maximum CPC). These considerations suggest to use a properly shifted and scaled logistic function. Starting from the saturation clicks,

$$n^{\text{sat}}(b) = \underbrace{k(1 + e^{-ac})}_{\text{scale factor}} \left(\underbrace{\frac{1}{1 + e^{-a(x-c)}}}_{\text{logistic function}} - \underbrace{\frac{1}{1 + e^{ac}}}_{\text{vert. shift}} \right). \tag{6}$$

An example of this function is shown in Fig. 1a. The term in parentheses in the scale factor has the goal of providing a meaningful k, which is the *saturation value*, i.e., the maximum number of clicks one can expect when setting a very high bid. The coefficients a and c have the meaning of an inverted length scale and of a horizontal shift. In order to give them a more intuitive meaning (since we will have to place priors on them), we can link them to the elbows of the curve, which can be identified as the maximum and minimum of the second derivative of the function. The left elbow can be interpreted as the threshold below which the bid yields a negligible number of clicks, while above the right elbow the function effectively saturates. For a standard logistic function (with $a = c = 1$) such elbow points are: $x_\pm = \log(2 \pm \sqrt{3})$. For a general logistic function, the elbows

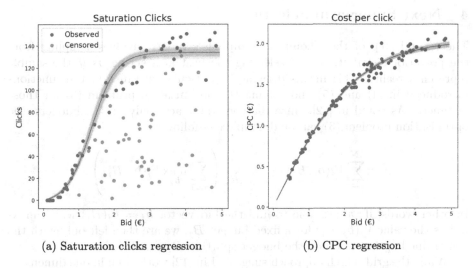

(a) Saturation clicks regression   (b) CPC regression

Fig. 1. Bayesian regression for models (6) and (7). Data have been simulated as in Sect. 5. Orange dots denote *censored* quantities: the number of saturation clicks is greater or equal than the observed number (see Sect. 3).

b_- and b_+ are linked to the parameters a and c via: $a = \frac{x_+ - x_-}{b_+ - b_-}$, $c = b_+ - \frac{x_+}{a}$. Switching to the CPC function φ (Fig. 1b), the additional hypothesis of being linear near the origin suggests the same functional form, with $c = 0$:

$$\varphi(b) = 2\kappa \left(\frac{1}{1 + e^{-\alpha x}} - \frac{1}{2} \right). \tag{7}$$

Here κ is the maximum CPC which can be paid, and the same considerations connecting α with elbows apply.

This model has the following advantages: *i*) lower variance (with a small bias increase), *ii*) closed form functions, *iii*) forces monotonicity, which helps optimisation to choose the next action, *iv*) transparent hyperparameters make it easy to elicit priors, and finally *v*) parametric Bayesian regression easily accommodates *censoring*. In particular, we are here trying to infer the relationship between bid and saturation clicks, which are often only partially known: in these cases, all we know is that saturation clicks are greater or equal than observed clicks, and one needs a principled way to take these data into account, without introducing systematic bias. In the bandit model one can assume *non-informative censoring*, thus only a simple change in the likelihood is needed [11].

Contextual bandits have been mostly studied in the linear [2,6], generalized linear [7] and kernelised domain [17,20]. More recently, deep neural networks have been explored for the regression step [15]; their expressive power is, however, balanced by the large need of data. To the best of our knowledge, this is the first work which uses full-fledged Bayesian regression on a parametric function which is not (generalized) linear.

4   Next Super-arm Selection

The second step of the Thompson sampling strategy involves sampling from the posterior distribution, and selecting the best arm acting *as if* the sample represents reality. This means drawing a particular instance of the functions introduced in (4) and (5) and solving the optimisation problem (3) for those instances. As noted in [12], since the constraint acts only on the budgets, the optimisation problem (3) can be decoupled as follows:

$$\max \sum_{j=1}^{N} V_j(\boldsymbol{b}_j, B_j) = \max_{B_1, \dots, B_N} \left(\sum_{j=1}^{N} \max_{\boldsymbol{b}_j} V_j(\boldsymbol{b}_j, B_j) \right).$$

In other words, if we are able to find the bid vector $\boldsymbol{b}_j = \boldsymbol{b}_j(B_j)$ which maximises the value $V_j(\boldsymbol{b}_j, B_j)$ for a fixed budget B_j, we are then left only with the constrained optimisation on the budget splitting $B_1, \dots, B_N$.

While the grid search approach suggested in [12] works well in one dimension (i.e. for single-ad group campaigns), it scales badly with increasing dimensionality. If a GP regression model is employed, owing to the non-monotonic nature of extracted samples, one must recur to *global* methods, as opposed to local ones. On the other hand, employing the monotonic functions (6) and (7), the function $V_j(\boldsymbol{b}_j, B_j)$ with fixed budget was empirically found to have only one local maximum, which is also global (see Fig. 2a). Therefore, optimisation is amenable to local methods: when applicable, these are both faster and more reliable. We will here describe how such methods can be applied in practice. Starting from (5), $V_j(\boldsymbol{b}_j, B_j)$ can be rewritten as a piecewise function:

$$V_j(\boldsymbol{b}_j, B_j) = \begin{cases} V_j^{\text{sat}}(\boldsymbol{b}_j) & \text{if } c_j^{\text{sat}}(\boldsymbol{b}_j) \leq B_j \\ B_j \frac{V_j^{\text{sat}}(\boldsymbol{b}_j)}{c_j^{\text{sat}}(\boldsymbol{b}_j)} & \text{if } c_j^{\text{sat}}(\boldsymbol{b}_j) \geq B_j \end{cases} \tag{8}$$

Note that, on the boundary $\{c^{\text{sat}}(\boldsymbol{b}_j) = B_j\}$ between the two regions, the functions coincide. On the other hand, traversing the boundary the gradient changes abruptly, thus hindering the direct application of gradient-based optimisation methods. We will see, however, that a constrained optimisation on just one region is sufficient. First we note that, if both regions are non-empty, the global maximum of $V_j(\boldsymbol{b}_j, B_j)$ is given by the maximum between the two maxima of the function on the two regions. Moreover, every directional derivative of $V_j^{\text{sat}}(\boldsymbol{b}_j)$ (sum of monotonic single-variable functions) is strictly positive. If the boundary $\{c^{\text{sat}}(\boldsymbol{b}_j) = B_j\}$ is not empty, the maximum of $V_j^{\text{sat}}(\boldsymbol{b}_j)$ then lies on said boundary.

This, in turn, means that the maximum over the region $\{c^{\text{sat}}(\boldsymbol{b}_j) \leq B_j\}$ is less than or equal to the maximum over the region $\{c^{\text{sat}}(\boldsymbol{b}_j) \geq B_j\}$, i.e. that, if the latter region is not empty, it suffices to search the maximum there.

Up to now we dealt with finding the optimal bids for a campaign given the budget, thus finding a function $\boldsymbol{b}_j = \boldsymbol{b}_j(B_j)$. We must now solve the following optimisation problem:

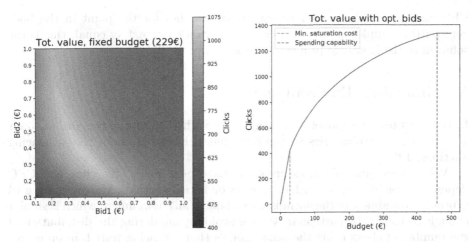

(a) Total value function for fixed budget

(b) Total value function varying budget, selecting optimal bids for each budget

Fig. 2. Thompson sample of the total value function $V_j(\boldsymbol{b}_j, B_j)$ of a campaign with two ad groups (click values v_{jk} are set to 1 for simplicity).

$$\max \sum_{j=1}^{N} V_j(\boldsymbol{b}_j(B_j), B_j) \quad \text{s.t.} \sum_j B_j \leq B. \qquad (9)$$

The terms $V_j(\boldsymbol{b}_j(B_j), B_j)$ in the sum are single-argument functions that depend only on the budget of the campaign. For budgets B_j greater than the *spending capability* $c_j^{\text{sat}}(\overline{b}_{j1}, \dots \overline{b}_{jm_j})$ of the campaign, such functions become constant, as can be seen by (8). For budgets below the spending capabilities, the functions have been empirically found to be downwards concave (see Fig. 2b), in agreement with the law of diminishing returns.

This also means that this optimisation step is amenable to local gradient methods too. If, however, the optimisation over bids is performed with numerical methods, extra care must be taken in choosing the step size over budgets: small errors in the first step translate to a small noise in the function $V_j(\boldsymbol{b}_j(B_j), B_j)$. To control this issue, we developed an intuitive optimisation procedure which generalises the budget splitting strategy presented in [9] to the case of non-constant return on investment (i.e. non-linear functions):

1. Assign to all campaigns their spending capability
2. Find the *most expendable campaign*, which has the smallest discrete derivative of $V_j(\boldsymbol{b}_j(B_j), B_j)$
3. Subtract one unit from its assigned budget (e.g. one euro)
4. Repeat from step 2 until the total assigned budget is smaller than B

The procedure keeps subtracting budget from a campaign until the discrete derivative matches or becomes smaller than the discrete derivative of another.

This means that the procedure effectively searches for the point in the budget splitting simplex where the derivatives are approximately equal: this is the solution of the Lagrange problem corresponding to (9).

5   Simulation Environment

In order to test the model and compare it with the state-of-the-art, we built an environment which tries to capture what is disclosed about the ad placing auctions [10].

While some simplifying assumptions have been made, the click and CPC dependence on bids agrees with experience on actual auctions. Moreover, the goal is proving the ability of the optimiser to adapt to an environment which is similar enough to reality. In particular, we are avoiding simulating the distribution of the number of clicks using the same models that are being tested, in order not to introduce bias.

For every day, the number of searches compatible with an ad group is sampled from a Poisson distribution. Then, for each search we simulate an auction. The number of competing advertisers is again sampled from a Poisson distribution (with a different mean). The ads belonging to different advertisers are ranked according to the product of three quantities. The first is the bid: for competing advertisers, the bid is sampled from an exponential distribution. The second is a *static quality score*, which measures the intrinsic quality of the ad: it is sampled from a triangular distribution. The third is an *instantaneous quality score*, which measures the affinity between the single search and the ad group. It is modeled as an angle between vectors, which is extracted from a rescaled beta distribution; then the quality score is calculated as the scalar product between said vectors.

After the ads have been ranked, the first ones appear on the search engine result page: whether they are clicked or not is determined by a Bernoulli distribution. Then, in keeping with the meaning of the bid as maximum CPC, the advertiser that has received a click pays the minimum amount necessary to appear in that position.

The budget is then updated accordingly, until either available searches end or the budget is finished. A second Bernoulli distribution governs which clicks turn into contacts. What distinguishes various simulations are the parameters of the manifold of the probability distributions involved. To run a comparison between the parametric regression model introduced in Sect. 3 and the GP model introduced in [12], the latter needs some additional metric to extrapolate the saturation clicks of the day from the observed clicks, as stated in Sect. 1. We have chosen *lost impression share*, an estimate of the fraction of times the ad was eligible for appearing in a search, but did not due to limited budget. To capture the fact that it is inherently a noisy quantity, a convex combination of the actual lost impression share with random fractions was used, with varying coefficients.

The code of the simulation environment is available at https://github.com/MarcoGigli/sem-simulation.

6   Numerical Experiments

We run 40 experiments randomly drawing the parameters introduced in Sect. 5. The number of campaigns of the portfolio varied between 2 and 8 and, for each campaign, the number of ad groups varied between 1 and 4. For each parameter setting, both the parametric and the GP model optimised the total value for 100 virtual days. To compare performances, we evaluated the *regret* of using the GP model instead of the parametric one, $R_n = \sum_{t=1}^{n} \left(r_t^{\mathrm{par}} - r_t^{\mathrm{GP}} \right)$. Here r_t^{par} and r_t^{GP} represent the rewards received at day t using the parametric and GP model respectively. In particular, it is given by the number of contacts. We also calculated *relative regret* $\rho_n = \frac{\sum_{t=1}^{n} \left(r_t^{\mathrm{par}} - r_t^{\mathrm{GP}} \right)}{\sum_{t=1}^{n} r_t^{\mathrm{par}}}$ to meaningfully compare performances of experiments with different parameters.

In Fig. 3, the behavior in time of regret and relative regret is shown for a particular set of parameters, i.e. one of the 40 experiments. After a short time in which, due to random fluctuations, the GP model gathers more contacts, the relative regret quickly raises to 40%. Then, as both models are given more data, the relative difference in performance gradually tapers off, and converges to approximately 10%.

This example is typical (Fig. 4): at $n = 10$ days, only in two experiments the regret is negative, and in most cases the relative regret ranges from 20% to 60%. Fast convergence is especially important if a sliding window strategy is employed to retroactively take time dependence into account, as in [13]. At $n = 100$ days, the relative regrets are much less spread out and lower on average, but only in three cases they are negative. Of these three, two are very close to zero (i.e. the two methods are nearly equivalent) and in the third case the portfolio is composed of just two campaigns, with one ad group each: in this simple case, a more flexible model pays off in the long run.

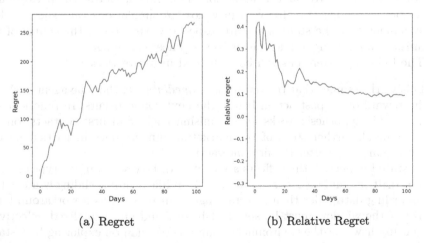

(a) Regret   (b) Relative Regret

Fig. 3. Time dependence of regret and relative regret suffered by the GP model when compared to the parametric model.

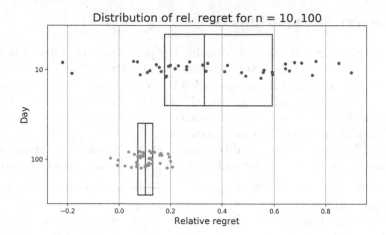

Fig. 4. Distribution of the relative regret at $n = 10$ days and $n = 100$ days for all experiments. The boxes show the first quartile, median and third quartile.

7   Conclusion and Next Steps

In this paper, we extended a state-of-the-art method for Search Engine Marketing optimisation to the multi-ad group domain, thus bridging a gap with application. Exploiting domain knowledge, we introduced a parametric Bayesian regression model to reduce the need of data with respect to GPs and to naturally account for censoring, further freeing up resources for both exploration and exploitation. Parameters are interpretable, hence allowing for the easy elicitation of priors. Benefiting from the properties of this model, we presented how the optimisation step in Thompson sampling can be carried out by local (as opposed to global) methods. We built a simulation environment to test the performance of competing models. Finally, we run a host of simulations that show a clear improvement over the state-of-the art, especially early on in the course of the optimisation, implying a much faster convergence on average.

The following extensions will be addressed in future works:

- This work assumes that the reward is *immediate*, i.e. that the agent is shown the reward of its past action before the next round occurs. In practical settings, this hypothesis works for optimising clicks and first contacts; on the other hand, further steps of the marketing funnel (sales in particular) can occur many days after the first interaction.
- As stated in Sect. 3, time effects are excluded; however, various types of concept drift can and do occur in practice [3]. While a simple sliding window (i.e. discarding data older than a certain age) can be used as a workaround, the size of the window must be specified beforehand and kept fixed, effectively ignoring how fast the environment changes. We plan on exploring how state-of-the-art non-stationary bandit techniques can be applied to the present approach.

References

1. Agrawal, S.: Recent advances in multiarmed bandits for sequential decision making. Oper. Res. Manage. Sci. Age Anal. 167–188 (2019)
2. Agrawal, S., Goyal, N.: Thompson sampling for contextual bandits with linear payoffs. In: ICML, pp. 127–135. PMLR (2013)
3. Cavenaghi, E., Sottocornola, G., Stella, F., Zanker, M.: Non stationary multi-armed bandit: empirical evaluation of a new concept drift-aware algorithm. Entropy **23**(3), 380 (2021)
4. Chapelle, O., Li, L.: An empirical evaluation of Thompson sampling. Adv. Neural Inf. Process. Syst. **24**, 2249–2257 (2011)
5. Chen, W., Wang, Y., Yuan, Y., Wang, Q.: Combinatorial multi-armed bandit and its extension to probabilistically triggered arms. JMLR **17**(50), 1–33 (2016)
6. Chu, W., Li, L., Reyzin, L., Schapire, R.: Contextual bandits with linear payoff functions. In: AISTATS 2011, pp. 208–214. PLMR (2011)
7. Filippi, S., Cappe, O., Garivier, A., Szepesvári, C.: Parametric bandits: the generalized linear case. In: NIPS, vol. 23, pp. 586–594 (2010)
8. Gammelli, D., Peled, I., Rodrigues, F., Pacino, D., Kurtaran, H.A., Pereira, F.C.: Estimating latent demand of shared mobility through censored Gaussian processes. Transp. Res. Part C Emerg. Technol. **120**, 102775 (2020)
9. Geyik, S.C., Saxena, A., Dasdan, A.: Multi-touch attribution based budget allocation in online advertising. In: Proceedings of the Eighth International Workshop on Data Mining for Online Advertising, pp. 1–9 (2014)
10. Google Ads: Google Ads Help (2021). https://support.google.com/google-ads, see: answer/1704396, answer/1722122, answer/2616012
11. Harrell, F.E.: Regression Modeling Strategies: with Applications to Linear Models, Logistic Regression, and Survival Analysis, vol. 608. Springer, New York (2001). https://doi.org/10.1007/978-1-4757-3462-1
12. Nuara, A., Trovò, F., Gatti, N., Restelli, M.: A combinatorial-bandit algorithm for the online joint bid/budget optimization of pay-per-click advertising campaigns. In: Thirty-Second AAAI Conference on Artificial Intelligence, April 2018
13. Nuara, A., Trovò, F., Gatti, N., Restelli, M.: Online joint bid/daily budget optimization of internet advertising campaigns. arXiv preprint arXiv:2003.01452 (2020)
14. PwC: IAB Internet advertising revenue report, Full year 2020 results, April 2021
15. Riquelme, C., Tucker, G., Snoek, J.: Deep Bayesian bandits showdown: An empirical comparison of Bayesian deep networks for Thompson sampling. arXiv:1802.09127 (2018)
16. Russo, D., Van Roy, B., Kazerouni, A., Osband, I., Wen, Z.: A tutorial on Thompson sampling. arXiv:1707.02038 (2017)
17. Srinivas, N., Krause, A., Kakade, S.M., Seeger, M.: Gaussian process optimization in the bandit setting: No regret and experimental design. arXiv:0912.3995 (2009)
18. Stan Development Team: Stan modeling language users guide and reference manual, version 2.28 (2019)
19. Thompson, W.R.: On the likelihood that one unknown probability exceeds another in view of the evidence of two samples. Biometrika **25**(3/4), 285–294 (1933)
20. Valko, M., Korda, N., Munos, R., Flaounas, I., Cristianini, N.: Finite-time analysis of kernelised contextual bandits. arXiv:1309.6869 (2013)
21. Varian, H.R., Harris, C.: The VCG auction in theory and practice. Am. Econ. Rev. **104**(5), 442–45 (2014)
22. Williams, C.K., Rasmussen, C.E.: Gaussian Processes for Machine Learning, vol. 2. MIT Press, Cambridge (2006)

User Incentive Based Bike-Sharing Dispatching Strategy

Bing Shi[1,2], Zhaoxiang Song[1], Xizi Huang[1], and Jianqiao Xu[3]($\boxtimes$)

[1] School of Computer Science and Artificial Intelligence,
Wuhan University of Technology, Wuhan 430070, China
[2] Shenzhen Research Institute of Wuhan University of Technology,
Shenzhen 518000, China
[3] Department of Information Security, Naval University of Engineering,
Wuhan 430033, China
xujianqiao321@163.com

Abstract. As a green and low-carbon transportation way, bike-sharing provides a lot of convenience in the daily traveling. However, after a period of usage, the distribution of shared bikes may not meet the traveling requirements of users. Shared bikes may converged in some areas while in other areas users have no bikes to ride. Therefore, the bike-sharing platform needs to dispatch the shared bike effectively to improve the user service ratio. Instead of using trucks to dispatch bikes, one possible way is to incentivize users to return shared bikes to the desired destination by subsidizing certain monetary, while ensuring users reach their original destinations within walking distance. However, the platform usually has a limited budget to incentivize users. In this paper, we design a bike-sharing dispatching strategy by incentivizing users to improve the user service ratio by taking into account the budget constraint, users' maximum walking distance, users' riding demands and dynamic changes of the distribution of shared bikes. The dispatching strategy consists of budget allocation and task allocation algorithms. In the budget allocation algorithm, we first predict the user's riding demand based on LSTM, so as to generate dispatching tasks, then we model the allocation of budgets for each time step as a Markov decision process, and then design a budget allocation algorithm based on the deep deterministic strategy gradient algorithm. In the task allocation algorithm, due to the budget constraint that makes it impossible to use the mainstream bipartite graph matching algorithm, we choose to use the greedy matching algorithm for the task allocation. Finally, we run experiments based on the Mobike dataset to evaluate our strategy against greedy budget algorithm, unlimited budget algorithm, and the truck hauling algorithm. The experimental results show that our shared bike dispatching strategy with user incentive can outperform other benchmark dispatching strategies and maximize the long-term user service rate with limited budgets.

Keywords: Bike-sharing dispatching · Demand prediction · User incentive · Deep reinforcement learning

J. Gama et al. (Eds.): PAKDD 2022, LNAI 13282, pp. 338–352, 2022.
https://doi.org/10.1007/978-3-031-05981-0_27

1   Introduction

As an economical, low-carbon and convenient traveling way, bike-sharing can solve the problem of "the last mile" in the public transportation system, and thus is growing rapidly around the world [1]. In the bike sharing systems, bikes can be parked in the public lands when riders reach their destinations. Under this situation, shared bikes may pile up in some areas, other riders may not find available bikes in the nearby area. At this moment, the bike-sharing platform needs to dispatch bikes regularly to specific locations in order to satisfy the riding demands (e.g. dispatching bikes to the exits of subway stations). However, the bikes are usually distributed anywhere within a specific geographical area, and thus it is inconvenient and costive to use trucks to convey them. Actually, in the real world, some riders might be willingly to dispatch the riding bike to the destination desired by the platform, which is close to its original destination within walking distance, in order to get subsidy from the platform, and then walk back to their original destinations. In this paper, we intend to analyze how the bike-sharing platform incentivize riders to dispatching shared bikes by subsidizing them efficiently.

Specifically, when incentivizing riders to dispatch bikes to the destinations, the platform should address several key challenges. Firstly, the bike-sharing platform usually has a limited budget to subsidize users, it need to allocate the budget to each time step and how much allocated in the current step may affect the allocation in the future steps. Therefore, the platform should allocate the total budget efficiently to subsidize users to dispatch bikes to increase service ratio over the whole time step. Secondly, the spatial and temporal riding demands of users are constantly changing. The platform needs to incentivize users to dispatch the bikes to meet the changing demand within the budget constraint of the current step. The platform also needs to predict the riding demands of each step.

The main contributions of this paper are as follows. We design an efficient user incentivized bike-sharing dispatching strategy to improve the user service ratio by taking into account the total budget constraint and the dynamic riding demands. The strategy consists of a budget allocation algorithm and a task allocation algorithm. We model the allocation of budget as a Markov Decision Process (MDP), and use a deep reinforcement learning algorithm to solve it. We then design a greedy based task allocation algorithm to maximize the number of served users in the current step given the allocated budget. We then experimentally evaluate the proposed dispatching strategy on the Mobike dataset against some typical benchmark approaches under different budgets and different initial supply of bikes. The experimental results show that our bike dispatching strategy under user incentives can achieve the best performance except for the approach with unlimited budget.

The rest of this paper is organized as follows. We introduce the related work in Sect. 2, we then describe the basic settings in Sect. 3, and introduce the proposed algorithms in Sect. 4. We provide experimental analysis in Sect. 5 and conclude the paper in Sect. 6.

2   Related Work

There exist a number of works analyzing bike-sharing dispatching strategies. Ghosh et al. proposed two methods to solve the truck marshalling and trailer routing problem [2]. Zheng et al. proposed an optimization model called DRRPVT (Dynamically Repositioning and Routing Problem with carrier Vehicles and bike Trailers), ton increase the overall profit of hired bikes and reduce the lost demand [3]. Li et al. proposed a method to alleviate the contradiction between supply and demand during peak riding periods by providing incentives for reverse-peak riding [4]. Svenja et al. proposed a user dispatching strategy based on price discounts and a manual dispatching strategy to reduce carbon emissions and platform costs [5].

There also exist a number of works by treating the shared bike dispatching as a crowdsourcing problem, i.e. incentivizing users to dispatch shared bikes to the desired destinations. Aeschbach et al. proposed four different strategies to involve users in the dispatching of shared bike [6]. Ban et al. proposed a simulation system to test the effect of different user parameters on the service ratio in a bike-sharing dispatching strategy with user participation [7]. Lin et al. incentivized users to ride to the area desired by the platform to rebalance distribution of bike [8]. Huang et al. used a sparse network to guide the bike dispatching of volunteers [9]. Duan et al. extended the work of [9] to facilitate user incentives and to combine origin and destination incentives in an adaptive manner [10].

The above works usually do not consider the overall budget constraints and users' maximum walking distance. In this paper, we will analyze how the platform allocates the whole budget into each time step to efficiently incentivize users of this step to dispatch bikes under the maximum walking distance constraints of users and the dynamic spatial and temporal demands of shared bikes.

3   Basic Settings

In this section, we introduce the basic settings of shared bike dispatching problem with user incentives. The bike-sharing platform dispatching bikes to the destinations by incentivizing users is a typical crowdsourcing problem. Figure 1 shows how the system works. mainly means that the bike-sharing platform crowdsources the dispatching task to users and gives them a certain monetary incentive to return the shared bikes to the appropriate area, so as to achieve the purpose of dispatching the shared bikes, as shown in Fig. 1. Location A is the original destination where the user wants to ride. However, there may exist shortage of bikes in the nearby area. Therefore, the platform can incentivize the user to ride the bike to the desired destination (location B) by giving the user some subsidy. Then the user can walk back from location B to A.

3.1   Symbols

The main mathematical symbols used in this paper are shown in Table 1 and we will explain separately in the following.

3.2 Region Settings

In this paper, we divide the overall area R into n small square regions which do not cross and overlap with each other, i.e. $R = \{r_1, r_2, \ldots, r_n\}$. We consider to maximize the whole long-term user service ratio, and we divided the whole time into T time steps, which is denoted as $\{t_1, t_2, \ldots, t_T\}$. We use $SU_{r_i}^t$ to represent the supply of shared bikes in region $r_i \epsilon R$ in the beginning of time step t, where $t = 0$ represents the initial time supply.

Fig. 1. Users incentive for bike-sharing dispatching

Table 1. Symbols

Parameter	Description
$R = \{r_1, r_2 \cdots r_n\}$	The whole area divided into n non-overlapping small regions
$\mathcal{T} = \{t_1, t_2 \cdots t_T\}$	Dividing time into T time steps
$SU_{r_i}^t$	The supply of shared bikes in region r_i at the beginning of time step t
$U_t = \{u_1, u_2 \ldots u_n\}$	The set of users arriving in time step t
r_{s_j}	The starting area of user u_j
r_{arr_j}	The original destination area of user u_j
r_{want_j}	Dispatch task area of user u_j
r_{act_j}	The actual return area of user u_j
r_{lack_j}	The lacking bike area of region j
d_{max_j}	Maximum walking distance of user u_j
B_U	Total budget over the whole time steps
X_j	Walking distance of user u_j

3.3 Users Setting

We use U to represent the set of all users, and $U_t = \{u_1, u_2, \ldots, u_{n_{u_t}}\}$ is the set of users who arrive the bike-sharing platform in time step t. $u_j = (r_{s_j}, r_{arr_j}, r_{want_j}, r_{act_j}, d_{max_{u_j}})$ is the information of the j_{th} user, where r_{s_j} is the starting area of user u_j, r_{arr_j} is the original destination area of user u_j; r_{want_j} is the area where the bike-sharing platform expect the user to return the bike, which is the dispatched destination; r_{act_j} represents the area where the user actually returns the shared bike. When the user accepts the dispatching task, $r_{act_j} = r_{want_j}$, on the contrary, $r_{act_j} \neq r_{arr_j}$; $d_{max_{u_j}}$ is the maximum walking distance that the user would like to walk, When the X_j is larger than $d_{max_{u_j}}$, the user will not accept the dispatching task.

3.4 Problem Formulation

We generate dispatching tasks for users by predicting the future bike shortage in each region. Specifically, we use Long Short-Term Memory (LSTM) network to predict the future riding demands based on historical data since it is a time series prediction problem[12]. $Pred^t = \{Pred_{r_1}^t, Pred_{r_2}^t, \ldots, Pred_{r_n}^t\}$ is the predicted number of users in region at time step t, where $Pred_{r_i}^t$ is the predicted number of users in region r_i in time step t. $Lack_{r_i}^t$ is the number of lacked bikes in region r_i at the beginning of time step t, which is:

$$
Lack_{r_i}^{t+1} = \begin{cases} 0, Pred_{r_i}^{t+1} \leq SU_{r_i}^t - \sum_{j,r_{s_j}=r_i}^{num_t} u_j \\ Pred_{r_i}^{t+1} - \left(SU_{r_i}^t - \sum_{j,r_{s_j}=r_i}^{num_t} u_j \right), Pred_{r_i}^{t+1} > SU_{r_i}^t - \sum_{j,r_{s_j}=r_i}^{num_t} u_j \end{cases} \tag{1}
$$

where $\sum_{j,r_{s_j}=r_i}^{num_t} u_j$ is the number of bikes ridden from region r_i in time step t. $SU_{r_i}^t - \sum_{j,r_{s_j}=r_i}^{num_t} u_j$ is the number of bikes that are ridden and not returned in time step t. After the dispatching tasks are generated, the platform needs to determines the matching between the tasks and users given the allocated budget.

The spatial and temporal demands of users are constantly changing and the number of shared bikes in each region is affected by the users' riding demands, $SU_{r_i}^t$ can be calculated as:

$$
SU_{ri}^{t+1} = SU_{r_i}^t - \sum_{j,r_{s_j}=r_i}^{num_t} u_j + \sum_{j,r_{act j}=r_i}^{num_t} u_j \tag{2}
$$

where $\sum_{j,r_{s_j}=r_i}^{num_t} u_j$ is the number of users departing from region r_i in time step t. $\sum_{j,r_{act j}=r_i}^{num_t} u_j$ is the number of users whose actual return area r_{act_j} is r_i in time step t. For users performing the dispatching task, after they return the shared bike to the dispatching region, they still need to walk back to their destination region, which will incur a walking cost, and the walking cost $cost_{u_j}$ can be calculated as:

$$cost_{u_j} = \begin{cases} \hbar X_j{}^2, X_j < d_{\max_{u_j}} \\ \infty, X_j > d_{\max_{u_j}} \end{cases} \tag{3}$$

where $\hbar$ is the user's walking cost parameter and X_j represents the distance walked by the user, i.e. the distance between the user's destination region r_{arr_j} and the user's dispatching task region r_{want_j}. Note that Eq. 3 can reflect the user's cost definition for walking distance [13]. When the X_j is greater than the $d_{max_{u_j}}$, the user's walking cost is extremely high and is not willing to perform the dispatching task. We use p_{u_j} to represent the subsidy of the bike-sharing platform incentivizing u_j in time step t, satisfying $p_{u_j}^t \geq cost_{u_j}^t$, that means, users are willing to perform dispatching tasks when the received subsidy is greater than the walking cost. We use $P = \{P_1, P_2, \ldots, P_T\}$ to represent the budget allocated by the platform at each time step. Note that $\sum_j^{U_t} p_{u_j}^t \leq P_t$, P_t and $\sum_t^T P_t \leq B_U$, where B_U is the whole budget. We need to maximize the long-term user service ratio, i.e. minimize the number of users who do not ride the bike.

$$\begin{cases} \text{remove }_{r_i}^t = \sum_{j,r_{s_j}=r_i}^{n_{u_t}} u_j - SU_{r_i}^t, & SU_{r_i}^t < \sum_{j,r_{s_j}=r_i}^{n_{u_t}} u_j \\ \text{remove }_{r_i}^t = 0, & SU_{r_i}^t > \sum_{j,r_{s_j}=r_i}^{n_{u_t}} u_j \end{cases} \tag{4}$$

4   Users Incentive Based Bike-sharing Dispatching Strategy

In this section, we introduce the user incentive based bike-sharing dispatching strategy, including budget allocation algorithm and task allocation algorithm.

4.1   Budget Allocation Algorithm

We use the LSTM network to predict the riding demands in each region based on the historical user usage data, and then compares it with the available bikes in each region to find the lacking number of bikes in each region. Then the platform needs to allocate the whole budget into each time step in order to maximize the long-term user service ratio. Allocating budget into each time slot is a sequential decision process. We model it as a Markov Decision Process (MDP), and use a deep reinforcement learning algorithm to design the budget allocation strategy.

Markov Decision Process. The budget allocation algorithm is modeled as MDP, which is denoted as a five-tuple (S, A, P, R, γ).

State is denoted as $s_t = (SU^t, Usernum^t, Lack^{t+1}, RB^t) \in S$, where SU^t is the set of shared bike supply in each region $r_i(r_i \in R)$ at the beginning of the current time step t, $Usernum^t$ is the number of users arriving in each region at the current time step t, $Lack^{t+1}$ is the set of lacking bikes in the future predicted according to the task generation algorithm, RB^t is the remaining budget for the current time step.

Action is denoted as $a_t = Budget^t \in A$, where $Budget^t$ is the budget allocated by the platform for the current time step t.

Return is denoted as $r_t = \sum_{r_i}^{R} remove_{r_i}^{t+1}$, $\sum_{r_i}^{R} remove_{r_i}^{t+1}$ is the number of users who did not ride to the bike in all regions R in time step $t + 1$.

p and γ is the state transfer probability and the discount factor respectively.

Budget Allocation Algorithm. In budget allocation, Since the action space for budget allocation is continuous, we use Deep Deterministic Policy Gradient (DDPG) reinforcement learning algorithm [14] to solve this problem.

The DDPG algorithm is based on the Actor-Critic framework, where actor policy network is a probability-based actor, and critic value network evaluates each action of actor through feedback from the environment and thus modifies the weights of actor. The whole algorithm converges through exploratory learning. The actor policy network consists of online policy network and target policy network. The actor online policy network outputs action according to the current state s, interacting with the environment under the current state-action to generate the state s_0 and reward r for the next time, and updates the policy network parameters θ^μ according to Eq. 5.

$$\nabla\theta^\mu|s_i \approx \frac{1}{Batch\_SIZE} \sum_i \nabla_a Q\left(s, a|\theta^Q\right)|_{s=s_i, a=\mu(s_i)} \theta_\mu \mu\left(s|\theta^\mu\right)|_{s_i} \quad (5)$$

Meanwhile, random noise $\mathcal{N}$ is added to the output action a_t of the Actor Online policy network, as shown in Eq. 6. The actions performed by the agent are changed from deterministic values to random values, thus increasing the randomness of the learning process and increasing the exploration capability.

$$a_t = \mu\left(s_t|\theta^\mu\right) + \mathcal{N}_t \quad (6)$$

where $\mathcal{N}_t$ decreases over time. The Actor Target policy network selects the optimal next action a_0 based on the next state s_0 sampled in the replay memory.

Critic value network also consists of online policy network and target policy network. Critic online value network updates the value network parameters θ^Q by minimizing the loss function, while the current $Q(s, a, \theta^Q)$ is:

$$L\left(\theta^Q\right) = \frac{1}{Batch\_Size} \sum_i \left(y_i - Q\left(s_t, a|\theta^Q\right)\right)^2 \quad (7)$$

Critic Target value network calculates the target Q value y_i as:

$$y_i = r + \gamma Q(s_{t+1}, a|\theta^{Q'}) \quad (8)$$

The target network of both actor and critic are updated by the online network parameters, can be calculated as:

$$\begin{cases} \theta^{\mu'} = \tau\theta^\mu + (1-\tau)\theta^{\mu'} \\ \theta^{Q'} = \tau\theta^Q + (1-\tau)\theta^{Q'} \end{cases} \quad (9)$$

where τ is the update factor and $\tau \ll 1$. The budget allocation algorithm is shown in Algorithm 1.

Algorithm 1: Budget Allocation Algorithm

---

 input : Initial supply of bikes SU^0 and U_t
 output: Budget allocated for each time step t
1 **initialization**;
2 initialize replay memory $\mathscr{D}$ to capacity N;
3 initialize actor and critic value function Q with random weights, are denoted as
   θ^μ and θ^Q respectively, meanwhile initialize actor and critic target function Q
   weights as $\theta^{\mu'} = \theta^\mu$ and $\theta^{Q'} = \theta^Q$ respectively;
4 **for** $episode = 1, M$ **do**
5 │   Initialized state $s_0 = \left(SU^0, Usernum^0, Lack^1, RB^0\right)$;
6 │   Select a random action a_t;
7 │   **for** $t = 1, T$ **do**
8 │   │   Calculate reward r_{t-1} according to Equation 4;
9 │   │   Get $Usernum^t$ according to the user information coming in the current
  │   │   time step and update s_t;
10 │   │   Calculate $Lack\,(t+1)$ for each area in the next time step according to
   │   │   the task generation algorithm and update s_t;
11 │   │   temps $= s_t$;
12 │   │   Store samples (temps, a_{t-1}, r_{t-1}, s_t) in the replay memory $\mathscr{D}$;
13 │   │   Select action a_t based on neural network and noise parameters;
14 │   │   Use the task allocation algorithm to allocate tasks to users and obtain
   │   │   the actual return area of them, while updating $RB\,(t)$ and s_{t+1};
15 │   │   Calculate SU^{t+1} and update s_{t+1} according to Equation 2;
16 │   │   **if** $\mathscr{D} \neq \emptyset$ **then**
17 │   │   │   Sample random $BATCH\_SIZE$ from $\mathscr{D}$, then update θ^μ and θ^Q
   │   │   │   according to equations 5, 7, 8;
18 │   │   │   Update $\theta^{\mu'}$ according to equations 9 and $\theta^{Q'}$;
19 │   │   **end**
20 │   │ $s_t = s_{t+1}$;
21 │   **end**
22 **end**
23 Output the $Budget^t$ in each time steps

4.2 Task Allocation Algorithm

After the dispatching task is generated and a budget is allocated to the current time step, the platform needs to allocate the dispatching task to users for dispatching under the budget constraint and the maximum walking distance of the user, to allocate as many users with the dispatching tasks as possible. The task allocation algorithm is shown in Algorithm 2.

Algorithm 2: Task Allocation Algorithm

---

    input : $Budget_t$, $Lack_{r_j}^{t+1}$ of each region r_j $(r_j \in R)$ in $t+1$ time steps,
               $U_t = \{u_1, u_2, \ldots, u_{n_{u_t}}\}$ where
               $u_i = \{r_{s_i}, r_{arr_i}, r_{want_i}, r_{act_i}, d_{max_i}\}$ (r_{want_i}, r_{act_i} are unknown)
    output: The dispatching policy of users $\pi_{u_j}(u_j \in U_t')$, a.k.a. r_{want_i} and r_{act_i}
               for each user

1 **for** *each* $u_i \in U_t$ **do**
2   **for** $Lack_{r_j}^{t+1}$ *of each region* r_j $(r_j \in R)$ **do**
3   **for** *each* $lack_j \in Lack_{r_j}^{t+1}(r_j \in R)$ **do**
4   Calculate the distance $distance_{ij}$ between $r_{arr_{u_i}}$ and r_{lack_j}, then
                store it in list $distance$;
5   **if** $distance_{ij} > d_{max_i}$ **then**
6                 $distance_{ij} = \infty$
7   **end**
8   **end**
9   **end**
10 **end**
11 **for** *each* $distance_{ij} \in distance$ **do**
12   **if** $distance_{ij}$ *is the minimum* **then**
13         $cost_i = k * (distance_{ij})^2$
14   Store the number of u_i and $lack_j$ in list $match$, a.k.a. add $(i, j, cost_i)$ to
            the $match$;
15         $r_{want_{u_i}} = r_{lack_j}$; Remove u_i from U_t; Remove $lack_j$ from $Lack_{r_j}^{t+1}$
16   **end**
17 **end**
18 **for** *each* $match_m \in match$ **do**
19   **if** $match_m[2] < Budget_t$ **then**
20   User i in $match_m$ perform dispatching tasks, a.k.a $r_{act_i} = r_{lack_j}$;
21         $Budget_t- = match_m[2]$
22   **end**
23 **end**
24 **for** *each rest* u_i **do**
25     $r_{act_{u_i}} = r_{arr_{u_i}}$
26 **end**
27 Output the dispatching policy $\pi_{u_j}(u_j \in U_t)$

Note that we cannot use the traditional bipartite matching algorithm to match users with dispatching tasks given the extra budget constraint. Since the traditional matching algorithm such as Kuhn-Munkres can make the perfect matching of maximizing the number of matching pairs [15], but ignore the dispatching subsidy of each pair. Therefore, when there exists a budget constraint for subsidizing the dispatching, the perfect matching of maximizing the matching numbers may not proceed when the subsidy is more than the budget. Therefore, maximization bipartite matching algorithm, such as Kuhn-Munkres cannot be applied in this situation. In this paper, we adopt a greedy based matching algorithm, which searches to find the user-task matching pair with the minimum subsidy until the budget is exhausted.

5   Experiment Analysis

In this section, we run experiments to evaluate the proposed user incentive dispatching strategy.

5.1   Experiment Setting

We evaluate the proposed dispatching strategy based on Mobike Dataset[1]. In this dataset, each data contains information of order ID, bike ID, user ID, starting time of riding, end time of riding, users' starting location and end location. The dataset includes one month of user usage data, with different characteristics of user demand profiles for weekdays and weekends, which is shown in Fig. 2. In this paper, we choose the user usage data from 7:00 *a.m.* to 8:00 *p.m.* of weekdays as the experimental data.

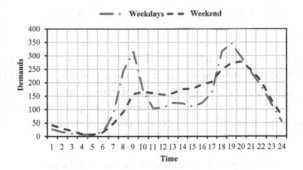

Fig. 2. Demands for weekdays and weekends

Furthermore, from the dataset we find that most of the users' riding time is about ten minutes, therefore, we set the length of time step as ten minutes. The platform will make the dispatching decision at the end of each time step given the collected information of this time step. Furthermore, the platform divides the whole area into several small squares, with side length 1 km, which is shown in Fig. 3.

---

[1] https://www.heywhale.com/mw/dataset/5eb6787e366f4d002d77c331/file.

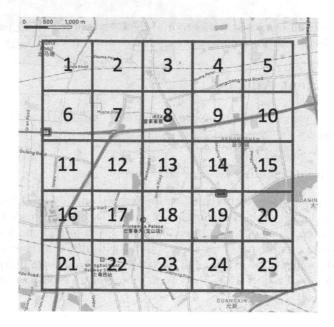

Fig. 3. Regional division and indexing

Other experimental parameters are shown in Table 2. Where the maximum walking distance of the user is set as a normal distribution with the mean of 1 km, that means the user has a certain probability of not wanting to perform the dispatching task. We set the user walking cost parameter $k = 2$. The total number of time steps T = 78, which covers the time from 7:00 *am* to 8:00 *pm*.

Table 2. Experimental Parameter Table

Parameter	Description
Number of regional divisions	$5 * 5$
Maximum walking distance of user	Normal distribution with a mean of 1 km
Walk cost parameters of user	2
Total number of time steps T	78
Time interval	10 min

5.2 Benchmark Algorithms

We evaluate the proposed dispatching strategy against some typical benchmark algorithms, which are introduced in the below. user incentives with unlimited budget, the dispatching strategy with greedy budget allocation, the dispatching strategy under truck hauling, and the situation without dispatching.

Unlimited Budget: This strategy is similar to our strategy. However, the platform has a unlimited budget and can use any money to incentivize users to dispatch bikes. We expect that this algorithm will have the best performance due to having an unlimited budget to incentivize users to complete the dispatching tasks.

Greedy Budget: Tong et al. showed that the greedy based strategy can perform well in the spatio-temporal crowdsourcing problem [16]. Therefore, we evaluate the proposed strategy against this greedy based user incentive dispatching strategy, where the platform allocates the budget into each time step in a greedy way to maximize the matching pairs of the current time step.

Truck Hauling: In this strategy, the platform uses trucks to haul bikes to the desired destinations by paying to them to improve the long-term user service ratio. The process is also a sequential decision process and thus can be modeled as a Markov decision process. We also use reinforcement learning to solve this problem.

Non-dispatch: The platform does not perform any dispatching operations.

5.3 Experimental Discussion

We run the experiments with different budget constraints and different initial supply of bikes. The experimental results of considering different budgets are shown in Fig. 4. Where the initial bike supply in each region is set to 5, and the budgets increased from 500 to 1000. We can find that as the budget is increased, the number of unserved users decreases. This is because given the increased budget, more users can be incentivize to dispatch bikes. As can be seen in Fig. 4,

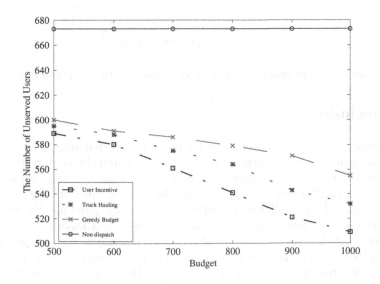

Fig. 4. The number of unserved users under different budgets

when the budget increases, the dispatching strategy under user incentives can reasonably allocate the budget to each time step, resulting in a higher long-term user service ratio, i.e. the lowest number of unserved users.

We now fix the budget as 1000 and increase the bikes in each region from 2 to 10. The experiment results are shown in Fig. 5. We can find that the number of unserved users for all the dispatching strategies decreases with the increased initial supply of shared bikes in the region. This is because when the supply of shared bikes in the region increases, the demand shortage of the platform decreases, and the number of unserved users decreases. We also find that the proposed strategy can outperform other strategies with the lower number of unserved users. Note that unlimited budget based strategy can perform best since it can provide infinite budget to incentivize users.

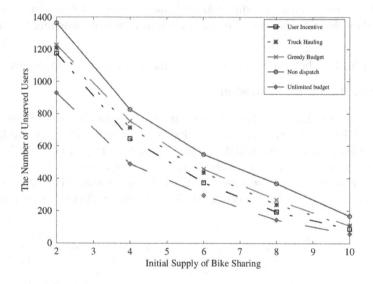

Fig. 5. The number of unserved users under different initial supply of bike sharing

6   Conclusion

In this paper, we propose a user incentive based bike dispatching strategy, where the bike-sharing platform subsidize the users to dispatch the bikes to the desired destination by taking into account the budget constraint and maximum walking distance of users. The user incentive based bike dispatching strategy consist of task generation, budget allocation and task allocation algorithms. We use LSTM to predict the dispatching task, use a deep reinforcement learning algorithm to divide the whole budget into each time step and then use a greedy based method to match users with tasks to maximize the number of served users of the current time step. We then run experiments to evaluate the proposed strategy. We find that our strategy can outperform these typical benchmark algorithms, i.e. can increase the user service ratio. The experimental results show that the performance of the dispatching strategy under user incentives in this paper is second only to that of the unlimit strategy. It shows that the dispatching strategy under user incentives is relevant, can effectively improve the long-term user service rate of the platform.

Acknowledgement. This paper was funded by the Shenzhen Fundamental Research Program (Grant No. JCYJ20190809175613332), the Humanity and Social Science Youth Research Foundation of Ministry of Education (Grant No. 19YJC790111) and the Philosophy and Social Science Post-Foundation of Ministry of Education (Grant No. 18JHQ060).

References

1. Demaio, P.: Bike-sharing: history, impacts, models of provision, and future. J. Public Transp. **12**(4), 41–56 (2009)
2. Ghosh, S., Varakantham, P., Adulyasak, Y., et al.: Dynamic repositioning to reduce lost demand in bike sharing systems. J. Artif. Intell. Res. **58**, 387–430 (2017)
3. Zheng, X., Tang, M., Liu, Y., et al.: Repositioning bikes with carrier vehicles and bike trailers in bike sharing systems. Appl. Sci. **11**(16), 7227 (2021)
4. Li, L., Shan, M.: Bidirectional incentive model for bicycle redistribution of a bicycle sharing system during rush hour. Sustainability **8**(12), 1299 (2016)
5. Svenja, R., Klaus, B.: A relocation strategy for Munich's bike sharing system: combining an operator-based and a user-based scheme. Transp. Res. Procedia **22**, 105–114 (2017)
6. Aeschbach, P., Zhang, X., Georghiou, A., et al.: Balancing bike sharing systems through customer cooperation - a case study on London's Barclays Cycle Hire. In: The 54th IEEE Conference on Decision and Control, pp. 4722–4727 (2015)
7. Ban, S., Hyun, K.H.: Designing a user participation-based bike rebalancing service. Sustainability **11**(8), 2396 (2019)
8. Ling, P., Cai, Q., Fang, Z., et al.: A deep reinforcement learning framework for rebalancing dockless bike sharing systems. AAAI Conf. Artif. Intell. **33**(01), 1393–1400 (2019)
9. Huang, J., Chou, M.C., Teo, C.P.: Bike-repositioning using volunteers: crowd sourcing with choice restriction. AAAI Conf. Artif. Intell. **35**(13), 11844–11852 (2021)
10. Duan, Y., Wu, J.: optimizing rebalance scheme for dock-less bike sharing systems with adaptive user incentive. In: The 20th IEEE International Conference on Mobile Data Management, pp. 176–181 (2019)
11. Caggiani, L., Camporeale, R., Marinelli, M., et al.: User satisfaction based model for resource allocation in bike-sharing systems. Transport Policy, pp. 117–126 (2019)
12. Sutskever, I., Vinyals, O., Le, Q.V.: Sequence to sequence learning with neural networks. Advances in Neural Information Processing Systems, pp. 3104–3112 (2014)
13. Singla, A., Santoni, M., Bartók, G., et al.: Incentivizing users for balancing bike sharing systems. In: The 29th AAAI Conference on Artificial Intelligence, pp. 723–729 (2015)
14. Lillicrap, T.P., Hunt, J.J., Pritzel, A., et al.: Continuous control with deep reinforcement learning. In: The 4th International Conference on Learning Representations (2016)

15. Munkres, J.: Algorithms for the assignment and transportation problems. J. Soc. Ind. Appl. Math. **5**(1), 32–38 (1957)
16. Tong, Y., She, J., Ding, B., et al.: Online minimum matching in real-time spatial data: experiments and analysis. VLDB Endowm. **9**(12), 1053–1064 (2016)

ALBIF: Active Learning with BandIt Feedbacks

Mudit Agarwal[(✉)] and Naresh Manwani

Machine Learning Lab, KCIS, International Institute of Information Technology
Hyderabad, Telangana, India
mudit.agarwal@research.iiit.ac.in, naresh.manwani@iiit.ac.in

Abstract. Online active learning algorithms reduce human labeling costs by querying only a subset of informative incoming instances from the data stream to update the classification model. Active learning for online multiclass classification under complete information has been well addressed; however, it remains unaddressed for the bandit setting. In this paper, we investigate online active learning techniques under the bandit feedback setting. We proposed an efficient algorithm for learning a multiclass classifier with bandit feedbacks under the active learning setting. The proposed algorithms enjoy a regret bound of the order $\mathcal{O}(\log T)$ in the active learning setting as well as in the standard (non-active) bandit feedbacks. We show the effectiveness of the proposed approach using extensive experiments on several benchmark datasets.

Keywords: Online learning · Active learning · Recommender system

1 Introduction

In many machine learning applications, the biggest challenge is to obtain labeled data. Traditional machine learning often neglects the cost of labeling data, which can sometimes be time-consuming, laborious, and expensive. This problem can be worse in the online learning setting. Online learning trains a classifier sequentially. At each round t, the learner receives an instance $\mathbf{x}^t$ from the environment and predicts a label $\hat{y}^t$. After that, it is often assumed that it will immediately obtain a true class label y^t from the environment, which helps calculate the loss and update the classifier if necessary. However, obtaining the feedback (*i.e.*, true class label, 1-bit bandit feedback, etc.) from the environment is often time-consuming and quite expensive in many real-life scenarios [15,20]. This led to the *active learning* setting in which the feedback is queried from the environment only when it is needed. Active learning for classification has been extensively studied in [9,13,20,22] under full information setting.

Some of the pioneering studies in online active learning under for binary classification are discussed in [8–10,13]. In [9], at each round t, the learner receives an instance $\mathbf{x}^t$, predicts a label $\hat{y}^t$, and then draws a Bernoulli random variable $Q^t \in \{0, 1\}$ which depends on the classification margin of instance $\mathbf{x}^t$. If $Q^t = 1$, the learner only asks for the true label from the environment and

J. Gama et al. (Eds.): PAKDD 2022, LNAI 13282, pp. 353–364, 2022.
https://doi.org/10.1007/978-3-031-05981-0_28

follows the standard Perceptron algorithm to update the weight vector. The idea behind this margin-based approach is to query instances near the hyperplane that are more challenging to classify correctly and thus more informative. On the other hand, [13] uses a confidence-based approach for querying the label. If the learner predicts the incoming instance with low confidence, then the probability of querying the true label increases and vice versa. Recently, [15] proposed an active learning algorithm by exploiting the second-order information, which enables each parameter of the model to be updated with a different and adaptive learning rate. In [20], a similar margin-based approach with a passive-aggressive setting is proposed for updating the weights.

All the above approaches deal with multiclass classification under a complete information setting. However, active learning under the bandit feedback setting is still unaddressed. In this paper, we investigate online active learning techniques under the bandit feedback setting. To the best of our knowledge, this is the first attempt in this direction.

Key Contribution of The Paper:

1. We proposed an efficient stochastic sub-gradient descent algorithm for learning a multiclass classifier under the active bandit feedback setting.
2. To the best of our knowledge, we are the first to propose an online active learning algorithm under the bandit feedback setting.
3. The proposed algorithm enjoys a regret bound of $\mathcal{O}(\log T)$.
4. We have validated the efficacy of the proposed algorithm with thorough experiments on various benchmark datasets.

2   Learning Multiclass Classifier with Bandit Feedback

Multiclass classifier is a function $g : \mathcal{X} \rightarrow [K]$ which takes an instance $\mathbf{x} \in \mathcal{X} \subseteq \mathbb{R}^d$ as input and outputs a label in the set $[K] = \{1, \ldots, K\}$. The simplest case is when function g is a linear classifier. In that case, g is modeled using a weight matrix $\mathbf{W} \in \mathbb{R}^{K \times d}$ as $g(\mathbf{x}^t) = \arg\max_{j \in [K]} (\mathbf{W}\mathbf{x}^t)_j$.

In the bandit feedback setting [18], the learner receives the signal indicating whether the predicted label is correct or incorrect. Here, the goal of the learner is to learn a classifier sequentially from the incoming instances. Let $\mathbf{W}^t$ be the classifier at the beginning of round t. At round t, the learner receives an instance $\mathbf{x}^t$ and finds $\hat{y}^t = \arg\max_{j \in [K]} (\mathbf{W}^t \mathbf{x}^t)_j$. Banditron [18] uses an exploitation-exploration scheme proposed in [3], to predict a label $\tilde{y}^t$. Banditron modifies the Perceptron updates to deal with the 1-bit bandit feedback. When the data is linearly separable, the expected number of mistakes made by Banditron is $\mathcal{O}(\sqrt{T})$ and $\mathcal{O}\left(T^{2/3}\right)$ for the general case. Another bandit algorithm, named Newtron [16], is based on the online Newton method. It uses a strongly convex objective function (adding regularization term with the loss function) and Follow-The-Regularized-Leader (FTRL) strategy to achieve $\mathcal{O}(\log T)$ regret bound in the best case and $\mathcal{O}\left(T^{2/3}\right)$ regret bound in the worst case. [26] extended the passive-aggressive online multiclass learning approach proposed in [11] in the bandit feedback setting. Crammer and Gentile [12] extended the Second-order Perceptron in the bandit feedback setting. It uses upper-confidence bounds (UCB) [2] based approach to handle exploration-exploitation and achieves regret bound of $\mathcal{O}\left(\sqrt{T}\log(T)\right)$. Beygelzimer et al. [6] proposed efficient algorithms under bandit feedback when the data is linearly separable by a margin of γ. They [6] achieve a near-optimal bound of $\mathcal{O}\left(K/\gamma\right)$ under strong linear separability.

Above discussed algorithms query the bandit feedbacks after every round. In many real-life scenarios, obtaining feedback is often time-consuming and can also be expensive [15,20]. This motivates us to come up with an active learning algorithm in the bandit feedback setting.

3 Proposed Approach: Active Learning in Bandit Feedback Setting

In this section, we introduce an active learning algorithm for multiclass classification under the bandit feedback, called *Active Learning with BandIt Feedback* (ALBIF). Active learning in bandit feedback setting is more challenging than active learning in full information setting. Besides querying the labels intelligently, the learner must discover a good classifier with only limited information (bandit feedback). ALBIF aims to reduce the number of queries for bandit feedback without adversely affecting the algorithm's performance.

In particular, at round t, ALBIF receives an unlabeled instance $\mathbf{x}^t$, calculates $\hat{y}^t = \arg\max_{j \in [K]} (\mathbf{W}^t \mathbf{x}^t)_j$, and computes its prediction margin by the current classifier, defined as $p^t = [(\mathbf{W}^t\mathbf{x}^t)_{\hat{y}^t} - \max_{r \neq \hat{y}^t}(\mathbf{W}^t\mathbf{x}^t)_r]_+$. Then the algorithm simulates a Bernoulli random variable $Q^t \sim \text{Bernoulli}\left((1 + \eta p^t)^{-1}\right)$ [4,20,24], where $\eta \geq 0$ is the smoothing parameter. If $Q^t = 0$, we don't query the bandit feedback, and we set $\mathbf{W}^{t+1} = \mathbf{W}^t$. If $Q^t = 1$, we define distribution P^t over the class labels as $P^t(r) = (1-\gamma)\mathbb{I}[r = \hat{y}^t] + \frac{\gamma}{K}$, $r \in [K]$. We sample label $\tilde{y}^t$ from P^t and bandit feedback $\mathbb{I}[\tilde{y}^t = y^t]$ is queried. Let $f(\mathbf{W}, (\mathbf{x}^t, y^t))$ be defined as follows.

$$f(\mathbf{W}, (\mathbf{x}^t, y^t)) = \frac{\lambda}{2}\|\mathbf{W}\|^2 + \left[1 + (1 - 2\mathbb{I}[\tilde{y}^t = y^t])(\mathbf{W}\mathbf{x}^t)_{\tilde{y}^t}\right]_+. \qquad (1)$$

We update the parameters in the following way.

$$\mathbf{W}^{t+\frac{1}{2}} = \mathbf{W}^t - n^t\nabla f(\mathbf{W}^t, (\mathbf{x}^t, y^t)); \qquad \mathbf{W}^{t+1} = \min\left\{1, \frac{1}{\sqrt{\lambda}\left\|\mathbf{W}^{t+\frac{1}{2}}\right\|}\right\}\mathbf{W}^{t+\frac{1}{2}}$$

where $\nabla f(\mathbf{W}^t, (\mathbf{x}^t, y^t))$ is gradient of $f(\mathbf{W}, (\mathbf{x}^t, y^t))$ evaluated at $\mathbf{W}^t$ given as below.

$$\nabla f(\mathbf{W}^t, (\mathbf{x}^t, y^t)) = \begin{cases} \lambda\mathbf{W}^t, & 1 - (\mathbf{W}^t\mathbf{x}^t)_{\tilde{y}^t} \leq 0, \tilde{y}^t = y^t \\ \lambda\mathbf{W}^t - \phi(\mathbf{x}^t, \tilde{y}^t), & 1 - (\mathbf{W}^t\mathbf{x}^t)_{\tilde{y}^t} > 0, \tilde{y}^t = y^t \\ \lambda\mathbf{W}^t + \phi(\mathbf{x}^t, \tilde{y}^t), & 1 + (\mathbf{W}^t\mathbf{x}^t)_{\tilde{y}^t} > 0, \tilde{y}^t \neq y^t \\ \lambda\mathbf{W}^t, & \text{otherwise} \end{cases} \qquad (2)$$

Algorithm 1. ALBIF: **A**ctive **L**eanring with **B**and**I**t **F**eedback

---

Input: $\eta, \lambda, \gamma \in [0,1]$

Initialization: Set $\mathbf{W}^1 = \mathbf{0}, \frac{1}{n_0} = \lambda$

  for $t = 1, 2, \cdots, T$ **do**

    Receive $\mathbf{x}^t \in \mathbb{R}^d$.

    Set $\hat{y}^t = \arg\max_{r \in [K]} (\mathbf{W}^t \mathbf{x}^t)_r$

    $p^t = [(\mathbf{W}^t \mathbf{x}^t)_{\hat{y}^t} - \max\limits_{r \in [K] \setminus \{\hat{y}^t\}} (\mathbf{W}^t \mathbf{x}^t)_r]_+$

    Draw a Bernoulli $Q^t \in \{0,1\}$ with parameter $\frac{1}{1+p^t\eta}$

    Set $\frac{1}{n^t} = \frac{1}{n^{t-1}} + \lambda Q^t$

    if $Q^t = 1$ **then**

      Set $P^t(r) = (1 - \gamma)\mathbb{I}[r = \hat{y}^t] + \frac{\gamma}{K}$, $r \in [K]$. Sample $\tilde{y}^t$ according to P^t.

      Query Bandit Feedback $\mathbb{I}[\tilde{y}^t = y^t]$

      Update weight matrix W as follows.

$$\mathbf{W}^{t+\frac{1}{2}} = \mathbf{W}^t - n^t \nabla f(\mathbf{W}^t, (\mathbf{x}^t, y^t)); \quad \mathbf{W}^{t+1} = \min\left\{1, \frac{1/\sqrt{\lambda}}{\left\|\mathbf{W}^{t+\frac{1}{2}}\right\|}\right\} \mathbf{W}^{t+\frac{1}{2}}$$

    else

      $\mathbf{W}^{t+1} = \mathbf{W}^t$

    end if

  end for

Here, matrix $\phi(\mathbf{x}^t, i) \in \mathbb{R}^{K \times d}$ is such that all its rows except i-th row are zero and i-th row is $\mathbf{x}^t$. While updating the weight matrix, we have a projection step that ensures that the set of admissible solutions are restricted to the ball of radius $\frac{1}{\sqrt{\lambda}}$. Complete details of the approach are given in the Algorithm 1.

4 Analysis

We first prove the following lemma, which upper bounds the gradient of the objective function at any round t of the proposed algorithm.

Lemma 1. *Let* $\|\mathbf{x}^t\| \leq R, \forall t \in [T]$ *and* $\nabla_\mathbf{W} f(\mathbf{W}^t, (\mathbf{x}^t, y^t))$ *denotes the gradient of* $f(\mathbf{W}, (\mathbf{x}^t, y^t))$ *at* $\mathbf{W}^t$. *Then,* $\|\nabla f(\mathbf{W}^t, (\mathbf{x}^t, y^t))\| \leq \sqrt{\lambda} + R$ *if we perform a projection step and* $\|\nabla f(\mathbf{W}^t, (\mathbf{x}^t, y^t))\| \leq 2R$ *otherwise.*

Proof. If we perform the projection step, then we know that the $\|\mathbf{W}^t\| \leq \frac{1}{\sqrt{\lambda}}$. Combining it with the fact that $\|\mathbf{x}^t\| \leq R$ and using the triangle inequality, we get $\|\nabla_\mathbf{W} f(\mathbf{W}^t, (\mathbf{x}^t, y^t))\| \leq \sqrt{\lambda} + R$. If we do not perform the projection step, then the update equation become $\mathbf{W}^{t+1} = (1 - \frac{1}{t})\mathbf{W}^t - \frac{1}{\lambda t}\mathbf{U}^t$, where $\mathbf{U}^t = [1 + (1 - 2\mathbb{I}[\tilde{y}^t = y^t])(\mathbf{W}^t \mathbf{x}^t)_{\tilde{y}^t} \geq 0]\mathbf{x}^t$. Therefore, the initial weight of each $\mathbf{U}^i$ is $\frac{1}{\lambda i}$ and then on rounds $j = i + 1, \cdots, t$ it will be multiplied by $1 - \frac{1}{j} = \frac{j-1}{j}$. Thus, the overall weight of $\mathbf{U}^i$ in $\mathbf{W}^{t+1}$ is $\frac{1}{\lambda i} \prod_{j=i+1}^t \frac{j-1}{j} = \frac{1}{\lambda t}$. Using the above fact, we can rewrite $\mathbf{W}^{t+1}$ as $\mathbf{W}^{t+1} = \frac{1}{\lambda t} \sum_{i=1}^t \mathbf{U}^i$. Thus, we can see that $\left\|\mathbf{W}^{t+1}\right\| \leq \frac{R}{\lambda}$ which implies that $\|\nabla_\mathbf{W} f(\mathbf{W}^t, (\mathbf{x}^t, y^t))\| \leq 2R$. $\quad\square$

Lemma 2. *Let n^t be the step size used is ALBIF at round t. Then,*

$$\sum_{t=1}^{T} \mathbb{E}[n^t] \le \frac{1 + \eta c}{\lambda}(1 + \ln T).$$

Proof. Expanding the definition of $\frac{1}{n^t}$ and taking the expectation on both sides.

$$\mathbb{E}\left[\frac{1}{n^t}\right] = \mathbb{E}\left[\frac{1}{n_0}\right] + \sum_{i=1}^{t} \lambda \mathbb{E}[Q_i]$$

$$\Rightarrow \mathbb{E}[n^t] = \frac{1}{\lambda(1 + \sum_{i=1}^{t} \mathbb{E}[Q_i])} \le \frac{1}{\sum_{i=1}^{t} \lambda \mathbb{E}[Q_i]} = \frac{1}{\sum_{i=1}^{t} \lambda \frac{1}{1+\eta p^i}}$$

We observe that $p^t \le c$ where $c = \frac{2R}{\sqrt{\lambda}}$ when we perform projection else $c = \frac{2R^2}{\lambda}$. Therefore, $\frac{1}{1+\eta p^i} \ge \frac{1}{1+\eta c}$. Plugging this in the above equation, we get $\mathbb{E}[n^t] \le \frac{1}{\sum_{i=1}^{t} \lambda \frac{1}{1+\eta c}} = \frac{1+\eta c}{\lambda \sum_{i=1}^{t} 1} = \frac{1+\eta c}{\lambda t}$. Summing up over all the values of t, we get the required upper bound as follows.

$$\sum_{t=1}^{T} \mathbb{E}[n^t] \le \sum_{t=1}^{T} \frac{1 + \eta c}{\lambda t} = \frac{1 + \eta c}{\lambda} \sum_{t=1}^{T} \frac{1}{t} \le \frac{1 + \eta c}{\lambda}(1 + \log T)$$

□

Equipped with the above two lemmas, we are now in the position to upper bound the regret of ALBIF. We use the following definition of the regret.

$$\mathcal{R}(T) = \mathbb{E}\left[\sum_{t=1}^{T}\{f(\mathbf{W}^t, (\mathbf{x}^t, y^t)) - \min_{\mathbf{W} \in \mathbb{B}} \sum_{t=1}^{T} f(\mathbf{W}, (\mathbf{x}^t, y^t))\}\right] \quad (3)$$

Theorem 1. *Let f be as defined in Eq. 1 and $\mathbb{B}$ be the Euclidean ball of radius $1/\sqrt{\lambda}$ if we perform a projection step and $\mathbb{B} = \mathbb{R}^n$ otherwise. Let $\mathbf{W}^1, \cdots, \mathbf{W}^{T+1}$ be the sequence of weight matrices generated by ALBIF. Let $\|\nabla f(\mathbf{W}^t, (\mathbf{x}^t, y^t))\| \le G$ where G is defined in Lemma 1, $T \ge 3$, regret bound achieved by ALBIF is*

$$\mathcal{R}(T) \le \frac{\lambda(1 + \eta c)}{2}\|\mathbf{W}^*\|^2 + \frac{G^2(1 + \eta c)^2}{2\lambda}(1 + \log T),$$

where $\mathbf{W}^* = \arg\min_{\mathbf{W} \in \mathbb{B}} \sum_{t=1}^{T} f(\mathbf{W}, (\mathbf{x}^t, y^t))$.

Proof. For any $\mathbf{W} \in \mathbb{B}$, we will bound the $\mathbb{E}[\langle \mathbf{W}^t - \mathbf{W}, \nabla f(\mathbf{W}^t, (\mathbf{x}^t, y^t))\rangle]$ from above and below, to prove the above theorem. Since f is a λ-strongly convex function, then from [7], we know that

$$\langle \mathbf{W}^t - \mathbf{W}, \nabla f(\mathbf{W}^t, (\mathbf{x}^t, y^t))\rangle \ge f(\mathbf{W}^t, (\mathbf{x}^t, y^t)) - f(\mathbf{W}, (\mathbf{x}^t, y^t)) + \frac{\lambda}{2}\|\mathbf{W}^t - \mathbf{W}\|^2$$

Taking the expectation on both sides gives us the following lower bound.

$$\mathbb{E}[\langle \mathbf{W}^t - \mathbf{W}, \nabla f(\mathbf{W}^t, (\mathbf{x}^t, y^t))\rangle] \geq \mathbb{E}[f(\mathbf{W}^t, (\mathbf{x}^t, y^t)) - f(\mathbf{W}, (\mathbf{x}^t, y^t))]$$
$$+ \frac{\lambda}{2}\mathbb{E}[\|\mathbf{W}^t - \mathbf{W}\|^2] \quad (4)$$

Next we derive an upper bound on $\mathbb{E}[Q^t]\mathbb{E}[\langle \mathbf{W}^t - \mathbf{W}, \nabla f(\mathbf{W}^t, (\mathbf{x}^t, y^t))\rangle]$. Since $\mathbf{W}^{t+1}$ is the projection of $\mathbf{W}^{t+\frac{1}{2}}$ onto $\mathbb{B}$ and $\mathbf{W} \in \mathbb{B}$, then we have $\left\|\mathbf{W}^{t+\frac{1}{2}} - \mathbf{W}\right\|^2 \geq \|\mathbf{W}^{t+1} - \mathbf{W}\|^2$. Therefore,

$$\|\mathbf{W}^t - \mathbf{W}\|^2 - \|\mathbf{W}^{t+1} - \mathbf{W}\|^2 \geq \|\mathbf{W}^t - \mathbf{W}\|^2 - \left\|\mathbf{W}^{t+\frac{1}{2}} - \mathbf{W}\right\|^2$$
$$= 2n^t\langle \mathbf{W}^t - \mathbf{W}, Q^t\nabla f(\mathbf{W}^t, (\mathbf{x}^t, y^t))\rangle - (n^t)^2\|\nabla f(\mathbf{W}^t, (\mathbf{x}^t, y^t))\|^2$$

Rearranging the above equation and taking the expectation on both sides, we get,

$$\mathbb{E}[Q^t]\mathbb{E}[\langle \mathbf{W}^t - \mathbf{W}, \nabla f(\mathbf{W}^t, (\mathbf{x}^t, y^t))\rangle]$$
$$\leq \mathbb{E}\left[\frac{\|\mathbf{W}^t - \mathbf{W}\|^2 - \|\mathbf{W}^{t+1} - \mathbf{W}\|^2}{2n^t}\right] + \mathbb{E}\left[\frac{n^t}{2}\|\nabla f(\mathbf{W}^t, (\mathbf{x}^t, y^t))\|^2\right]$$

Using the fact that $\|\nabla f(\mathbf{W}^t, (\mathbf{x}^t, y^t))\| \leq G$ and Eq. (4), we get

$$2\,\mathbb{E}[Q]\,\mathbb{E}[f(\mathbf{W}^t, (\mathbf{x}^t, y^t)) - f(\mathbf{W}, (\mathbf{x}^t, y^t))]]$$
$$\leq \mathbb{E}\left[\frac{\|\mathbf{W}^t - \mathbf{W}\|^2 - \|\mathbf{W}^{t+1} - \mathbf{W}\|^2}{n^t}\right] - \mathbb{E}\left[Q^t\lambda\|\mathbf{W}^t - \mathbf{W}\|^2\right] + G^2\mathbb{E}[n^t]$$
$$= \mathbb{E}\left[\frac{1}{n^t} - \lambda Q^t\right]\mathbb{E}[\|\mathbf{W}^t - \mathbf{W}\|^2] - \mathbb{E}\left[\frac{1}{n^t}\right]\mathbb{E}[\|\mathbf{W}^{t+1} - \mathbf{W}\|^2] + G^2\,\mathbb{E}[n^t]$$
$$\leq \mathbb{E}\left[\frac{1}{n^{t-1}}\right]\mathbb{E}[\|\mathbf{W}^t - \mathbf{W}\|^2] - \mathbb{E}\left[\frac{1}{n^t}\right]\mathbb{E}[\|\mathbf{W}^{t+1} - \mathbf{W}\|^2] + G^2\,\mathbb{E}[n^t]$$

Summing from $t = 1$ to T and using Lemma 2, we get

$$\mathbb{E}\left[\sum_{t=1}^{T}\{f(\mathbf{W}^t, (\mathbf{x}^t, y^t)) - f(\mathbf{W}, (\mathbf{x}^t, y^t))\}\right] \leq \frac{1}{2\mathbb{E}[Q^t]}\mathbb{E}\left[\frac{1}{n_0}\right]\|\mathbf{W}\|^2$$
$$- \mathbb{E}\left[\frac{1}{n^t}\right]\frac{\|\mathbf{W}^{T+1} - \mathbf{W}\|^2}{2\mathbb{E}[Q^t]} + \frac{G^2(1 + \eta c)}{2\lambda\mathbb{E}[Q^t]}(1 + \log T)$$

We know that $\mathbb{E}[Q^t] = P(Q^t = 1) \geq \frac{1}{1+\eta c}$, where $c = \frac{2R}{\sqrt{\lambda}}$ when we perform projection and $c = \frac{2R^2}{\lambda}$ if we do not perform projection. Using this, we get

$$\mathbb{E}\left[\sum_{t=1}^{T}\{f(\mathbf{W}^t, (\mathbf{x}^t, y^t)) - f(\mathbf{W}, (\mathbf{x}^t, y^t))\}\right] \leq \frac{\lambda(1 + \eta c)}{2}\|\mathbf{W}\|^2$$
$$+ \frac{G^2(1 + \eta c)^2}{2\lambda}(1 + \log T)$$

Let $W^* = \arg\min_{\mathbf{W} \in \mathbb{B}} \sum_{t=1}^{T} f(\mathbf{W}, (\mathbf{x}^t, y^t))$. Thus,

$$\mathcal{R}(T) \leq \frac{\lambda(1 + \eta c)}{2} \|\mathbf{W}^*\|^2 + \frac{G^2(1 + \eta c)^2}{2\lambda}(1 + \log T).$$

$\square$

Thus, the proposed approach achieves $\mathcal{O}(\log T)$ regret bound in the active learning setting with bandit feedbacks. Analyzing the above bounds, we see that as η increases, so does the regret bound. This happens because increasing η decreases the query's probability, resulting in fewer instances for which we get the bandit feedback. As the number of bandit feedback decreases, the error rate is bound to increase. We can realize standard bandit feedback (non-active) by using $\eta = 0$, i.e., $Q^t = 1 \ \forall t$. Regret bound of ALBIF in that case is $\mathcal{R}(T) \leq \frac{\lambda}{2}\|\mathbf{W}^*\|^2 + \frac{G^2(1+\log T)}{2\lambda}$.

5   Experiments

We empirically evaluate the proposed algorithm against the present state-of-the-art algorithms on synthetic and real-world data sets. We use CIFAR-10 [19], Fashion-MNIST [25], USPS [17], Abalone and Ecoli datasets from UCI repository [14]. To extract the feature of Fashion-MNIST, we use a four-layer convolutional neural network as described in [1]. Moreover, for CIFAR-10, we used a pre-trained VGG-16 [23] model to extract features. We also performed experimentation on synthetic datasets, which we called SynSep and SynNonSep. SynSep is a 9-class, 400-dimensional synthetic data set of size 10^5. While constructing SynSep, we ensure that the dataset is linearly separable. For more details about the dataset, one can refer to [18]. The idea behind SynSep is to generate a simple dataset simulating a text document. The coordinates represent different words in a small vocabulary of size 400. SynNonSep is constructed the same way as SynSep except that a 5% label noise is introduced, making the dataset non-separable.

Benchmark Algorithms: We compare the proposed algorithm with other bandit algorithms, which include Banditron [18], Bandit Passive Aggressive (BPA) [26] and, Second-Order Banditron Algorithm (SOBA) [5]. We also run ALBIF[1]. with $\eta = 0$, i.e., $Q^t = 1 \ \forall t$ to see how the proposed algorithm perform under (non-active) standard bandit setting. We use grid and line search to tune the best values of the hyper-parameters for ALBIF and other benchmarking algorithms. Table 1 shows the parameters used by ALBIF for different datasets.

Results: We compare the average error rate[2] of the proposed algorithm ALBIF and ALBIF($\eta = 0$) with other benchmark algorithms as shown in Fig. 1 and 2.

---

[1] The complete code for all the experiments can be found https://github.com/Mudit-1999/ALBIF-Active-Learning-With-BandIt-Feedbacks.

[2] Note that here averaging is done over 20 independent simulations of the algorithm.

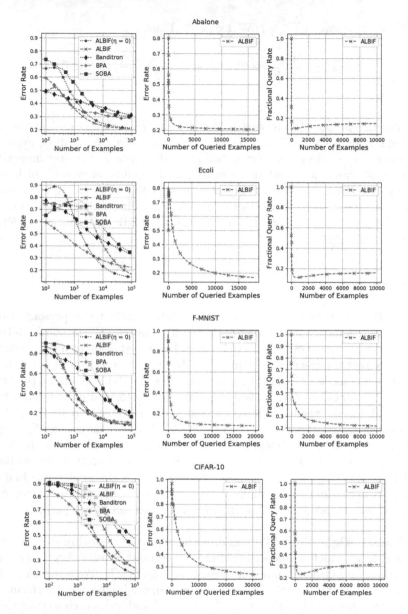

Fig. 1. Average error rates of ALBIF, ALBIF($\eta = 0$) and other benchmarking algorithms for (top to bottom) Abalone, Ecoli, F-MNIST and CIFAR-10 dataset.

The second column shows how the error rate of ALBIF decreases with the number of bandit feedback asked, while the third column shows the fraction of bandit feedbacks asked till round t. It shows that the query rate becomes constant after some time, *i.e.*, ALBIF needs to ask bandit feedback for only a small fraction of incoming instances over T rounds. Figure 1 and 2 shows that the proposed

Table 1. Parameter values used by the proposed algorithm ALBIF

Parameter	Dataset						
	USPS	F-MNIST	Syn-Sep	Syn-NonSep	Abalone	Ecoli	CIFAR-10
γ	0.007812	0.003906	0.003906	0.003906	0.003906	0.031250	0.007812
η	16.0	16.0	128.0	32.0	64.0	8.0	8.0
λ	0.125	0.25	32.0	8.0	0.031250	0.000977	2.0

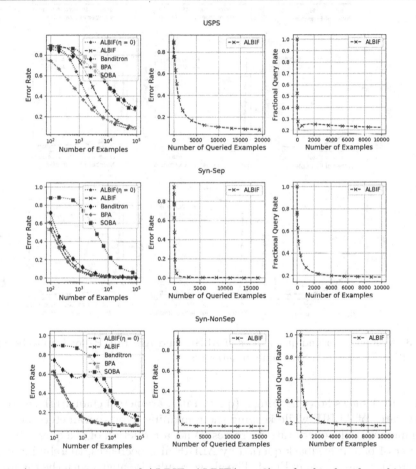

Fig. 2. Average error rates of ALBIF, ALBIF($\eta = 0$) and other benchmarking algorithms for (top to bottom) USPS, Syn-Sep and Syn-NonSep dataset.

algorithm ALBIF can learn a good classifier, beating the state-of-the-art algorithms. The query rate of ALBIF is less than 20% for Abalone, Ecoli, SynSep, and SynNonSep, while for USPS, it is approximately 20%. For complex datasets like Fashion-MNIST and CIFAR-10, ALBIF queries bandit feedback for only 20% and 32% of the incoming instances, respectively.

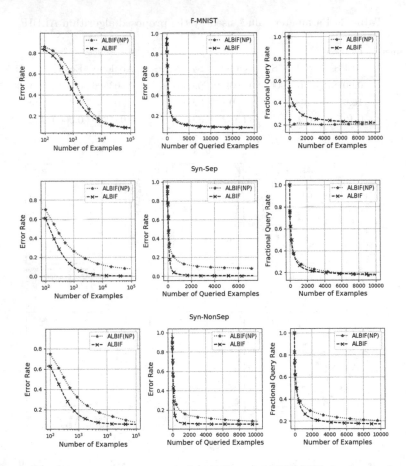

Fig. 3. Average error and query rates of ALBIF and ALBIF(NP) for (top to bottom): F-MNIST, Syn-Sep and Syn-NonSep dataset

Figure 4 shows how changing η (query rate parameter) changes the error and query rates of ALBIF while keeping the remaining hyper-parameters the same. As η increases, the query rate decreases. With a reduction of the query rate, the number of bandit feedback queried by ALBIF also decreases, thereby increasing the error rate, as shown in Fig. 4. The above result also aligns with the theoretical bound obtained in Theorem 1. We also empirically evaluated the performance of the proposed algorithm without the optional projection step, denoted by the acronym ALBIF(NP) in Fig. 3. We compare the performance of ALBIF(NP) with ALBIF on F-MNIST, SynSep, and SynNonSep datasets. The first column of Fig. 3 compares the average error rate of ALBIF(NP) with ALBIF. We see that after 10^5 iteration, the error rate of both algorithms converges approximately to the same value while querying approximately the same number of instances. From this, we can conclude that both algorithms are performing comparably.

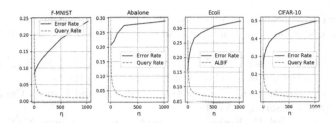

Fig. 4. Average error and query rates of ALBIF against parameter's value η for (left to right): F-MNIST, Abalone, Ecoli and CIFAR-10 dataset

6 Conclusion and Future Work

We proposed an efficient active learning algorithm (ALBIF) for multiclass classification in the bandit feedback setting. The proposed algorithm theoretically achieves regret bound of the order of $\mathcal{O}(\log T)$. Under the standard (non-active) bandit feedback setting, the proposed algorithm has a regret bound of the order of $\mathcal{O}(\log T)$, which is better than the other state-of-the-art algorithms. We also show the effectiveness of the proposed approach using extensive experiments on several benchmark datasets. Experimental results show that by querying the bandit feedback for only a tiny fraction of incoming instances, ALBIF outperforms state-of-the-art bandit algorithms. Thus, saving a lot of time and resources spent on obtaining the bandit feedback. Addressing other challenging online learning tasks, such as the issue of concept drift [21], would be an interesting direction for the future work.

References

1. Agarwal, M., Manwani, N.: Learning multiclass classifier under noisy bandit feedback. In: PAKDD, pp. 448–460. Springer (2021). https://doi.org/10.1007/978-3-030-75765-6_36
2. Auer, P., Cesa-Bianchi, N., Fischer, P.: Finite-time analysis of the multiarmed bandit problem. Mach. Learn. **47**(2), 235–256 (2002)
3. Auer, P., Cesa-Bianchi, N., Freund, Y., Schapire, R.E.: The nonstochastic multi-armed bandit problem. SIAM J. Comput. **32**(1), 48–77 (2002)
4. Balcan, M.-F., Broder, A., Zhang, T.: Margin based active learning. In: Bshouty, N.H., Gentile, C. (eds.) COLT 2007. LNCS (LNAI), vol. 4539, pp. 35–50. Springer, Heidelberg (2007). https://doi.org/10.1007/978-3-540-72927-3_5
5. Beygelzimer, A., Orabona, F., Zhang, C.: Efficient online bandit multiclass learning with $\tilde{O}(\sqrt{T})$ regret. In: ICML 2017, pp. 488–497. PMLR (2017)
6. Beygelzimer, A., Pál, D., Szorenyi, B., Thiruvenkatachari, D., Wei, C.-Y., Zhang, C.: Bandit multiclass linear classification: efficient algorithms for the separable case. In: ICML 2019, pp. 624–633. PMLR (2019)
7. Boyd, S., Boyd, S.P., Vandenberghe, L.: Convex Optimization. Cambridge University Press, Cambridge (2004)

8. Cavallanti, G., Cesa-Bianchi, N., Gentile, C.: Linear classification and selective sampling under low noise conditions. Adv. Neural. Inf. Process. Syst. **21**, 249–256 (2008)

9. Cesa-Bianchi, N., Gentile, C., Zaniboni, L., Warmuth, M.: Worst-case analysis of selective sampling for linear classification. JMLR **7**(7) (2006)

10. Cesa-Bianchi, N., Lugosi, G.: Prediction, Learning, and Games. Cambridge University Press, Cambridge (2006)

11. Crammer, K., Dekel, O., Keshet, J., Shalev-Shwartz, S., Singer, Y.: Online passive aggressive algorithms (2006)

12. Crammer, K., Gentile, C.: Multiclass classification with bandit feedback using adaptive regularization. Mach. Learn. **90**(3), 347–383 (2012). https://doi.org/10.1007/s10994-012-5321-8

13. Dasgupta, S., Kalai, A.T., Tauman, A.: Analysis of perceptron-based active learning. J. Mach. Learn. Res. **10**(2) (2009)

14. Dua, D., Graff, C.: UCI machine learning repository (2017)

15. Hao, S., Zhao, P., Lu, J., Hoi, S.CH., Miao, C., Zhang, C.: Soal: second-order online active learning. In: ICDM 2016, pp. 931–936 (2016)

16. Hazan, E., Kale, S.: Newtron: an efficient bandit algorithm for online multiclass prediction. In: NIPS, vol. 11, pp. 891–899. Citeseer (2011)

17. Hull, J.J.: A database for handwritten text recognition research. IEEE Trans. Pattern Anal. Mach. Intell. **16**(5), 550–554 (1994)

18. Kakade, S.M., Shalev-Shwartz, S., Tewari, A.: Efficient bandit algorithms for online multiclass prediction. In: ICML 2008, pp. 440–447 (2008)

19. Krizhevsky, A., Hinton, G., et al.: Learning multiple layers of features from tiny images (2009)

20. Lu, J., Zhao, P., Hoi, S.C.: Online passive-aggressive active learning. Mach. Learn. **103**(2), 141–183 (2016)

21. Minku, L.L., Yao, X.: Ddd: a new ensemble approach for dealing with concept drift. IEEE TKDE **24**(4), 619–633 (2011)

22. Sculley, D.: Online active learning methods for fast label-efficient spam filtering. In: CEAS, vol. 7, p. 143 (2007)

23. Simonyan, K., Zisserman, A.: Very deep convolutional networks for large-scale image recognition. arXiv:1409.1556 (2014)

24. Tong, S., Koller, D.: Support vector machine active learning with applications to text classification. JMLR **2**(Nov), 45–66 (2001)

25. Xiao, H., Rasul, K., Vollgraf, R.: Fashion-mnist: a novel image dataset for benchmarking machine learning algorithms. arXiv:1708.07747 (2017)

26. Zhong, H., Daucé, E.: Passive-aggressive bounds in bandit feedback classification. In: Proceedings of the ECMLPKDD, pp. 255–264 (2015)

A Novel Protein Interface Prediction Framework via Hybrid Attention Mechanism

Haifang Wu[1,2,3], Shujie Luo[1,2,3], Weizhong Zhao[1,2,3,4,5(✉)],
Xingpeng Jiang[1,2,3(✉)], and Tingting He[1,2,3]

[1] Hubei Provincial Key Laboratory of Artificial Intelligence and Smart Learning,
Central China Normal University, Wuhan 430079, Hubei, People's Republic of China
{wzzhao,xpjiang}@ccnu.edu.cn
[2] School of Computer, Central China Normal University,
Wuhan 430079, Hubei, People's Republic of China
[3] National Language Resources Monitoring and Research Center for Network Media,
Central China Normal University, Wuhan 430079, Hubei, People's Republic of China
[4] Guangxi Key Laboratory of Trusted Software, Guilin University of Electronic
Technology, Guilin 541004, People's Republic of China
[5] Guangxi Key Lab of Multi-source Information Mining and Security,
Guangxi Normal University, Guilin 541004, People's Republic of China

Abstract. Protein interface prediction is fundamental to understand the hidden principles of many living activities. Although many
approaches to the task of protein interface prediction have been proposed,
most of existing methods fail to make full use of the available sequence
information and structure information. To address the challenge, we propose a deep learning-based end-to-end framework for protein interface
prediction, in which a hybrid attention mechanism is utilized to take
into account the semantic associations and complementary effect between
both sequence and structure information. More specifically, a cross-modal
attention is built to capture the semantic associations between sequence
representations and structure representations for proteins. In addition,
a type-level attention is introduced to model the different contributions
of sequence and structure information for predicting protein interaction
interface. Experimental results on three commonly used datasets demonstrate the effectiveness of the proposed method.

Keywords: Protein interface prediction · Sequence information ·
Structure information · Hybrid attention mechanism

1 Introduction

Proteins perform a variety of functions in cells, covering almost all aspects of
life activities, such as signal transmission and material transportation [21]. The
realization of most protein functions depends on protein interactions, and protein
interactions occur through an interface, which is represented as combinations of

© The Author(s), under exclusive license to Springer Nature Switzerland AG 2022
J. Gama et al. (Eds.): PAKDD 2022, LNAI 13282, pp. 365–378, 2022.
https://doi.org/10.1007/978-3-031-05981-0_29

amino acids from each involved protein. Therefore, protein interface prediction is of great significance for understanding protein functions [3], disease mechanism [14], drug design [7], etc.

Protein-protein interactions generate protein complexes, and the interface of a protein complex is composed of a series of residue pairs. Thereafter, the task of protein interface prediction is to predict the set of residue pairs which form the interface of protein interactions in the given protein complex. For the task of protein interface prediction, the traditional experimental method is a time-consuming and expensive process, which involves crystallization of protein complex, X-ray crystallography or nuclear magnetic resonance imaging [24]. With the continuous growth of protein sequence data and structure data, many computational methods for predicting protein interface have been proposed based on different strategies [6,9]. Based on existing results on the already known protein complexes from the Protein Data Bank (PDB) [4] (which is the most popular protein database globally), it has shown that there exist differences in physical or chemical properties between protein interface regions and non-interface surface regions [25]. More specifically, compared with non-interface surface regions, interface regions usually have more hydrophobic residues and fewer hydrophilic residues, and more evolutionary conservation [25]. Therefore, it is feasible to build effective prediction systems based on computational methods by using meaningful features of proteins in given protein complexes [5].

Generally, existing computational approaches to protein interface prediction can be classified into two groups: *partner-independent* prediction and *partner-specific* prediction [6]. The partner-independent prediction methods predict typically whether an amino acid in the given protein will be involved in an interaction interface with any other proteins, while the latter approaches predict whether a pair of residues (each from a different protein) participate in an interaction interface. Due to using the interaction information in protein complexes, partner-specific prediction approaches have derived relatively better prediction performance [19], which is also the focus of interest in this study.

For traditional machine learning based approaches to partner-specific prediction, the features employed for protein representations are typically obtained via a procedure called "feature engineering", on which machine learning methods are trained to predict the interface for new protein complexes. PAIRpred [1] is the most representative traditional machine learning based system for protein interface prediction. Generally speaking, PAIRpred utilizes protein sequence information (such as evolution-related information) and structure information (such as solvent accessibility and residue depth) to construct features of proteins. And then a SVM classifier is trained on the obtained features for protein interface prediction. Although traditional machine learning based methods are simple and with clear interpretation, the feature engineering needs often expert knowledge in biological or biochemical fields. In addition, employing directly the derived features usually fails to make full use of the sequence and structure information, leading to undesirable performance for predicting protein interface.

Deep learning (*a.k.a.* "representation learning"), which is able to learn automatically meaningful features for applications, has shown excellent performance in various fields such as image [11], speech [15], and natural language processing [23]. Inspired by the success, recent studies try to utilize deep learning technology to automatically learn useful features of amino acids, which are then used for predicting the protein interface. Raphael et al. [20] proposed an end-to-end interface prediction method called SASNet, in which based on information from PDB, each amino acid of a protein is represented as a 4D grid. Then a deep neural network framework is developed to learn the hidden representations of proteins, and to predict the protein interaction interface accordingly. The main drawback to SASNet lies in that the topological structure information, which is important to determine the inherent properties of amino acids and proteins, is ignored by the 4D representation of proteins. In order to utilize the topological information of protein structures, Fout et al. [8] proposed a graph convolutional neural network (GCN) based framework for the task of protein interface prediction. The GCN based framework first represents each protein as a graph, in which nodes are amino acids and edges are the relationships between amino acids, and then GCN is applied to learning features on which the prediction model is trained accordingly.

For the task of protein interface prediction, although existing approaches (both traditional machine learning based methods and deep learning based methods) utilize mainly sequence and/or structure information of proteins, we argue that sequence and structure information of proteins are not fully employed in existing approaches. More specifically, most of existing approaches usually concatenate directly the derived sequence features and structural features, which ignore possible semantic associations between these two types of information. In addition, both sequence and structure information are generally treated equally to represent proteins, which fails to distinguish the different contributions of sequence and structural features for predicting protein interface in complexes.

According to the above observations, this paper proposes a novel deep learning based end-to-end framework via hybrid attention mechanism for the task of protein interface prediction. More specifically, a cross-modal attention mechanism is introduced to learn representations of proteins based on sequence and structure information, in which the semantic associations between both types of information are effectively utilized. Moreover, a type-level attention is employed to model the different roles played by sequence and structural features for predicting protein interface. With the hybrid attention mechanism, the available sequence and structure information for proteins can be fully utilized to learn meaningful representations for the task of protein interface prediction. Extensive experiments are conducted on PDB datasets, which demonstrate the effectiveness of the proposed method in this study.

Note that we use residue and amino acid interchangeably in the whole paper.

2 Problem Formulation

In this compact section, we first formally define the task of protein interface prediction to make it more convenient to present the details of our framework in Sect. 3. Then, we give the overview of our proposed solution.

Protein Interface Prediction. A protein complex is composed of two proteins, which are known as ligand protein and receptor protein, and denoted by X and Y, respectively. Without loss of generality, assume that X consists of L_l amino acids and Y consists of L_r amino acids. Thereafter, there are total $L_l \times L_r$ amino acid pairs which are candidate interaction interfaces.

Formally, given a protein complex (X, Y), the task of protein interface prediction is to derive the predicted interaction interfaces (denoted by $O \in \{0,1\}^{L_l \times L_r}$) involving in the complex. For each $i \in \{1, 2, \ldots, L_l\}$ and $j \in \{1, 2, \ldots, L_r\}$, the predicted $O_{i,j} = 1$ indicates that the pair of amino acids (x_i, y_j) forms an interaction interface, while $O_{i,j} = 0$ denotes (x_i, y_j) does not form an interaction interface.

Solution Overview. In real applications, it is readily to obtain structure information and sequence information for proteins. Therefore, in this study, each protein is represented by both structure information and sequence information. As the discussions in previous sections, both information is complementary to represent the whole aspects of proteins, and a specific mechanism is required to make full use of the available information for the task of protein interface prediction. Accordingly, a cross-modal attention mechanism is introduced to model the semantic associations between sequence and structure information. Moreover, a type-level attention is utilized to distinguish the contributions of sequence and structure information to predicting protein interface. With the hybrid attention mechanism, our framework is able to derive significant performance. The next section will detail the proposed framework more specifically.

3 Methods

The proposed framework in this study is shown in Fig. 1. Generally, it consists of three modules, including preprocessing module, hybrid attention module, and prediction module. Given each protein complex (X, Y), the prediction module will derive the predicted protein interface O. It is worth noting that preprocessing module and hybrid attention module are shared for ligand protein X and receptor protein Y. Thereafter, without loss of generality, the ligand protein X will be used to present the processing details in the two shared modules.

Given the protein X, the initial structure information (denoted by $X^{<1>}$) is calculated from its structure file (which is typically a protein pdb file), while the initial sequence information (denoted by $X^{<2>}$) is calculated from its sequence file (which is typically a protein fasta file). In the preprocessing module, $X^{<1>}$ and $X^{<2>}$, each of which is represented as a matrix, will input into a convolution neural network (CNN) respectively, and the local hidden features will be captured. The representations derived by the two CNNs are denoted by $X'^{<1>}$ and

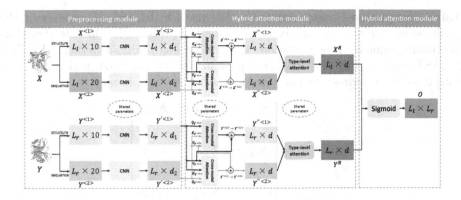

Fig. 1. The framework of our proposed method.

$X'^{<2>}$, respectively. In the hybrid attention module, the cross-modal attention mechanism is first utilized to transform $X'^{<1>}$ and $X'^{<2>}$ into new representations $X''^{<1>}$ and $X''^{<2>}$ respectively, which takes into account the semantic associations between structure information and sequence information. Then a type-level attention mechanism is utilized to integrate $X''^{<1>}$ and $X''^{<2>}$ into the final representation of the ligand protein denoted by X^R, which considers different weight of structure information and sequence information for predicting the interaction interface. With the final representations of ligand and receptor X^R and Y^R, in the prediction module, a sigmoid function is used to derive the predicted interaction interface.

The details of implementation in each module are presented as follows.

3.1   Preprocessing Module

Given a protein X, the initial features are derived from sequence and structure information, and then the initial features are transformed into new representations by CNNs.

Initial Structure-Based Features. The initial structure features are generated by using STRIDE [10], MSMS [18], and PSAIA [17]. More specifically, hydrogen bond energy and amino acid backbone torsional angle are employed by STRIDE to derive the relative accessible surface area (rASA) of amino acids, which reflects the fraction of a residue that is exposed to a potential solvent. MSMS is utilized to generate residue depth (RD), which is described as the minimum distance from the residue to the protein surface. PSAIA is used to calculate the protrusion index (CX) and hydrophobicity of amino acids. Note that CX indicates the degree to which a sphere (radius 10Å) centered at a non-hydrogen atom is not filled with other atoms, and hydrophobicity indicates the tendency for an amino acid to avoid water. Finally, each amino acid is denoted by a vector with 10 dimensions, i.e. $X^{<1>} \in R^{L_l \times 10}$.

Initial Sequence-Based Features. We use PSI-BLAST [2] to extract the position-specific scoring matrix (PSSM) for the protein X, which is viewed as the initial sequence representation $X^{<2>}$. By using multiple iterations of PSI-BLAST, evolutionary information of proteins can be captured, and the derived initial sequence representation is denoted by $X^{<2>}$ as follows.

$$X^{<2>} = \begin{bmatrix} S_{1,1} & \cdots & S_{1,j} & \cdots & S_{1,20} \\ \vdots & \vdots & \vdots & \vdots \\ S_{i,1} & \cdots & S_{i,j} & \cdots & S_{i,20} \\ \vdots & \vdots & \vdots & \vdots \\ S_{L_l,1} & \cdots & S_{L_l,j} & \cdots & S_{L_l,20} \end{bmatrix}$$

where 20 denotes the total number of types of standard amino acids. The i-th row of $X^{<2>}$ denotes the probability of the i-th amino acid in sequence of protein X mutating into other types of amino acids during the evolutionary process. It's clear that $X^{<2>} \in R^{L_l \times 20}$.

Convolutional Neural Networks. For $X^{<1>}$ and $X^{<2>}$, we use two convolutional neural networks (CNNs) respectively to capture the local hidden features for structure and sequence information. Note that the two CNNs have similar architecture but with different training parameters. Therefore, without loss of generality, we will take the structure-based representation (i.e. $X^{<1>}$) as an example to describe the implementation of CNN.

Taking $X^{<1>}$ as input, the CNN layer consists of F feature maps and a max-pooling layer. Formally, the convolution operation is defined as follows.

$$Conv(X^{<1>}) = (C^1, \ldots, C^f, \ldots, C^F)$$

where C^f denotes the output of the f-th feature map. Then, the max-pooling operation is used to derive the new representation $X'^{<1>}$ which captures the local context information in structure information. Formally, given i and j, the entry $X'^{<1>}_{i,j}$ is derived as follows.

$$X'^{<1>}_{i,j} = max(\{C^1{}_{i,j}, \ldots, C^f{}_{i,j}, \ldots, C^F{}_{i,j}\})$$

3.2 Hybrid Attention Module

The hybrid attention module, which consists of a cross-modal attention layer and a type-level attention layer, integrates $X'^{<1>}$ and $X'^{<2>}$ into a final representation denoted by X^R.

Cross-modal Attention. Firstly, for structure representation $X'^{<1>}$, the Query, Key and Value matrices are defined as $Q_{X'^{<1>}} = X'^{<1>} \cdot W_Q^{<1>}$, $K_{X'^{<1>}} = X'^{<1>} \cdot W_K^{<1>}$, and $V_{X'^{<1>}} = X'^{<1>} \cdot W_V^{<1>}$, respectively. Similarly, the Query, Key and Value matrices for sequence representation $X'^{<2>}$

are derived as $Q_{X'^{<2>}} = X'^{<2>} \cdot W_Q^{<2>}$, $K_{X'^{<2>}} = X'^{<2>} \cdot W_K^{<2>}$, and $V_{X'^{<2>}} = X'^{<2>} \cdot W_V^{<2>}$, respectively.

Then, by considering the sequence information, the intermediate representation of structure information is derived by a scaled dot-product attention mechanism formally as follows.

$$Att_{X'^{<2>} \to X'^{<1>}} = softmax(\frac{Q_{X'^{<1>}} \cdot K_{X'^{<2>}}^{\top}}{\sqrt{d}}) \cdot V_{X'^{<2>}} \tag{1}$$

From Eq. (1), we can find that the similarity between the structure information (i.e. $Q_{X'^{<1>}}$) and the sequence information (i.e. $K_{X'^{<2>}}$) are computed. Then a normalized weight is obtained by the *softmax* function. Finally, the intermediate representation of structure information is obtained by the weighted average of the sequence information $V_{X'^{<2>}}$.

With a residual connection, the final representation of the structure information is derived by an element-wise addition operation (denoted by $\oplus$) as follows.

$$X''^{<1>} = Att_{X'^{<2>} \to X'^{<1>}} \oplus V_{X'^{<1>}} \tag{2}$$

With the same treatment, the final representation of the sequence information $X''^{<2>}$ is derived as follows.

$$X''^{<2>} = Att_{X'^{<1>} \to X'^{<2>}} \oplus V_{X'^{<2>}} \tag{3}$$

Type-Level Attention. In the type-level attention layer, the unnormalized weight vector of structure information is calculated based on multiplication of $X''^{<1>}$ and a context vector (which is the parameter of the whole framework to be trained). And the unnormalized weight vector of sequence information is calculated in the same way.

Then, a *softmax* function is applied on the two unnormalized vectors to obtain the normalized weight vectors denoted by $\alpha^{<1>}$ and $\alpha^{<2>}$ for structure and sequence representations, respectively.

It's worth noting that for each $i \in \{1, 2, \ldots, L_l\}$, $\alpha_i^{<1>}$ and $\alpha_i^{<2>}$ denote respectively the weight of structure information and sequence information for the i-th amino acid in protein X. It holds that $\alpha_i^{<1>} + \alpha_i^{<2>} = 1$. Thereafter, for the i-th amino acid in protein X, the final representation (which is denoted by X_i^R) will be derived as follows.

$$X_i^R = \alpha_i^{<1>} \cdot X''^{<1>} + \alpha_i^{<2>} \cdot X''^{<2>} \tag{4}$$

Based on the above description, we can find that the cross-modal attention layer models the semantic associations between structure and sequence representations, while the type-level attention layer models the different weights of structure and sequence information for the final representation of the given protein. Therefore, the hybrid attention module makes full use of the available structure and sequence information to learn the representations of proteins, which will be beneficial for the following prediction task.

3.3 Prediction Module

Given the final representation X^R and Y^R, the *sigmoid* function is used to derive the prediction for each pair of amino acids.

Formally, given each $i \in \{1, 2, \ldots, L_l\}$ and each $j \in \{1, 2, \ldots, L_r\}$, the output of prediction module is denoted by $O_{i,j}$, which is the predicted probability of forming the interaction interface for the i-th amino acid from protein X and the j-th amino acid from protein Y.

3.4 Training and Testing

The above subsections describe the processing procedure for one complex. In order to explain our idea more clearly, we provide the training and testing phases in the subsection.

In the training phase, assume the input mini-batch consists of N samples, which are expressed as $\{(X_1, Y_1), \ldots, (X_n, Y_n), \ldots, (X_N, Y_N)\}$. The ground-truth labels for N samples are denoted by $\{O_1^{True}, \ldots, O_n^{True}, \ldots, O_N^{True}\}$, where O_n^{True} is labels for the n-th complex (X_n, Y_n). The entry $O_{n(i,j)}^{True} = 1$ means the i-th amino acid from protein X_n and the j-th amino acid from protein Y_n form an interaction interface. Otherwise the amino acid pair does not form an interaction interface.

As described above, the predicted results by our model are denoted by $\{(O_1, \ldots, O_n, \ldots, O_N)\}$. The cross-entropy loss function can be defined as follows.

$$Loss = -(\sum_{n=1}^{N} \sum_{i=1}^{L_l^n} \sum_{j=1}^{L_r^n} O_{n(i,j)}^{True} \log O_{n(i,j)} + (1 - O_{n(i,j)}^{True}) \log(1 - O_{n(i,j)})) \quad (5)$$

where L_l^n and L_r^n are the length of the ligand protein and the receptor protein in the n-th complex.

Through minimizing the loss function, all parameters in our model are trained in an end-to-end manner. Once the training process is complete, the obtained model can be used directly for testing, in which the output of the prediction module is the predicted interaction interfaces for the given complex.

The proposed model is evaluated extensively on PDB datasets, and the details of experiments and results analysis will be presented in the following section.

4 Experiments

4.1 Datasets

In experiments, we use three datasets, DB3 [12], DB4 [13] and DB5 [22], to evaluate the proposed method. Note that these three datasets are different versions of the public Protein Data Bank (PDB) [4], and DB3 and DB4 are the subsets of DB5. Totally, the number of protein complexes are 230, 175, and 127 in DB5, DB4, and DB3, respectively.

Following the treatment in [1], in each complex, each pair of amino acids with the distance between any non-hydrogen atoms less than or equal to $6\mathring{A}$ will be annotated as an interaction interface, i.e. a positive sample. Since most pairs of residues are negative samples (i.e. not interaction interface), we down-sample the negative samples to keep the ratio of positive and negative samples as bout 1:10. In addition, to train and evaluate the model, each dataset is divided further into training set, validation set and testing set. The detailed statistics of datasets are presented in Table 1.

Table 1. Data statistics for three datasets.

Dataset	Train	Validation	Test	Positive samples
DB5	140	35	55	20,857
DB4	105	35	35	16,004
DB3	77	25	25	12,335

4.2 Experimental Setup

Hyper-parameter Setting. For performance comparison, three hyper-parameters, i.e. dimension of hidden representations d, number of the feature maps F, and dropout rate in CNN layers are set as 100, 16, and 0.6, respectively. From the result of hyper-parameter sensitivity analysis, this setting of hyper-parameters derives the best performance, and is used for performance comparison in the following results. In the training process, the mini-batch size, the maximum number of epochs, and the learning rate are set as 8, 10, and 0.001, respectively.

In addition, to set an appropriate value of sequence length in our implementation, we calculate the distribution of sequence lengths (i.e. number of amino acids in protein sequences) of proteins in PDB, which is illustrated in Fig. 2. Figure 2 shows clearly that although the range of sequence length is [29, 2128], about 98% of proteins have less than 800 residues. Thereafter, we set the length of protein sequence in our model as 800, and truncate the sequence over 800 while padding accordingly the value zero for sequence less than 800.

Baselines and Evaluation Metrics. To evaluate our framework, we select several representative methods as baselines, including the machine learning-based method PAIRpred [1], the CNN-based method SASNet [20], and the GCN-based method NEA [8]. It's worth noting that since the source code is not available, HOPI [16] is not selected as a baseline here. The proposed model in this study is denoted by HAM (the abbreviation for Hybrid Attention Mechanism) in the following analysis.

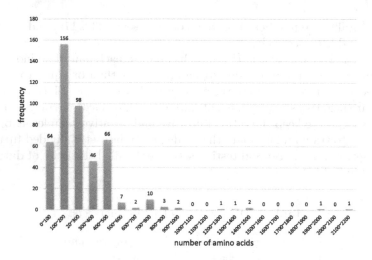

Fig. 2. Histogram of protein sequence length distribution.

In experiments, the Receiver Operating Characteristic (ROC) curve is computed for each complex. Then, the median AUC (MedAUC) for all the complexes in the test sets is used for performance comparison, since it can avoid the influence of too large or too small proteins on the performance for the whole dataset [16].

4.3   Results and Analysis

Overall Performance. Table 2 depicts the overall performance of our proposed model HAM on all datasets, by comparing with selected baselines. According to the evaluation metric MedAUC, HAM performs stably better than all the baselines. For SASNet, which leverages 3D spatial information of amino acids at the atom level, might ignore the structure information which is crucial for interaction interface prediction. As for baselines, PAIRpred and NEA, both of them concatenate directly sequence representation and structure representation, which ignore possible semantic associations between these two types of information. By utilizing the hybrid attention mechanism, our proposed method (i.e. HAM) is able to capture the relevant and complementary information between structure and sequence information, deriving the best performance on the task of protein interface prediction.

Ablation Study. In this section, we conduct an ablation study to investigate the contributions of the cross-modal attention and the type-level attention to the performance improvement in our framework. The comparison results are shown in Table 3. Note that in Table 3 "CMA" denotes the variant of our model with cross-modal attention only, in which we just concatenate the structure representation and sequence representation as the final protein representation for interaction interface prediction. "TLA" denotes the variant of our model with

Table 2. Comparison results on overall performance.

Methods	DB5	DB4	DB3
PAIRpred	0.809	0.781	0.774
SASNet	0.876	0.866	0.862
NEA	0.876	0.884	0.881
HAM	**0.892**	**0.887**	**0.882**

type-level attention only, in which we apply the type-level attention on the initial representations of structure and sequence (i.e. the output of the "Preprocessing Module"), and the derivation is used as the final protein representation for inter-action interface prediction.

Table 3. Results of ablation test.

Methods	DB5	DB4	DB3
CMA	0.871	0.869	0.875
TLA	0.659	0.632	0.633
HAM	**0.892**	**0.887**	**0.882**

As shown in Table 3, the performance of method CMA is clearly better than TLA, indicating that the cross-modal attention mechanism is more important for learning the protein representations than that of the type-level attention mechanism. Generally, HMA outperforms the best among the three models, suggesting that both cross-modal attention mechanism and type-level attention mechanism are beneficial for learning meaningful representations of proteins. For the task of protein interface prediction, both mechanisms are required to derive desirable performance.

Hyper-parameter Sensitivity Analysis. In this section, we investigate the impact of hyper-parameters in the proposed model on the performance of protein interface prediction on DB5. More specifically, we report the sensitivity analysis of three important hyper-parameters in the proposed model, including dimension of hidden representations d, number of the feature maps F, and dropout rate in CNN layers. The detailed experimental results are presented in Fig. 3. Note that each sub-figure demonstrates the results derived by the proposed model with a fixed dropout rate (i.e. dropout rate $\in \{0.2, 0.4, 0.6, 0.8\}$).

Dropout Rate. We can find that the performance (i.e. MedAUC) is improved as the dropout rate changing from 0.2 to 0.6, and it derives the best MedAUC 0.892. However, as we use a bigger dropout rate (i.e. 0.8), the performance of the proposed method goes down a little bit (as shown in Fig. 3(d)). That is the

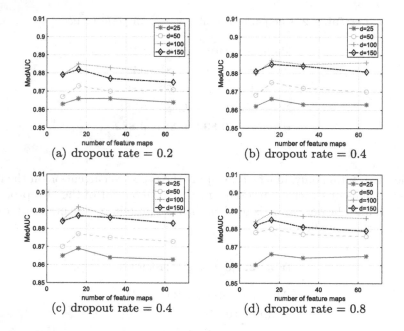

Fig. 3. Hyper-parameter sensitivity analysis of the proposed method.

reason why we present the results obtained by our model with dropout rate 0.6 in previous two subsections.

Number of the Feature Maps F. Generally, our model with setting $F = 16$ derives the best results by comparing with other settings of F. The possible reason is that the training samples are really limited in our experiments, and a more complex model is liable to over-fitting.

Dimension of Hidden Representations d. From Fig. 3, we can find that when keeping the dropout rate and F unchanged, the model with a bigger dimension of hidden representations (100 or 150) performs better than its counterpart with a smaller size of dimension (25 or 50). The results indicate that hidden representation with a bigger dimension can generally encode more meaningful information for protein interface prediction. Considering a complex model is liable to over-fitting, we set $d = 100$ in our model for performance comparison in the previous two subsections.

5   Conclusions

In this paper, we have proposed a novel deep learning-based end-to-end framework via a hybrid attention mechanism for the task of protein interface prediction. More specifically, a cross-modal attention mechanism is utilized to learn representations of proteins based on sequence and structure information, which can capture the semantic associations between both types of information. In addition, a type-level attention mechanism is employed to model the different roles played by sequence-based and structure-based features for predicting protein interface. The extensive experiments are conducted on three commonly used datasets, the results of which demonstrate that both cross-modal attention mechanism and type-level attention mechanism are beneficial for predicting protein interaction interface.

Acknowledgment. The work is partially supported by the National Natural Science Foundation of China (No. 61532008, No. 61872157, and No. 61932008), the Wuhan Science and Technology Program (2019010701011392), the Key Research and Development Program of Hubei Province (2020BAB017), the Fundamental Research Funds for the Central Universities (CCNU19TD004), the Research Fund of Guangxi Key Lab of Multi-source Information Mining & Security (MIMS19-02) and the Guangxi Key Laboratory of Trusted Software (kx201905). Authors are grateful to the anonymous reviewers for helpful comments.

References

1. Minhas, A.F.U.A., Geiss, B.J., Ben-Hur, A.: Pairpred: partner-specific prediction of interacting residues from sequence and structure. Prot. Struct. Funct. Bioinform. **82**(7), 1142–1155 (2014)
2. Altschul, S.F., et al.: Gapped BLAST and PSI-BLAST: a new generation of protein database search programs. Nucleic Acids Res. **25**(17), 3389–3402 (1997)
3. Bartlett, G.J., Annabel, E.T., Thornton, J.M.: Inferring Protein Function from Structure, Chap. 19, pp. 387–407. Wiley (2003)
4. Berman, H.M., et al.: The protein data bank. Nucl. Acids Res. **28**(1), 235–242 (2000)
5. Dai, B., Bailey-Kellogg, C.: Protein interaction interface region prediction by geometric deep learning. Bioinformatics (2021)
6. Esmaielbeiki, R., Krawczyk, K., Knapp, B., Nebel, J.C., Deane, C.M.: Progress and challenges in predicting protein interfaces. Brief. Bioinform. **17**(1), 117–131 (2015)
7. Fauman, E.B., Hopkins, A.L., Groom, C.R.: Structural Bioinformatics in Drug Discovery, Chap. 23, pp. 477–497. Wiley (2003)
8. Fout, A.M.: Protein interface prediction using graph convolutional networks. Ph.D. thesis, Colorado State University (2017)
9. Frappier, V., Keating, A.E.: Data-driven computational protein design. Curr. Opin. Struct. Biol. **69**, 63–69 (2021). (engineering and Design Membranes)
10. Frishman, D., Argos, P.: Knowledge-based protein secondary structure assignment. Prot. Struct. Funct. Bioinform. **23**(4), 566–579 (1995)
11. Gupta, A., et al.: Deep learning in image cytometry: a review. Cytom. A **95**(4), 366–380 (2019)
12. Hwang, H., Pierce, B., Mintseris, J., Janin, J., Weng, Z.: Protein-protein docking benchmark version 3.0. Prot. Struct. Funct. Bioinform. **73**(3), 705–709 (2008)
13. Hwang, H., Vreven, T., Janin, J., Weng, Z.: Protein-protein docking benchmark version 4.0. Prot. Struct. Funct. Bioinform. **78**(15), 3111–3114 (2010)

14. Jubb, H.C., Pandurangan, A.P., Turner, M.A., Ochoa-Montaño, B., Blundell, T.L., Ascher, D.B.: Mutations at protein-protein interfaces: small changes over big surfaces have large impacts on human health. Prog. Biophys. Molec. Biol. **128**, 3–13 (2017). (exploring mechanisms in biology: simulations and experiments come together)
15. Kumar, A., Verma, S., Mangla, H.: A survey of deep learning techniques in speech recognition. In: 2018 International Conference on Advances in Computing, Communication Control and Networking (ICACCCN), pp. 179–185. IEEE (2018)
16. Liu, Y., Yuan, H., Cai, L., Ji, S.: Deep learning of high-order interactions for protein interface prediction. In: Proceedings of the 26th ACM SIGKDD International Conference on Knowledge Discovery and Data Mining, pp. 679–687 (2020)
17. Mihel, J., Sikić, M., Tomić, S., Jeren, B., Vlahovicek, K.: Psaia - protein structure and interaction analyzer. BMC Struct. Biol. **8**, 21 (2008)
18. Sanner, M.F., Olson, A.J., Spehner, J.C.: Reduced surface: an efficient way to compute molecular surfaces. Biopolymers **38**(3), 305–320 (1996)
19. Shandar, A., Kenji, M., Deane, C.M.: Partner-aware prediction of interacting residues in protein-protein complexes from sequence data. PLoS ONE **6**(12), e29104 (2011)
20. Townshend, R., Bedi, R., Suriana, P., Dror, R.: End-to-end learning on 3d protein structure for interface prediction. Adv. Neural. Inf. Process. Syst. **32**, 15642–15651 (2019)
21. Urbanc, B.: Protein actions: principles and modeling. In: Bahar, I., Jernigan, R.l., Dill, K.A. (eds.) Garland science. Taylor and Francis group, 1st ed. 09 Feb 2017, ISBN: 9780815341772. (Journal of Biological Physics 43(4), 585-589 (2017))
22. Vreven, T., et al.: Updates to the integrated protein-protein interaction benchmarks: Docking benchmark version 5 and affinity benchmark version 2. J. Mol. Biol. **427**(19), 3031–3041 (2015)
23. Xie, Y., Le, L., Zhou, Y., Raghavan, V.V.: Chapter 10 - deep learning for natural language processing. In: Gudivada, V.N., Rao, C. (eds.) Computational Analysis and Understanding of Natural Languages: Principles, Methods and Applications, Handbook of Statistics, vol. 38, pp. 317–328. Elsevier (2018)
24. Xue, L.C., Dobbs, D., Bonvin, A.M., Honavar, V.: Computational prediction of protein interfaces: A review of data driven methods. FEBS Lett. **589**(23), 3516–3526 (2015)
25. Yan, C., Wu, F., Jernigan, R.L., Dobbs, D., Honavar, V.: Characterization of protein-protein interfaces. Protein. J. **27**(1), 59–70 (2008)

Author Index

Printed in the United States
by Baker & Taylor Publisher Services

Printed in the United States
by Baker & Taylor Publisher Services